Cognitive Unconscious and Human Rationality

Cognitive Unconscious and Human Rationality

edited by Laura Macchi, Maria Bagassi, and Riccardo Viale

The MIT Press
Cambridge, Massachusetts
London, England

This book was set in Stone Sans Std and Stone Serif Std by Toppan Best-set Premedia Limited. Printed and bound in the United States of America.

Library of Congress Cataloging-in-Publication Data

Names: Macchi, Laura, 1961– editor. | Bagassi, Maria, editor. | Viale, Riccardo, editor.
Title: Cognitive unconscious and human rationality / Laura Macchi, Maria Bagassi, and Riccardo Viale, eds. ; foreword by Keith Frankish.
Description: Cambridge, MA : MIT Press, [2015] | Includes bibliographical references and index.
Identifiers: LCCN 2015038315 | ISBN 9780262034081 (hardcover : alk. paper)
Subjects: LCSH: Reasoning (Psychology) | Cognition. | Subconsciousness.
Classification: LCC BF442 .C64 2015 | DDC 154.2–dc23 LC record available at http://lccn.loc.gov/2015038315

10 9 8 7 6 5 4 3 2 1

Contents

Foreword: Cooking Eggs on a Toaster

Keith Frankish

If you want to understand a device and evaluate its performance, you need to know what it is for. If you think that toasters were designed for cooking eggs, then you will be puzzled by their features and conclude that they are not very effective. With ingenuity, you might be able to cook an egg on a toaster, but it would be tricky and might even be dangerous. There is a strong case for thinking that we have been making a similar mistake about human reasoning.

Traditionally, psychologists thought of reasoning as an abstract, formal process. In the case of deductive reasoning most researchers adopted a model based on classical logic, on which the aim was to construct formally valid, truth-preserving arguments from accepted premises. The model reasoner, we might say, was René Descartes, secluded from the world, effortfully trying to build up a body of secure knowledge from certain foundations by a series of unshakeable deductions. Accordingly, experimental work focused heavily on assessing people's competence with the logical connectives, especially conditionals. In the study of judgment and decision making, similar assumptions were made. A rational agent was thought of as one who applies formal procedures to determine the correct judgment or choice, given his or her beliefs, expectations, and preferences.

If there is one thing that has been learned from experimental work on thinking and reasoning over the last four decades, it is that humans are not very good at this kind of thing. Although we can learn to apply formal rules of inference and choice, it takes discipline and practice, and we have to fight against strong natural tendencies to think in other ways. Our minds, it seems, are wired to be sensitive to content and context and to employ heuristics rather than formal decision procedures.

A moral often drawn from this is that we are naturally irrational. Our attempts to reason correctly are thwarted by innate dispositions to take shortcuts and to be swayed by irrelevant factors. These shortcuts and biases may be necessary, given limitations of time and cognitive resources, but they are suboptimal all the same. This view was often incorporated into the "dual-process" theories of reasoning developed since the 1980s. Such theories distinguish two types of reasoning process: implicit (Type 1) processes, which are fast, effortless, and nonconscious, and explicit (Type 2) processes, which are slow, effortful, and conscious. Typically, biases and fallacies are seen as the product of implicit processing, whereas logical responses are seen as the product of explicit processing. (This is not to say that all explicit thought is rule based; many dual-process theorists argue that explicit thought is also crucially involved in hypothetical thinking.)

From this logic-based, formalist perspective, then, it looks as if the human reasoning system is not well adapted to its job, succeeding at best only with effort. But maybe that perspective is the wrong one. Perhaps we have been judging human reasoning against an inappropriate standard, like someone who thinks that a toaster is for cooking eggs. Indeed, a new perspective has been emerging, drawing on models of reasoning that assign crucial roles to context, content, and pragmatics in addition to formal features.

A major aspect of this change has been a paradigm shift in the way psychologists think about everyday deductive reasoning, where the classical model has been widely replaced with a probabilistic one on which premises are assigned graded probabilities and the aim of argument is to preserve probability rather than truth. This model represents deductive reasoning as sensitive to the reasoner's degrees of confidence as well as to the logical structure of the premises, and it offers new normative standards based on probabilistic notions of validity and coherence.

This "new paradigm" reflects the type of problems we face in everyday life, which typically involve reasoning from qualified beliefs rather than accepted premises, and experimental evidence suggests that it is a better fit to actual human performance. In addition, it offers a common framework for thinking about deductive reasoning, judgment, decision making, and belief updating, and it allows for the development of decision-theoretic approaches to reasoning, in which considerations of utility and obligation play important roles—as they do in real-life problem solving. Unlike Descartes, everyday reasoners are not seeking timeless certainties but timely solutions to particular problems, and their reasoning activities are shaped by practical goals as well as epistemic ones.

Another important development has been Mercier and Sperber's proposal that human reasoning is not designed for solitary enquiry but for public argumentation—the production and evaluation of arguments intended to persuade others.[1] In this view the model reasoner is not the philosopher by his stove but the lawyer in court. Again, this changes how we think about some supposed biases, in particular confirmation bias (the tendency to look only for evidence that confirms one's views). This is a weakness in solitary enquiry but a useful trait in public argumentation and one that is self-correcting in group enquiry. (There is an interesting link here with the work of Giuseppe Mosconi, to whom this volume is dedicated. Mosconi held that there is a language-involving form of thought, *thinking-speaking*, which is employed in both public argumentation and private reasoning and which is shaped by communicative and rhetorical principles as well as logical ones.)

There are complementary trends in judgment and decision making, where researchers in the traditions of bounded and ecological rationality argue that the heuristics we intuitively rely on are not suboptimal compromises but often the best solutions to particular context-specific problems. Again, it is argued, the traditional view mistook specific intelligent responses for general fallacies.

Thus, from this new perspective, many factors that were formerly seen as biases or shortcuts appear as legitimate influences and rational procedures. Of course, it does not follow that humans are optimally rational (not all toasters are good ones), or even that it is helpful to set abstract standards of rationality. But the change of perspective yields a new understanding of the strengths and weaknesses of human reasoning, and it is changing the questions psychologists ask and the interpretations they offer.

The new perspective also has important implications for dual-process theory. The developments mentioned do not undermine the dual-process approach: there remains a strong case for thinking that there are two types of mechanism underpinning human reasoning competence. But these developments do undermine the simple equation of implicit processing with bias and explicit processing with normative correctness. And in doing so, they invite us to rethink the function and powers of the implicit mind. There has been a tendency to think that implicit processing is simply a quick-and-dirty substitute for explicit thought, and that if explicit thought is engaged, then it is desirable that it should override implicit responses. From the new perspective, however, many implicit responses appear as intelligent solutions to context-specific problems and may be more successful than those generated by the application of formal rules.

At the same time, researchers are increasingly coming to appreciate the powers of the implicit mind and the crucial role that implicit processes play in enabling and supporting explicit thought—in initiating it, supplying content to it, monitoring it, and shaping the way it is conducted. Indeed, the implicit mind is increasingly looking like the engine room of cognition, and the explicit mind like a fragile superstructure. This approach is also shedding light on creative processes, such as insight problem solving and artistic activity. Rather than arising from some mysterious faculty of intuition, these can be understood as the result of complex implicit processing of various kinds, including, for example, a systematic search for relevance.

This is not to deny the value of explicitly applying logical rules and decision procedures. In modern life there are many situations where we need to do just that. But it is important to recognize that the apparatus we use to do it did not develop for that purpose. The better we understand the functions of the human reasoning system and its powers and limitations, the better placed we shall be to tweak it to do the more artificial jobs we sometimes require of it. If you must cook an egg on a toaster, then it helps to understand exactly how a toaster works.

The essays in this volume are written by researchers centrally involved in the trends mentioned above—the new paradigm, bounded rationality, dual-process theory, and the study of implicit cognition. They vividly illustrate the advantages of the new perspective, the insights it offers, and the new questions it poses. This is an exciting time in the psychology of thinking and reasoning, and this volume offers a stimulating encounter with the latest work in the field.[2]

Heraklion, Crete
May 2015

Notes

1. Mercier, H. & Sperber, D. (2011). Why do humans reason? Arguments for an argumentative theory. *Behavioral and Brain Sciences, 34,* 57–111.

2. Thanks to Maria Bagassi, Shira Elqayam, Laura Macchi, and David Over for their advice on this piece.

Introduction

Laura Macchi, Maria Bagassi, and Riccardo Viale

The aim of this volume is to give voice to the contemporary debate within the psychology of thought that has great impact on creativity, decision making, and economic behavior. The volume intends to promote new perspectives on a holistic view of mind, with in-depth and dialectical examination of the most promising theories and approaches to the study of thought, discussed and compared with the more recent experimental evidence.

The debate starts from a discussion on the paradigm shift in the psychology of reasoning, which goes beyond the idealized logic of certainty to the much more realistic *logic of uncertainty*. Both the deductive and probabilistic paradigms appear to respond to fundamental issues. They address the yearning for certainty and the need to come to terms with the impossibility of knowing the world, falling back on a knowledge only probable, more able to explain how human beings reason, make decisions, deal with the new in the real world.

A shadow area of thought, that is, implicit thought, is then reconsidered. The book focuses on the role of implicit, unconscious thinking in creativity and insight problem solving, the interaction of intuition and analytic thinking in decision making, and the relationship between communicative heuristics and thought, dealing with the eclipse of the conception of mind informed by extrapsychological theoretical models toward a genuinely psychological conception of rationality. A more complex idea of rationality emerges that is no longer limited to conscious, explicit thought but is able to exploit the intentional implicit level, resulting in the ability to contextualize and to move creatively within the complexity of reality.

Creativity may well represent a crucial issue at this stage of the debate for theories on thinking by challenging them from several points of view. Traditionally, our ability to face the novel was attributed to analytical-reflective thought, a process that is slow, sequential, effortful, and characterized by consciousness. But it is not possible to ascribe creativity only to reflective, conscious thinking since it implies incubation and unconscious, implicit thinking.

This general, fundamental issue will be investigated by exploring crucial questions, which are challenges for any theory of thought.

The first section, "Bounded Rationality Updated," addresses a new conception of human rationality, adopting a psychological approach rather than extrapsychological normative paradigms. To cope with all the uncertainty in the real world, thinking must consider probabilities, implicit presuppositions or degrees of belief.

This new conception is explained by Gerd Gigerenzer in terms of *ecological rationality,* which is functionalist and adaptive to the environment; as *grounded rationality* by Shira Elqayam, taking epistemic contexts into account; and as rationality informed by the fundamental interpretative function of thinking by Maria Bagassi and Laura Macchi.

Gigerenzer reestablishes the original meaning Simon gave to human rationality in terms of *bounded rationality*. Having pointed out its authentic psychological nature, he launches into a series of significant theoretical reflections that show how human rationality is able to adapt to a reality characterized by uncertainty in which "there is no single best answer, but there are several reasonable ways to behave in a rational way." The concept of *ecological rationality* explains how this inferential miracle can be possible.

Elqayam considers the question of whether psychology of reasoning needs an alternative norm at all (probabilistic rather than logical deductive norm) to be crucial. She proposes a descriptive approach to reasoning, decision making, and rationality in place of the normativist paradigm, in other words, focusing on reasoning as it is rather than as it ought to be. She suggests an extension of bounded rationality, *grounded rationality,* taking epistemic contexts into account, which offers a productive stance for the new paradigm.

Bagassi and Macchi propose a conception of mind bounded by the qualitative constraint of relevance at conscious and unconscious levels. The core of this conception is an *interpretative function,* as adaptive characteristic of the human cognitive system. This perspective is supported by evidence from the authors' research on insight problem solving, which they consider a privileged route to understanding what kind of special unconscious thought produces the solution. During incubation, in the absence of conscious control, relevance constraint allows multilayered thinking to discover a new interpretation of the data that finally offers an exit from the impasse. The authors speculate that the creative act of restucturing implies a form of high-level unconscious thought, the *unconscious analytic* thought.

The second section, "The Paradigm Shift: The Debate," is dedicated to essays on the paradigm shift in the psychology of reasoning. According to David Over, abandoning the normative model of classic logic implies a Bayesian and probabilistic approach, which takes into account the beliefs and knowledge of individuals in an attempt to integrate reasoning research much more closely with the study of subjective probability and utility in judgment and decision making. He considers the study of conditionals, which are fundamental to human reasoning, to be productive in assessing the difference between the old and new paradigms. Over's claim is that the psychology of reasoning should provide a theory of how people perform inferences from their beliefs and change their beliefs in a dynamic process of reasoning. In fact, little can be learned about everyday or scientific reasoning by simply studying inferences from fixed assumptions. Arbitrary assumptions do not help people make rational decisions, and absolute certainty cannot usually be found, assumed, or presupposed in the real world. He claims that to be rational, people must make inferences from their beliefs and judge how confident they should be in the conclusions that they infer from their mostly uncertain beliefs. Fixed epistemic states of mind cannot be assumed, but degrees of beliefs must be updated over time.

Mike Oaksford argues that in everyday reasoning in the real world there is invariably a trade-off between achieving our goals in a timely way and the search for truth. In other words, although our everyday reasoning is practically rational, it may not always fulfill the demands of theoretical rationality. The author shows that the recent Bayesian approaches to explicit conscious reasoning are directly related to the approaches to unconscious inference, which are proposed in perception and action by the Bayesian brain hypothesis. That is, in unconscious inference, there is always a trade-off between our epistemic and our practical goals, between exploration to reduce ignorance and exploitation to get what we want. Similar constraints are also operative in human reasoning at the conscious verbal level.

The ensuing new paradigm poses many new conceptual and methodological questions, which are addressed by Jean Baratgin and Guy Politzer. First and foremost, reasoning cannot be reduced to a neutral mechanism independent of the individual's motivation, preferences, and objectives. Moreover, individual subjective responses (judgments and decisions) may be different but nevertheless coherent as they are based on different states of knowledge. The various factors at the origin of the subjectivity of beliefs, such as cultural background, should also be taken into account. The objectives of the new paradigm meet the foundation of the Bayesian subjective theory and, more precisely, the de Finettian theory.

Linden Ball and Edward Stupple focus on *belief biases* as a key finding for dualistic and probabilistic theories on reasoning, involving experimental manipulations, neuroscientific investigations, and the analyses of individual differences. They argue that dual-process theories of belief-bias effects, which were established until a decade ago, no longer appear to be sustainable. According to these authors, the cognitive system is well adapted to engage in reflective processing only in those situations where there is a conflict between the outputs of intuitive heuristics and logical intuitions. In this way, highly constrained cognitive resources associated with reflective processing can be deployed efficiently to those reasoning situations that are characterized by uncertainty.

Finally, the critical approach of Steven Sloman and Aron Barbey raises doubts about the adequacy of the new probabilistic paradigm, as well as about the deductive paradigm, by considering the deterministic constraint of causality as the attempt of human thought to find explanations. They paint a sensible portrait of human cognition, determining a set of basic assumptions made by the cognitive system. To make inferences, the cognitive system needs some direction, some driving force or principles of operation. Human beings allow for probability despite the belief that there are deterministic laws, and the use of probability is justified in terms of uncertainty. So, even though we believe that all events are determined by their causes, we also know that they are impossible to predict because we are necessarily ignorant. Probability arises from ignorance of initial conditions or relevant variables or governing laws.

In the third section, "Epistemic Context in Reasoning," Hugo Mercier, Pierre Bonnier, and Emmanuel Trouche-Raymond deal with the argumentative function of reasoning, given its interactional nature, considering the argumentative-persuasive goal to be the core feature of thought rather than the true validity of the logical approach to the study of thinking. The study of argumentative discussion, with the interactions it entails, offers a genuine understanding of the strengths of reasoning and gives an account for the (supposed) failures.

This form of *interactional rationality*, as it could be called, has also been developed by Denis Hilton in the field of deontic reasoning. He argues that in addition to rule content, we need to consider rule-use in order to understand the meanings of deontic conditionals, as deontic rule contents can be used in both performative and constative (indicative) uses. Then, he suggests that an understanding of the interests at stake is necessary for the proper interpretation of deontic verbs when used in the context of institutional deontics, in which rules, whose contents are conventionally understood as rights or duties, can be used in "unconventional ways."

Jean-François Bonnefon and Eric Billaut have exercised the same attention to language use and comprehension to capture the individual differences in reasoning. The psychology of reasoning has focused on a rather narrow band of such characteristics (linked to the ability and disposition to engage effortful mental processing) and has not addressed other personality traits in any depth. The authors, instead, have assumed that an individual linguistic style would be linked to that individual reasoning style, and they suggest that different personality traits may predispose reasoners to endorse different premises and conclusions.

In the fourth section, "Implicit Thought: Its Role in Reasoning and in Creativity," a shadow area of thought is reconsidered, traditionally believed to be responsible for automatisms and often for errors of reasoning. In the *component* view of Tilmann Betsch and colleagues, intuition becomes central as dualistic visions of thought and logical-deductive paradigm are left behind. It is capable of rapidly processing multiple pieces of information through a parallel process of information integration, unconstrained by the amount of encoded information and cognitive capacity (Tilmann Betsch, Johannes Ritter, Anna Lang, and Stefanie Lindow).

Ron Sun investigates implicit processes, treating them as an integral part of human thinking in a variety of settings, and their interactions with explicit processes. Using the CLARION cognitive architecture, he shows that frequently the more implicit a process is, the faster it is but, also, that intuition can be very slow, especially when explicit thought emerges from implicit intuition.

Valerie A. Thompson, Nicole H. Therriault, and Ian R. Newman have investigated the basis of metacognitive judgments (FOR), which predict the extent of the subsequent analytic engagement. Metareasoning, like other metacognitive processes, operates in the background, exercising implicit control over explicit reasoning processes. Thus, explicit analytic thinking may be initiated or terminated in response to implicit monitoring cues such as fluency or familiarity.

Kenneth Gilhooly has explored the role of unconscious thought in creative processes, particularly in the incubation phase that precedes the response in divergent thought tasks. Through a review of the theories of incubation and of related empirical evidence, the author argues that unconscious thinking seems a viable mechanism for incubation in the form of implicit associative processes based on spreading activation. He also explores the question of what form unconscious work might take and the idea that unconscious work might be a subliminal version of conscious work.

Riccardo Viale analyzes some of the neurocognitive procedural characteristics that bound the process of creativity. Among the various expressions of creativity, figurative art is preferred. It seems to involve more parts of the brain and psychological functions than other forms of creativity. In particular, the chapter analyzes the tendency to assign responsibility for creative thought to the right hemisphere and tries to understand why this is. The author explains how a structure of the brain known as the *default mode network* produces one of the characteristics of human creativity: that of being enhanced, above all, during periods when the mind wanders. Both results highlight a particular dynamic of creative thought linked to the unconscious spreading activation of neural networks. This neurobiological finding seems to correspond to the results of psychological research on incubation. In conditions where the mind wanders or is engaged in other tasks, there is an empirically controlled increase in creative capacity with regard to the target problem. This appears to be caused by an unconscious spreading activation of semantic networks. Last, the characteristics of bounded creativity linked to the implicit and unconscious dynamic of neurocognitive and perceptual limits are highlighted, also by examining the emotional and visual dimension of creativity in figurative art. In this concluding paragraph, Viale shows how the unique structure of the brain has constrained the emergence of particular artistic styles and expressions.

The last section, "Closing Thoughts," includes a selection of excerpts from the works of Giuseppe Mosconi (1931–2009), whose innovative and pioneering works are an unexpected forerunner to the debate on reasoning. His philosophical, *psycho-rhetorical* perspective on the study of thought would have made him the ideal interlocutor in the current debate, establishing possible foundations for a holistic view on thought and providing experimental evidence of "psycho-logical" rationality. His approach refuted identification of the study of thought with that of logical-deductive reasoning, considering the idealized logical paradigm a "distillate" of human reasoning, neither able to be a descriptive nor a normative model of how we reason. His conception of mind involves interaction of thought with language in a unitary cognitive activity, *thinking-speaking*, leading him to suggest an approach to the study of thinking based on a direct reference to conversational rules rather than on those of logic. In this view, biases and errors are not considered as constitutive of human cognitive system, but as the effect of misleading communication.

Hence, what emerges across the contributions of the book is the potential of a cognitive unconscious, which, in addition to the task of ensuring automatic and effortless behaviors, appears to underlie almost every aspect of our conscious existence, promoting a creative and imaginative dimension capable of redescribing reality. In this case our mind, finally free from the quantitative limits of the consciousness, fluctuates between the conscious and unconscious levels in a wide-range analytical search. Paradoxically, it is due to these constraints that analytical thought dives, continuing research in depth, beyond the control of consciousness, to reemerge with a creative act.

I Bounded Rationality Updated

1 Rationality without Optimization: Bounded Rationality

Gerd Gigerenzer

The study of bounded rationality is not the study of optimization in relation to task environments.

—Herbert A. Simon (1991)

What is bounded rationality? The answer to this question seems obvious. After all, numerous books, articles, and encyclopedia entries have been written and conferences held on this topic ever since Herbert A. Simon coined the term in the mid-20th century. Yet when Kenneth Arrow, Daniel Kahneman, and Reinhard Selten—all three, like Simon, winners of the Bank of Sweden Prize in Economic Sciences in Memory of Alfred Nobel—write about bounded rationality, each of them has a strikingly different concept in mind. Nonetheless, each links his own particular concept to the same person, Herbert Simon, so that there appears to be a single definition of bounded rationality. In this chapter I explain the three different faces of this concept and what Herbert Simon actually meant. Let us begin with Simon himself.

How do people make decisions in a world of uncertainty? In the mid-1930s young Herbert Simon, fresh from a class on price theory at the University of Chicago, tried to apply the perspective of utility maximization to budget decision problems in his native Milwaukee recreation department. To his surprise, he learned that managers did not compare the marginal utility of a proposed expenditure with its marginal costs. Instead, they simply added incremental changes to last year's budget, engaged in habits and bargaining, or voted on the basis of their identification with organizations. Simon concluded that the framework of utility maximization "was hopeless" (Simon, 1988, p. 286). This discrepancy between theory and reality marked the beginning of what he later called the study of *bounded rationality*:

Now I had a new research problem: How do human beings reason when the conditions for rationality postulated by the model of neoclassical economics are not met? (Simon, 1989, p. 377)

Let us call this *Simon's question*. Budget problems are not the only ones for which Simon's question must be posed. It is equally relevant for financial investment, personnel selection, strategic decisions, and in general, in situations where the future cannot be known with certainty and surprises happen. What Simon means is that in such situations the neoclassical theory is of little help either *as a model of how decisions are actually made* or *as a prescription for how to make better decisions*. The hypothesis that managers behave *as if* they maximized their expected utility is an essentially different claim (see below). What are these conditions for rationality to which Simon refers?

There are two main approaches to defining rationality in economics and beyond (Sen, 1987). The first emphasizes internal *consistency*. In other words, choices must obey certain rules such as "if *x* is chosen from a set containing *y*, then *y* will not be chosen from any set containing *x*," which is known as the *Weak Axiom of Revealed Preferences* (Samuelson, 1938). A variety of further conditions of internal consistency have been proposed in the literature. The second approach emphasizes the reasoned pursuit of *self-interest*. The view that human beings relentlessly maximize self-interest is often attributed to Adam Smith, although Smith considered many other motivations as well; nevertheless, self-interest has shaped the understanding of individual motivation in economics for a long time.

These two approaches are not the same; a person can be consistent without maximizing self-interest by maximizing something else. As a consequence, some consider the maximization of self-interest too narrow a conception of rationality, and various proposals have been made to extend the motivational structure of *Homo economicus* to "other-regarding" motivations (e.g., kinship, reciprocal altruism, kindness, politeness) while maintaining the maximization framework. To others, the consistency framework is too general; for instance, acting consistently against one's own interests would be an odd conception of rationality.

What Is Radical about Simon's Question?

Simon's question is more radical than the question of how to add other-regarding motives into the utility function. He argued that most of the time, motivated by either self-interest or other-regarding interest, humans do not maximize at all but instead "satisfice" (Simon, 1955, 1979). Why would humans not optimize? As a polymath with a strong interest in artificial intelligence, Simon was aware that most important problems are computationally intractable, that is, no computer or mind can calculate the optimal course of action. This was not a new insight. Jimmy Savage, the father of modern Bayesian decision theory, had earlier pointed out how "utterly ridiculous" it would be to "plan a picnic or to play a game of chess in accordance with the principle," that is, with the expected utility maximization framework (Savage, 1954, p. 16). In chess the insurmountable problem is tractability: although the game is well defined, the optimal sequence of moves cannot be calculated by a human or machine. For planning a picnic, the problem is even deeper. Here all possible moves (what can happen) cannot be known ahead, and surprises may occur; thus, the problem space is ill defined.

In the following, I use the term *uncertainty* to refer to problems that are well defined but computationally intractable or ill defined, that is, where some alternatives, con-sequences, or probability distributions are either not known or knowable. My use of the term extends its earlier use by Knight (1921), who focused mainly on unknown probability distributions. In an uncertain world, by definition, we cannot calculate the "optimal" course of action (although, after the fact, many tend to exhibit supernatural powers of hindsight). We can now see that Simon's question involves two objectives.

The research program of bounded rationality is about:

1. Decision making under *uncertainty*—as opposed to programs about calculable *risk* only—that is, situations in which optimizing (determining the *best* course of action) is not feasible for the reasons mentioned above.
2. The process by which individuals and institutions actually reach decisions—as opposed to *as-if* models—including their search, stopping, and decision rules.

In Simon's words, the program's first objective is to model "rationality without opti-mization" (1990, p. 8). Note that the reason is not people's irrationality but the very nature of ill-defined or intractable problems. As we will see, some psychologists missed this point, blaming the human mind for not being able to optimize all the time. The second objective is to model the decision processes rather than construct as-if models, as do neoclassical economics and parts of psychology as well. Ever since Milton Fried-man declared that the realism of the assumptions is of little relevance in economic models and that only the quality of the predictions matters, the as-if doctrine has suc-cessfully eliminated concern with psychological processes.

The reader might ask: What is so radical about Simon's question? Let me consider two groups of researchers whom Simon addressed, economists and psychologists. Many neoclassical economists would consider a model that does not involve optimiza-tion as something alien to their field. As we will see, even behavioral economists have continued on this path. The resulting optimizing models are typically as-if. Rational-ity without some form of optimization appears to be unthinkable. Similarly, many psychologists define rational behavior by the laws of probability or logic and confront their experimental subjects with problems of risk rather of uncertainty. Standards of this kind of research are choices among lotteries, Bayesian problems, trolley problems, and experimental games in which the optimal course of action can be calculated. Both the problems and the theories of rationality are reduced to situations of calculable risks.

As a consequence, research devoted to Simon's question has been the exception rather than the rule.

Why Is There No General Theory of Bounded Rationality?

In *Administrative Behavior* (Simon, 1947), bounded rationality still referred to everything outside of full rationality or optimizing. Later, Simon (1955, 1956) became more specific by proposing concepts such as aspiration levels, recognition, limited search, and satisficing. Yet in the half-century after these beginnings, relatively little conceptual and empirical work has been done. As a consequence, we lack an overarching theory of bounded rationality. One reason is that Simon's question has been systematically avoided by misrepresenting the study of bounded rationality as either the study of optimization or deviations from optimization. Simon himself would be the first to admit that he left us with an unfinished project and that his own thinking developed gradually from an optimizing version of satisficing to the clear statement of his question in later years. In an email dated September 2, 1999, he wrote:

Dear Gerd,

I have never thought of either bounded rationality or satisficing as precisely defined technical terms, but rather as signals to economists that they needed to pay attention to reality, and a suggestion of some ways in which they might. ...

I guess a major reason for my using somewhat vague terms—like bounded rationality—is that I did not want to give the impression that I thought I had "solved" the problem of creating an empirically grounded theory of economic phenomena. ... There still lies before us an enormous job of studying the actual decision processes that take place in corporations and other economic settings. ... End of sermon—which you and Reinhard don't need. I am preaching to believers.

Cordially,

Herb

In signing off, he refers to the work of Reinhard Selten, who devoted part of his professional life to answering Simon's question (e.g., Gigerenzer & Selten, 2001; Selten, 1998, 2001). Selten's work on decision making under uncertainty, together with the work of my research group and others around the world (e.g., Gigerenzer, Todd, & the ABC Research Group, 1999; Todd, Gigerenzer, & the ABC Research Group, 2012), indicates that the tide is changing and that we are finally beginning to know how to answer Simon's question, including its extension to a normative one (see below). In this chapter, I do not cover the hundreds of experiments and analyses that have been performed to answer Simon's question in the last two decades (for an overview, see Gigerenzer, Hertwig, & Pachur, 2011; Hertwig, Hoffrage, & the ABC Research Group, 2013). Instead, I focus on the conceptual issues crucial to Simon's question.

I argue that there are three quite different programs of bounded rationality, two of them explicitly ignoring Simon's question. I then explain how these programs differ from what Simon had in mind and thus provide one answer to the question why there is no general theory of bounded rationality.

Three Programs of "Bounded Rationality"

The first two programs represent what most economists and psychologists, respectively, think bounded rationality is about. Each of these two programs ignores Simon's question by assuming that the conditions of rationality are met. Nevertheless, both misleadingly refer to Simon's concept of bounded rationality. In that way, the revolutionary potential of Simon's question has been largely defused.

As-If Rationality: Optimization under Constraints   Most neoclassical economists argue that bounded rationality is nothing but full rationality, taking costs of information and deliberation into account. As Arrow (2004, p. 48) asserts:

... boundedly rational procedures are in fact fully optimal procedures when one takes account of the cost of computation in addition to the benefits and costs inherent in the problem as originally posed.

As an aside, this statement is from Arrow's contribution to the memorial volume for the late Simon. In this view, bounded rationality is nothing but constrained optimization in disguise. Stigler's (1961) model of the purchase of a second-hand car is a classic example, where the buyer is assumed to stop search when the costs of further search exceed its benefits. But in Simon's own words that is by no means what it is about: "The study of bounded rationality is not the study of optimization in relation to task environments" (Simon, 1991).

In personal conversation Simon once wryly told me that he wanted to sue authors who misused his concept to describe another form of optimization (Gigerenzer, 2004). In doing so they ignored both objectives of Simon's bounded rationality program: to study behavior under uncertainty and the actual decision processes themselves.

Moreover, the optimization-under-constraints version of bounded rationality leads to a paradoxical result that provides a good reason not to pursue this program any further. The more realistic constraints are introduced into an optimization model, the more unrealistic knowledge and computations it needs to assume. Accordingly, in his *Bounded Rationality in Macroeconomics*, Sargent (1993, p. 22) describes bounded rationality as a "research program to build models populated by agents who behave like working economists or econometricians." As a consequence, the program of incorporating realistic constraints creates highly complex models that are unattractive compared to the easier models of full rationality, leaving us back at square one.

Deviations from Optimization: The Cognitive Illusions Program   Whereas most economists argue that bounded rationality is about rational behavior, most psychologists argue that it is about systematic biases and other forms of irrational behavior. The *cognitive illusions program*, also called the *heuristics-and-biases program*, takes this stance. Although the two interpretations appear to be diametrically opposed, a closer look shows that they differ only in their descriptive claims about people's rationality. In fact, the heuristics-and-biases program accepts the economic optimizing models as the universal norm for people's reasoning. In Kahneman's (2003, p. 1449) words:

Our research attempted to obtain a map of bounded rationality, by exploring the systematic biases that separate the beliefs that people have and the choices they make from the optimal beliefs and choices assumed in rational agent models.

In this view, optimization is descriptively incorrect as a model of human behavior but remains normatively correct. Accordingly, the experimental studies conducted in this program rely on content-free principles of logic, probability theory, or expected utility maximization in situations where the optimal belief or choice is known—or assumed to be known. Any discrepancy between human judgment and the proposed norm is then blamed on the humans rather than the models.

Let us compare the two programs. As-if rationality ignores both research goals in Simon's question. Instead, it models situations in which the optimal answer can be calculated and favors as-if models rather than models of the decision process. Although the cognitive illusions program also ignores Simon's first objective, it does accept his second objective, repeatedly stating its aim to uncover the underlying cognitive processes, called heuristics. However, unlike what Simon had in mind, its proponents propose no testable formal models of heuristics. With the exception of the earlier seminal work by Tversky (1972) on elimination by aspects, virtually no models of heuristics have been specified algorithmically. Instead, vague labels such as availability and representativeness were introduced after the fact (Gigerenzer, 1996).

By now it should be clear that Kahneman's use of the term *bounded rationality* is not Simon's. As Simon (1985, p. 297) said, "Bounded rationality is not irrationality." One explanation for this discrepancy is that Kahneman and Tversky never set out to find an answer to Simon's question. As Lopes (1992) pointed out, they adopted the fashionable term *bounded rationality* as an afterthought and as an acknowledgment to a distinguished figure. This hypothesis is supported by the fact that although bounded rationality is mentioned in the preface of Kahneman, Slovic, and Tversky's 1982 anthology, their influential papers reprinted in the same anthology neither mention it nor cite Simon at all.

Whereas economists have tried to reduce bounded rationality to neoclassical economics, Kahneman, Tversky, and their followers have tried to reduce it to their study of cognitive illusions. In this way, Simon's question has been abandoned (Callebaut, 2007; Gigerenzer, 2004), and the role of psychology has been reduced to documenting errors. A distinguished economist made this point crystal clear, concluding a discussion with the dictum "either reasoning is *rational* or it's *psychological*" (Gigerenzer, 2000, p. vii). By contrast, the study of ecological rationality dispenses with this supposed opposition between the rational and the psychological. To analyze ecological rationality, one needs to understand the heuristics that people use in the first place.

Ecological Rationality: The Adaptive Toolbox   A third perspective, the program of ecological rationality (Gigerenzer et al., 1999; Gigerenzer & Selten, 2001), addresses both objectives in Simon's question. Looking beyond expected utility maximization and Bayesian decision theory, it seeks to answer the question of how people make rational decisions under uncertainty, that is, when optimization is out of reach (Gigerenzer & Selten, 2001, p. 4):

Models of bounded rationality describe how a judgment or decision is reached (that is, the heuristic processes or proximal mechanisms) rather than merely the outcome of the decision, and they describe the class of environments in which these heuristics will succeed or fail.

This program has both a descriptive and a normative goal. Its descriptive goal is to model the heuristics, their building blocks, and the core cognitive capacities these exploit. This is called the study of the *adaptive toolbox* of an individual or an institution. Although the heuristics-and-biases program has made a first step in this direction, the program goes one step further and provides mathematical and computational models for heuristics that are testable. The normative study of heuristics addresses the question when people *should* rely on a simple heuristic rather than on a competing, more complex strategy. It is called the study of the ecological rationality of heuristics.

The Adaptive Toolbox

Simon had proposed the satisficing heuristic as a model of decision making under uncertainty. The task is to choose a satisfactory alternative when the set of alternatives is not known in advance. That kind of uncertainty is typical for many problems, such as whom to marry, where to invest one's money, or which house to buy. Here, the "optimal" stopping rule—the point in time where the costs of further search exceed its benefits—cannot be calculated, and a reduction to optimization of constraints would mean creating an illusion of certainty. The basic version of satisficing with a fixed aspiration level has three building blocks: a search rule, a stopping rule, and a decision rule (Simon, 1955):

1. Set an aspiration level α and sequentially search for objects.
2. Stop search after the first object x is found with $x \geq \alpha$.
3. Choose object x.

By adapting building blocks to the problem at hand, one can modify heuristics. For instance, aspiration levels can be constant over time or adjusted to reflect a learning process (Selten, 1998, 2001). By replacing the stopping rule (2) with (2′), we can incorporate such a learning process and attain an adjustable aspiration level:

2′. Stop search after the first object is found with $x \geq \alpha$. If no such object is found within a time interval β, lower α by a factor γ, and continue search.

Thus, the basic version has one individual parameter (the aspiration level α), whereas the adjustable version has three (aspiration level α, time interval β, and size of adjustment γ). A study of the pricing strategies for used BMWs (models 320 and 730) by 748 dealers selling 24,482 cars showed that 95% of these dealers used a satisficing heuristic: 15% relied on fixed aspiration levels, and 80% relied on an adjustable aspiration level (Artinger & Gigerenzer, 2014). Typically, α was set above the mean market price and adjusted downward after around 1 month. Simon (1955) referred to pricing in the housing market, but given the lack of research on satisficing, we do not know whether housing prices, like used car prices, are also shaped by aspiration levels. Yet that is also quite likely. For instance, Levitt and Syverson (2008) report that when selling on their own behalf, real estate agents take considerably longer until they sell than when selling for their clients. Satisficing has also been investigated as an alternative to optimizing models in mate choice (Todd & Miller, 1999) and to explain aggregate age-at-marriage patterns (Todd, Billari, & Simao, 2005).

Research on the adaptive toolbox has identified a number of other heuristics, many of which can be decomposed into the three common building blocks:

1. *Search* rules specify what information is considered and in which order or direction information search proceeds.
2. *Stopping* rules specify when the search for information is terminated.
3. *Decision* rules specify how the final decision is reached.

Although there are many heuristics, they consist of a small number of building blocks, similar to how the chemical elements in the periodic table are built from a small number of particles.

Ecological Rationality

The traditional explanation for why people rely on heuristics is that heuristics reduce effort at the cost of accuracy. This hypothesis is called the *accuracy–effort trade-off*. In this view, heuristics are always second-best—in terms of accuracy, not effort. One of the first discoveries in our research was that this accuracy–effort trade-off is not generally true in situations of uncertainty (Goldstein & Gigerenzer, 2002; Gigerenzer & Brighton, 2009). For illustration, across 20 real-world problems, a simple sequential heuristic called take-the-best made on average more accurate predictions than did a multiple regression, and with less effort (Czerlinski, Gigerenzer, & Goldstein, 1999). We called the reversal of the accuracy–effort trade-off a *less-is-more effect*; determining the exact conditions under which less-is-more occurs is one part of the normative study of ecological rationality.

The normative study of heuristics is new. It was unthinkable as long as researchers believed in the generality of the accuracy–effort trade-off: that logic and probability theory are always optimal and heuristics always second best. This may be correct in situations of risk but not in those of uncertainty. The study of ecological rationality is a normative extension of Simon's descriptive question:

Ecological Rationality Question: How *Should* Human Beings Reason in Situations of Uncertainty?

To answer this question, it is no longer sufficient to rely on optimization. But what are the alternatives? Simon provided a clue with his scissors analogy: "Human rational behavior (and the rational behavior of all physical symbol systems) is shaped by a scissors whose two blades are the structure of task environments and the computational capabilities of the actor" (Simon, 1990, p. 7). By looking at only one blade, cognition, it is impossible to understand when and why a behavior succeeds or fails. As a consequence, in situations under uncertainty, rationality can be defined as the "match" between a heuristic and the structure of the environment: "A heuristic is ecologically rational to the degree it is adapted to the structure of the environment" (Gigerenzer & Todd, 1999, p. 13).

The study of the ecological rationality of a heuristic is conducted with the help of mathematical analysis and computer simulation. It aims at identifying the environmental structures that a heuristic exploits. These structures include:

1. Predictability: how well a criterion can be predicted.
2. Sample size: number of observations (relative to number of cues).
3. Number N of alternatives.
4. Redundancy: the correlation between cues.
5. Variability in weights: the distribution of the cue weights (e.g., skewed or uniform).

For instance, (1) the smaller the sample size, (2) the higher the redundancy, and (3) the higher the variability in weights are, the more accurate the predictions tend to be of sequential search heuristics such as take-the-best in comparison to linear regression models (Martignon & Hoffrage, 2002). Given these insights, normative statements can be made about when a heuristic will likely be better at making predictions than an alternative strategy such as a weighted linear model. The conditions of the ecological rationality of some heuristics have been well identified (e.g., Baucells, Carrasco, & Hogarth, 2008; Hogarth & Karelaia, 2006, 2007; Katsikopoulos, 2013; Katsikopoulos & Martignon, 2006; Simşek, 2013).

A complementary analysis of the ecological rationality of heuristics is based on the bias-variance dilemma (Geman, Bienenstock, & Doursat, 1992). In earlier research on cognitive illusions, the term *bias* refers to ignoring part of the information, as in the *base rate fallacy*. This can be captured in the equation:

Error = bias + ε   (1.1)

where ε is an irreducible random error. In the heuristics-and-biases view, if the bias is eliminated, good inferences are obtained. That is true in a world of known risks. In an uncertain world, however, there are three sources of errors:

Error = bias^2 + variance + ε   (1.2)

where *error* is the prediction error (the sum of squared error) and *bias* refers to a systematic deviation between a model and the true state, as in equation 1.1. To define the meaning of *variance*, consider 100 people who rely on the same strategy but have different samples of observations from the same population. Because of sampling error, the 100 inferences may not be the same. Across samples, *variance* is the expected squared deviation around their mean; *bias* is the difference between the mean prediction and the true state of nature. Thus, to minimize total error, a certain amount of bias is needed to counteract the error due to oversensitivity to sampling noise (see Gigerenzer, 2016, for formal definitions).

Variance decreases with increasing sample size but also with simpler strategies that have fewer free parameters (and less flexible functional forms) (Pitt, Myung, & Zhang, 2002). Thus, a cognitive system needs to draw a balance between being biased and flexible (variance) rather than simply trying to eliminate bias. Heuristics can be fast, frugal, and accurate by exploiting the structure of information in environments, by being robust, and by striking a good balance between bias and variance. This bias-variance dilemma helps to clarify the rationality of simple heuristics and why less can be more (Brighton & Gigerenzer, 2008).

Intuitive Design

Alongside Simon's question and its normative extension to the study of ecological rationality, there is a third question: How can one apply the answers to both questions to design heuristics that improve expert decision making? The key idea is to use the research on the adaptive toolbox to identify classes of strategies that are intuitive to the human mind. For instance, unlike logistic regressions and similar models, which are alien to the thinking of, say, most doctors and lawyers, sequential search heuristics such as take-the-best and fast-and-frugal trees are easy to understand and learn. The research on ecological rationality is then used to determine what kind of heuristic is likely to work for the problem at hand. For instance, fast-and-frugal trees that have been designed for coronary care unit allocations are intuitive for doctors, enable fast decisions, and are reported to make fewer errors in predicting heart attacks than standard logistic regression models used in heart attack diagnosis (Green & Mehr, 1997; Martignon, Vitouch, Takezawa, & Forster, 2003). Similarly, predicting bank vulnerability concerns the realm of uncertainty, not of calculable risk, and standard probability models such as value-at-risk have failed to predict every financial crisis and prevented none. My research group is currently cooperating with the Bank of England to develop simple heuristics for a safer world of finance, both for regulation and for predicting bank failures (Aikman et al., 2014; Neth, Meder, Kothiyal, & Gigerenzer, 2014). Under uncertainty, simplicity and bias are not signs of cognitive limitations but are vital to making better inferences.

Three Programs of Behavioral Economics

These three visions of bounded rationality have shaped the research questions posed in quite a few research programs, including social psychology, neuroeconomics, and behavioral economics.

The term *behavioral economics* appears to have been first used by George Katona (1951, 1980) and by Herbert Simon (Sent, 2004). Behavioral economics as we know it today, however, began in 1984, funded by the Alfred P. Sloan Foundation. In its early stages it was understood as an advancement of Simon's question (Heukelom, 2014). Yet Kahneman, Tversky, and Thaler eventually took over and abandoned Simon's revolutionary ideas in favor of a program with better promise to be accepted by neoclassical economists. Contrary to Simon, Kahneman and Tversky argued that there was nothing wrong with expected utility maximization; there was only something wrong with us humans. These researchers eventually determined the agenda of the emerging field of behavioral economics, to the dismay not only of Simon but notably of Vernon Smith. Smith (2003) put the term *ecological rationality* in the title of his Nobel lecture, drawing an explicit link to our work on the adaptive toolbox, and discussed this relation in his subsequent book (Smith, 2007). The history of the struggle over behavioral economics is told by Heukelom (2014). My point is this: as a closer look reveals, not one but three different kinds of behavioral economics exist, corresponding to the three versions of bounded rationality.

As-If Behavioral Economics

Behavioral economists typically justify their approach with the claim that it "increases the explanatory power of economics by providing it with more realistic psychological foundations" (Camerer & Loewenstein, 2004, p. 3). It is thus surprising that the first version of behavioral economics, which I call *as-if behavioral economics*, appears to be indistinguishable from neoclassical economics in its reliance on as-if models (Berg & Gigerenzer, 2010). The program of as-if behavioral economics is to keep the as-if framework of expected utility maximization untouched, and just add a few new "psychological" parameters to it.

Fehr and Schmidt's (1999) *inequity aversion theory* provides an example of as-if behavioral economics. It accounts for the observation that people care about others as well as themselves by adding two free parameters to the expected utility maximization theory, assuming that individuals behave *as if* they maximized a weighted function of various utilities. *Prospect theory* (Kahneman & Tversky, 1979; Tversky & Kahneman, 1992) similarly introduces additional adjustable parameters to the expected utility equation, including nonlinear transformations of probabilities and utilities, resulting in calculations that few would maintain are realistic models of people's decision processes. Laibson's (1997) model of impulsiveness consists, in essence, of adding a parameter to the neoclassical model of maximizing an exponentially weighted sum of instantaneous utilities. The decision maker is assumed to make an exhaustive search of all consumption sequences, compute the weighted sum of utility terms for each of these, and choose the one with the highest weighted utility.

These three prominent theories illustrate the first kind of behavioral economics: as-if models with parameters added, which allow a better fit, not necessarily better predictions (Brandstätter, Gigerenzer, & Hertwig, 2006). This program serves as a "repair program" of neoclassical economics (Berg & Gigerenzer, 2010; Gigerenzer, 2008, p. 90; Güth, 2008). It ignores both objectives formulated in Simon's question.

Anomalies Program

The second program of behavioral economics is based on the cognitive illusions program. In Camerer's (1995) words, "limits on computational ability force people to use simplified procedures or 'heuristics' that cause systematic mistakes (biases) in problem solving, judgment, and choice. The roots of this approach are in Simon's (1955) …" (p. 588). More bluntly, Diamond (2008) states: "Behavioral economics is the identification of circumstances where people are making 'mistakes.'" Thaler (1991, p. 138) argues that the major contribution of behavioral economics has been the discovery of a collection of cognitive illusions completely analogous to visual illusions. Similarly, Ariely (2008) argues that visual illusions such as Shepard's two-tables illusion show the deficiencies of our brain. Neither Thaler nor Ariely seems to realize that, without useful biases, we would perceive only what is on the retina. To borrow a phrase from Jerome Bruner, cognition has to go beyond the information given. In order to infer a three-dimensional world from a retinal image, the perceptual system makes intelligent bets. Every intelligent system has to make bets and therefore necessarily makes "good" errors (Gigerenzer, 2008). A system that makes no errors is simply not intelligent.

The anomalies program inherited two problems from the cognitive illusions program: questionable norms for good reasoning and a focus on vague labels instead of precise models of heuristics (e.g., Camerer, 1998; Conlisk, 1996; Kahneman, 2011). What is considered a "cognitive error" is not always so. Consider the first three items in Camerer's list of cognitive errors.

The first apparent illusion is called *miscalibration*. In a typical study people answer a series of trivia questions and provide a confidence judgment that each of their answers is correct. Then the average percentage correct is plotted against each confidence category. Typically, if confidence is 100%, only about 80% of the answers are correct; if confidence is 90%, only 70% are correct; and so on. This discrepancy between confidence and percentage correct, a so-called "miscalibration," has been attributed to systematic mental or motivational flaws in people's minds (e.g., Camerer, 1998; Lichtenstein, Fischhoff, & Phillips, 1982). A closer inspection shows that these "miscalibration" curves look like regression to the mean, and several experiments have shown that this is indeed largely the case. Regression to the mean is a consequence of *unsystematic* error, and if one plots the data the other way around, one gets the other regression curve, which now looks like *underconfidence*. Regression to the mean is generated by unsystematic error, not by a systematic error such as overconfidence. Thus, the problem is largely a systematic error *by researchers*, not one of *ordinary people* (Dawes & Mulford, 1996; Erev, Wallsten, & Budescu, 1994; Gigerenzer, Fiedler, & Olsson, 2012).

The second error on the list is that the confidence intervals people construct are too small. In this research, people are asked to estimate the 95% confidence interval for, say, the length of the river Nile. A person might respond, "between 2,000 and 8,000 kilometers." The typical result reported is that in more than 5% of the cases the true estimate is outside the confidence interval. This phenomenon is attributed to overconfidence. Again, a bit of statistical thinking by the researchers would have helped. Juslin, Winman, and Hansson (2007) noted that when inferences are drawn from samples, statistical theory implies that a systematic error will appear in variance and therefore in the production of confidence intervals, whereas the mean (probability) is an unbiased estimate. In experiments they showed that when people are asked to provide the probability (mean) that the true value lies in the interval, the error largely disappears. In our example that means asking the question, "What is the probability that the length of the river Nile is between 2,000 and 8,000 kilometers?"

The third form of "overconfidence" listed is the observation that average confidence judgment is higher than average percentage correct. Long ago, we (Gigerenzer, Hoffrage, & Kleinbölting, 1991) showed that this difference disappears when experimenters use randomly sampled questions rather than selecting questions with surprising answers (such as that New York is south of Rome). After reviewing 135 studies, Juslin, Winman, and Olsson (2000) found that overconfidence was observed when questions were selected but was on average about zero with random sampling. The bias thus lies in the biased sampling of researchers, not simply in people's minds.

I could continue to review the list of so-called errors, but the point should be sufficiently clear. Quite a few "biases" actually appear to originate from researchers' lack of statistical thinking rather than, as claimed, from people's systematic errors (Gigerenzer, 2001; Gigerenzer, Fiedler, & Olsson, 2012). This is not to say that biases never exist. For instance, most of the research reported in the previous paragraphs and elsewhere contradicting the heuristics-and-biases story is surprisingly absent in the behavioral economics literature from Camerer (1998) to Kahneman (2011). That is known as a confirmation bias.

Toward an Ecological Rationality Program of Behavioral Economics

The third program of behavioral economics is an elaboration of Simon's question. The goal is to enrich economic theory by progressing from as-if models to process models while simultaneously moving from labels for heuristics to testable models and from logical rationality to ecological rationality. As a first step there is a need to develop process models that address systematic violations of expected utility maximization. Tversky's (1972) elimination by aspects described such a process model and was a move away from optimizing models in the very spirit of Simon's question. Unfortunately, he abandoned this program when joining Kahneman and turned to the repair program. Meanwhile, a quite successful process model has been proposed—the priority heuristic (Brandstätter, Gigerenzer, & Hertwig, 2006), which has been shown to outperform cumulative prospect theory in predicting choices between gambles. Moreover, without a single free parameter, it logically implies many of the so-called anomalies, including the Allais paradox and the fourfold pattern of risk attitude (Katsikopoulos & Gigerenzer, 2008). The second goal is to derive better norms, that is, norms that can guide decision making under uncertainty. The adaptation of the ecological rationality program to behavioral economics is practically nonexistent at the moment. As a consequence, behavioral economics bears a much greater similarity to neoclassical economics than to its original goal of a realistic account of decision processes (Berg & Gigerenzer, 2010).

Rationality in the Plural

It is sometimes said that there is only one way to be rational but several ways to be irrational. The only rational way is optimization. I hope to have made clear that this statement is much too narrow. In situations of uncertainty there is no single best answer, but there are several reasonable ways to behave in a rational way. The idea that there is one way to be rational is itself an illusion or, if you like, one of the many ways to be irrational. Simon gave us an idea of how to fare better when confronted with uncertainty as opposed to calculable risk.

References

Aikman, D., Galesic, M., Gigerenzer, G., Kapadia, S., Katsikopoulos, K., Kothiyal, A., et al. (2014, May 2). *Taking uncertainty seriously: Simplicity versus complexity in financial regulation.* Bank of England Financial Stability Paper No. 2028. Available at SSRN: http://ssrn.com/abstract=2432137. doi: 10.2139/ssrn.2432137

Ariely, D. (2008). *Predictably irrational: The hidden forces that shape our decisions.* New York: HarperCollins.

Arrow, K. J. (2004). Is bounded rationality unboundedly rational? Some ruminations. In M. Augier & J. G. March (Eds.), *Models of a man: Essays in memory of Herbert A. Simon* (pp. 47–55). Cambridge, MA: MIT Press.

Artinger, F., & Gigerenzer, G. (2014). Aspiration-adaptation, price setting, and the used-car market. Unpublished manuscript.

Baucells, M., Carrasco, J. A., & Hogarth, R. M. (2008). Cumulative dominance and heuristic performance in binary multi-attribute choice. *Operations Research, 56,* 1289–1304.

Berg, N., & Gigerenzer, G. (2010). As-if behavioral economics: Neoclassical economics in disguise? *History of Economic Ideas, 18,* 133–165. doi:10.1400/140334.

Brandstätter, E., Gigerenzer, G., & Hertwig, R. (2006). The priority heuristic: Making choices without trade-offs. *Psychological Review, 113,* 409–432. doi:10.1037/0033-295X.113.2.409.

Brighton, H., & Gigerenzer, G. (2008). Bayesian brains and cognitive mechanisms: Harmony or dissonance? In N. Chater & M. Oaksford (Eds.), *The probabilistic mind: Prospects for Bayesian cognitive science* (pp. 189–208). New York: Oxford University Press.

Callebaut, W. (2007). Simon's silent revolution. *Biological Theory, 2,* 76–86. doi:10.1162/biot.2007.2.1.76.

Camerer, C. (1995). Individual decision making. In J. H. Kagel & A. E. Roth (Eds.), *The handbook of experimental economics* (pp. 587–703). Princeton, NJ: Princeton University Press.

Camerer, C. F. (1998). Bounded rationality in individual decision making. *Experimental Economics, 1,* 163–183. doi:10.1007/BF01669302.

Camerer, C., & Loewenstein, G. F. (2004). *Behavioral economics: Past, present, future.* Princeton, NJ: Princeton University Press.

Conlisk, J. (1996). Why bounded rationality? *Journal of Economic Literature, 34,* 669–700.

Czerlinski, J., Gigerenzer, G., & Goldstein, D. G. (1999). How good are simple heuristics? In G. Gigerenzer, P. M. Todd, & the ABC Research Group, *Simple heuristics that make us smart* (pp. 97–118). New York: Oxford University Press.

Dawes, R. M., & Mulford, M. (1996). The false consensus effect and overconfidence: Flaws in judgment, or flaws in how we study judgment? *Organizational Behavior and Human Decision Processes, 65,* 201–211. doi:10.1006/obhd.1996.0020.

Diamond, P. (2008). Behavioral economics. *Journal of Public Economics, 92*(8), 1858–1862.

Erev, I., Wallsten, T. S., & Budescu, D. V. (1994). Simultaneous over- and underconfidence: The role of error in judgment processes. *Psychological Review, 101,* 519–527. doi:10.1037/0033-295X.101.3.519.

Fehr, E., & Schmidt, K. (1999). A theory of fairness, competition, and cooperation. *Quarterly Journal of Economics, 114,* 817–868. doi:10.1162/003355399556151.

Geman, S. E., Bienenstock, E., & Doursat, R. (1992). Neural networks and the bias/variance dilemma. *Neural Computation, 4,* 1–58. doi:10.1162/neco.1992.4.1.1.

Gigerenzer, G. (1996). On narrow norms and vague heuristics: A reply to Kahneman and Tversky (1996). *Psychological Review, 103,* 592–596. doi:10.1037/0033-295X.103.3.592.

Gigerenzer, G. (2000). *Adaptive thinking: Rationality in the real world.* New York: Oxford University Press.

Gigerenzer, G. (2001). Content-blind norms, no norms, or good norms? A reply to Vranas. *Cognition, 81,* 93–103. doi:10.1016/S0010-0277(00)00135-9.

Gigerenzer, G. (2004). Striking a blow for sanity in theories of rationality. In M. Augier & J. G. March (Eds.), *Models of a man: Essays in honor of Herbert A. Simon* (pp. 389–409). Cambridge, MA: MIT Press.

Gigerenzer, G. (2008). *Rationality for mortals.* New York: Oxford University Press.

Gigerenzer, G. (2016). Towards a rational theory of heuristics. In R. Frantz & L. Marsh (Eds.), *Minds, models, and milieu. Commemorating the centenary of Herbert Simon's birth.* Basingstoke, UK: Palgrave Macmillan.

Gigerenzer, G., & Brighton, H. (2009). *Homo heuristicus*: Why biased minds make better inferences. *Topics in Cognitive Science, 1,* 107–143. doi:10.1111/j.1756-8765.2008.01006.x.

Gigerenzer, G., Fiedler, K., & Olsson, H. (2012). Rethinking cognitive biases as environmental consequences. In P. M. Todd, G. Gigerenzer, & the ABC Research Group, *Ecological rationality: Intelligence in the world* (pp. 80–110). New York: Oxford University Press.

Gigerenzer, G., Hertwig, R., & Pachur, T. (Eds.). (2011). *Heuristics: The foundations of adaptive behavior.* New York: Oxford University Press.

Gigerenzer, G., Hoffrage, U., & Kleinbölting, H. (1991). Probabilistic mental models: A Brunswikian theory of confidence. *Psychological Review, 98,* 506–528. doi:10.1037/0033-295X.98.4.506.

Gigerenzer, G., & Selten, R. (Eds.). (2001). *Bounded rationality: The adaptive toolbox.* Cambridge, MA: MIT Press.

Gigerenzer, G., & Todd, P. M. (1999). Fast and frugal heuristics: The adaptive toolbox. In G. Gigerenzer, P. M. Todd, & the ABC Research Group, *Simple heuristics that make us smart* (pp. 3–34). New York: Oxford University Press.

Gigerenzer, G., Todd, P. M., & the ABC Research Group. (1999). *Simple heuristics that make us smart*. New York: Oxford University Press.

Goldstein, D. G., & Gigerenzer, G. (2002). Models of ecological rationality: The recognition heuristic. *Psychological Review, 109*, 75–90. doi:10.1037/0033-295X.109.1.75.

Green, L. A., & Mehr, D. R. (1997). What alters physicians' decisions to admit to the coronary care unit? *Journal of Family Practice, 45*, 219–226.

Güth, W. (2008). (Non)behavioral economics: A programmatic assessment. *Zeitschrift für Psychologie. Journal of Psychology, 216*(4), 244.

Hertwig, R., Hoffrage, U., & the ABC Research Group. (2013). *Simple heuristics in a social world*. New York: Oxford University Press.

Heukelom, F. (2014). *Behavioral economics: A history*. Cambridge, UK: Cambridge University Press.

Hogarth, R. M., & Karelaia, N. (2006). "Take-the-best" and other simple strategies: Why and when they work "well" with binary cues. *Theory and Decision, 61*, 205–249. doi:10.1007/s11238-006-9000-8.

Hogarth, R. M., & Karelaia, N. (2007). Heuristic and linear models of judgment: Matching rules and environments. *Psychological Review, 114*, 733–758. doi:10.1037/0033-295X.114.3.733.

Juslin, P., Winman, A., & Hansson, P. (2007). The naïve intuitive statistician: A naïve sampling model of intuitive confidence intervals. *Psychological Review, 114*, 678–703. doi:10.1037/0033-295X.114.3.678.

Juslin, P., Winman, A., & Olsson, H. (2000). Naive empiricism and dogmatism in confidence research: A critical examination of the hard-easy effect. *Psychological Review, 107*, 384–396. doi:10.1037/0033-295X.107.2.384.

Kahneman, D. (2003). Maps of bounded rationality: A perspective on intuitive judgment and choice. In T. Frangsmyr (Ed.), *Les Prix Nobel: The Nobel Prizes 2002* (pp. 1449–1489). Stockholm: Nobel Foundation.

Kahneman, D. (2011). *Thinking fast and slow*. London: Allen Lane.

Kahneman, D., Slovic, P., & Tversky, A. (Eds.). (1982). *Judgment under uncertainty: Heuristics and biases*. Cambridge, UK: Cambridge University Press.

Kahneman, D., & Tversky, A. (1979). Prospect theory: An analysis of decision under risk. *Econometrica, 47*, 263–291. doi:10.2307/1914185.

Katsikopoulos, K. (2013). Why do simple heuristics perform well in choices with binary attributes? *Decision Analysis, 10*, 327–340.

Katsikopoulos, K. V., & Gigerenzer, G. (2008). One-reason decision-making: Modeling violations of expected utility theory. *Journal of Risk and Uncertainty, 37*, 35–56. doi:10.1007/s11166-008-9042-0.

Katsikopoulos, K. V., & Martignon, L. (2006). Naive heuristics for paired comparisons: Some results on their relative accuracy. *Journal of Mathematical Psychology, 50,* 488–494. doi:10.1016/j.jmp.2006.06.001.

Katona, G. (1951). *Psychological analysis of economic behavior.* New York: McGraw-Hill.

Katona, G. (1980). *Essays on behavioral economics.* Ann Arbor: Survey Research Centers, Institute for Social Research, The University of Michigan.

Knight, F. (1921). *Risk, uncertainty and profit* (Vol. XXXI). Boston: Houghton Mifflin.

Laibson, D. (1997). Golden eggs and hyperbolic discounting. *Quarterly Journal of Economics, 112*(2), 443–477.

Levitt, S. D., & Syverson, C. (2008). Market distortions when agents are better informed: The value of information in real estate transactions. *Review of Economics and Statistics, 90*(4), 599–611. doi:10.1162/rest.90.4.599.

Lichtenstein, S., Fischhoff, B., & Phillips, L. D. (1982). Calibration of probabilities: The state of the art to 1980. In D. Kahneman, P. Slovic, & A. Tversky (Eds.), *Judgment under uncertainty: Heuristics and biases* (pp. 306–334). Cambridge, UK: Cambridge University Press.

Lopes, L. L. (1992). Three misleading assumptions in the customary rhetoric of the bias literature. *Theory & Psychology, 2,* 231–236. doi:10.1177/0959354392022010.

Martignon, L., & Hoffrage, U. (2002). Fast, frugal, and fit: Lexicographic heuristics for paired comparison. *Theory and Decision, 52,* 29–71. doi:10.1023/A:1015516217425.

Martignon, L., Vitouch, O., Takezawa, M., & Forster, M. R. (2003). Naive and yet enlightened: From natural frequencies to fast and frugal decision trees. In D. Hardman & L. Macchi (Eds.), *Thinking: Psychological perspectives on reasoning, judgment and decision making* (pp. 189–211). Chichester, UK: Wiley.

Neth, H., Meder, B., Kothiyal, A., & Gigerenzer, G. (2014). *Homo heuristicus* in the financial world: From risk management to managing uncertainty. *Journal of Risk Management in Financial Institutions, 7*(2), 134–144.

Pitt, M. A., Myung, I. J., & Zhang, S. (2002). Toward a method of selecting among computational models of cognition. *Psychological Review, 109,* 472–491.

Samuelson, P. A. (1938). A note on the pure theory of consumers' behavior. *Economica, 5,* 61–71.

Sargent, T. J. (1993). *Bounded rationality in macroeconomics.* New York: Oxford University Press.

Savage, L. J. (1954). *The foundations of statistics.* New York: Wiley.

Selten, R. (1998). Aspiration adaptation theory. *Journal of Mathematical Psychology, 42,* 191–214.

Selten, R. (2001). What is bounded rationality? In G. Gigerenzer & R. Selten (Eds.), *Bounded rationality: The adaptive toolbox* (pp. 13–36). Cambridge, MA: MIT Press.

Sen, A. (1987). Rational behaviour. In J. Eatwell, M. Milgate, P. Newman, & I. Erev (Eds.), *The new Palgrave dictionary of economics* (Vol. 4, pp. 68–76). London: Macmillan.

Sent, E.-M. (2004). Behavioral economics: How psychology made its (limited) way back into economics. *History of Political Economy, 36*(4), 735–760.

Simon, H. A. (1947). *Administrative behavior: A study of decision-making processes in administrative organizations.* New York: Macmillan.

Simon, H. A. (1955). A behavioral model of rational choice. *Quarterly Journal of Economics, 69,* 99–118. doi:10.2307/1884852.

Simon, H. A. (1956). Rational choice and the structure of the environment. *Psychological Review, 63,* 129–138. doi:10.1037/h0042769.

Simon, H. A. (1979). Rational decision making in business organizations. *American Economic Review, 69,* 493–513.

Simon, H. A. (1985). Human nature in politics: The dialogue of psychology and political science. *American Political Science Review, 79,* 293–304.

Simon, H. A. (1988). Nobel laureate Simon "looks back": A low-frequency mode. *Public Administration Quarterly, 12,* 275–300.

Simon, H. A. (1989). The scientist as problem solver. In D. Klahr & K. Kotovsky (Eds.), *Complex information processing: The impact of Herbert A. Simon* (pp. 375–398). Hillsdale, NJ: Lawrence Erlbaum Associates.

Simon, H. A. (1990). Invariants of human behavior. *Annual Review of Psychology, 41,* 1–19. doi:10.1146/annurev.ps.41.020190.000245.

Simon, H. A. (1991). Cognitive architectures and rational analysis: Comment. In K. V. Lehn (Ed.), *Architectures for intelligence* (pp. 25–39). Hillsdale, NJ: Lawrence Erlbaum Associates.

Şimşek, Ö. (2013). Linear decision rule as aspiration for simple decision heuristics. *Advances in Neural Information Processing Systems, 26,* 2904–2912.

Smith, V. L. (2003). Constructivist and ecological rationality in economics. *American Economic Review, 93,* 465–508. doi:10.1257/000282803322156954.

Smith, V. L. (2007). *Rationality in economics: Constructivist and ecological forms.* Cambridge: Cambridge University Press.

Stigler, G. J. (1961). The economics of information. *Journal of Political Economy, 69,* 213–225. doi:10.1086/258464.

Thaler, R. H. (1991). *Quasi rational economics.* New York: Russell Sage Foundation.

Todd, P. M., Billari, F. C., & Simao, J. (2005). Aggregate age-at-marriage patterns from individual mate-search heuristics. *Demography, 42,* 559–574.

Todd, P. M., Gigerenzer, G., & the ABC Reseach Group. (2012). *Ecological rationality: Intelligence in the world*. New York: Oxford University Press.

Todd, P. M., & Miller, G. F. (1999). From pride and prejudice to persuasion: Satisficing in mate search. In G. Gigerenzer, P. M. Todd, & the ABC Research Group, *Simple heuristics that make us smart* (pp. 287–308). New York: Oxford University Press.

Tversky, A. (1972). Elimination by aspects: A theory of choice. *Psychological Review, 79*, 281–299. doi:10.1037/h0032955.

Tversky, A., & Kahneman, D. (1992). Advances in prospect theory: Cumulative representation of uncertainty. *Journal of Risk and Uncertainty, 5*, 297–323. doi:10.1007/BF00122574.

2 Grounded Rationality and the New Paradigm Psychology of Reasoning

Shira Elqayam

When I ask my students what it means to be rational, there is always someone who suggests that to be rational is to be logical. This position is not all that different from that of the traditional paradigm in thinking and reasoning, variously dubbed *logicist* (Oaksford & Chater, 1991), *binary* (Over, 2009), or *the deduction paradigm* (Evans, 2002). The traditional paradigm regarded human reasoning as an offshoot of classical extensional logic, the sort of logic taught in textbooks, in which there are only two truth values—true or false. Although the traditional paradigm came in many shades and guises, the one common denominator was the status of logic as a normative system. People were considered rational precisely to the degree with which their reasoning and thinking agreed with the standards of classical logic.[1]

In recent times the field has moved on from this idea, resulting in what has been dubbed *the new paradigm* (Manktelow, Over, & Elqayam, 2011; Over, 2009; for a brief introduction see Elqayam & Over, 2013, and the contributions to the special issue, and Over, chapter 4, this volume). The new paradigm is a family of theories sharing some common ideas and working postulates. The hard core is a move away from classical logic and logicism and toward a more decision-theoretic view of human reasoning. Put simply, new-paradigm researchers acknowledge that when we reason, what we do is make decisions. From this follow a number of shared postulates such as the importance of the two key factors in decision making—probability and utility. Despite the growing consensus on a number of key issues, though, there are also a number of unresolved issues and disputes within the new paradigm (Evans, 2012). One of the central ones is this. Granted that classical logic is no longer considered a viable normative system, should it be replaced by another normative system—one that is more suitable to the conceptualization of reasoning as decision-theoretic—or is psychology of reasoning better off moving away from normative research questions altogether? My aims in this chapter are to present and defend the latter option.

The rest of this chapter is structured as follows. I set the stage with a brief introduction to the new paradigm, highlighting the shared views and values of the field. I then proceed to the two opposing approaches to the normativity question within the new paradigm—normativist Bayesianism versus descriptivism. From there I move on to describe grounded rationality as a special case of descriptivism. Last, I argue that, as the new paradigm moves toward a richer conception of rationality, grounded rationality provides a particularly attractive option.

New Paradigm Psychology of Reasoning

Although precursors to the new paradigm date back to the 1980s and 1990s (e.g., Evans, 1989; Oaksford & Chater, 1998; Manktelow & Over, 1991), it is mostly this century that has witnessed a full-blown paradigm shift with (arguably) truly Kuhnian dimensions (Evans & Over, 2004; Oaksford & Chater, 2007; Stanovich, 2004, to name but a few). The new paradigm has been variably dubbed *probabilistic*, *Bayesian*, and *decision-theoretic*. The last label is probably the broadest and provides the best fit with the range of theories under this shared umbrella. The idea is that reasoning is a branch of decision making and shares a research toolbox with this discipline—normative systems, theories, and experimental paradigms. The appellation "Bayesian" derives from the classic formal and normative system for decision making, although the term covers a range of rather different approaches (Elqayam & Evans, 2013; Jones & Love, 2011).

The same parameters that shape decision making, then, also shape reasoning. Reasoning is affected by how probable an event is—or, more often, by how probable a *conditional* event is. The new paradigm maintains the interest in conditionals—sentences of the form "*if p then q*"—introduced in the early days of the traditional paradigm, but where the traditional paradigm focused on truth and falsity, the new paradigm replaces this with a focus on conditional probability, the probability of *q* given *p*, or $P(q|p)$. The often-cited Ramsey test (Ramsey, 1931) serves as a philosophical as well as psychological starting point for many new paradigm approaches. The idea is that when we judge a conditional, if *p* then *q*, we do so by adding *p* hypothetically to our knowledge and judging *q* in this context. To take a classic example (Over, Hadjichristidis, Evans, Handley, & Sloman, 2007), if we need to evaluate the sentence "If global warming continues, London will be flooded," we do this by hypothetically supposing that global warming continues and evaluating the probability that London will be flooded under this supposition. From this it follows that the probability of a conditional is the conditional probability; that is, the probability of the global warming conditional is the probability that London will be flooded given that global warming continues. This can be expressed as $P(if\ p\ then\ q) = P(q|p)$, or, in our case, P(If global warming continues, London will be flooded) = P(London will be flooded | global warming continues). This relationship is so important that it has been dubbed simply *the Equation* (Edgington, 1995). The overwhelming psychological evidence shows that untutored reasoners generally conform to the Equation in their judgments (e.g., Douven & Verbrugge, 2010;

Evans, Handley, & Over, 2003; Fugard, Pfeifer, Mayerhofer, & Kleiter, 2011; Oberauer & Wilhelm, 2003; Over et al., 2007).

In the old paradigm, where premises were considered either true or false, validity was a matter of truth preservation: a classically valid inference is one in which, if the premises are true, the conclusion is necessarily true as well. But in everyday life we normally hold beliefs with a degree of uncertainty. We might be pretty confident that global warming will continue within the next five years, but not so sure that we can assign it a probability of 1. In the new paradigm (Chater & Oaksford, 1999; Oaksford & Chater, 2007; Oaksford, Chater, & Larkin, 2000; Over, 2009; Pfeifer & Kleiter, 2010), the focus has therefore shifted to Adams's (1998) notion of *probabilistic validity*, or *p*-validity. Here the idea is that validity is determined by the comparing the probability of the premise(s) to that of the conclusion. An argument is *p*-valid if and only if it cannot coherently[2] increase uncertainty: to put it informally (Elqayam & Over, 2012, 2013), a *p*-valid argument cannot coherently lead from a believable (roughly, probable) premise to an unbelievable (roughly, improbable) conclusion. Tversky and Kahneman's (1983) conjunction fallacy is sometimes evoked as an illustration (Gilio & Over, 2012; Elqayam & Over, 2012; Evans & Over, 2013; Over, chapter 4, this volume): given that Linda is a feminist, one cannot coherently infer that Linda is both a feminist and a banker, because the probability of Linda being both a banker and a feminist cannot be higher than the probability of Linda being a feminist, whether she is a banker or not. Unlike the logicist notion of validity, which is truth-preserving, *p*-validity preserves confidence—a much more psychological notion (Evans & Over, 2013).

Although probabilism is a key feature, there is more to the new paradigm than just probability; it is also affected by how desirable or undesirable we find various events—"utility" in the technical terminology of decision theory. In particular, *deontic thinking*—thinking about rules and regulations—is strongly related to the costs and benefits of actions and their outcomes (Manktelow & Over, 1991; Oaksford & Chater, 1994). For example, *ceteris paribus*, conditional promises are considered more effective when the cost to the agent is low and the benefit is high (Evans, Neilens, Handley, & Over, 2008): "If you do your homework tonight, I will buy you that bike you wanted" is more effective than "If you do your homework every night, I'll buy you a chocolate." Similarly, people tend to generate new normative rules to direct behavior when the outcome is desirable: Given a premise such as "If Martin uses Fontignani olive oil, his recipe will taste better," people tend to infer that Martin should use the Fontignani olive oil (Elqayam, Thompson, Wilkinson, Evans, & Over, 2015). Typically social pragmatics is thrown into the pot (e.g., Bonnefon, 2009, 2013), not least because cost and benefit often depend on social interactions; for example, if a mother tells her child "If you tidy your room, then you may go out to play," the utility for the mother is in the tidy room, whereas the utility for the child is being allowed to play (Manktelow & Over, 1991). Such cases fall under the general heading of *utility conditionals*

(Bonnefon, 2009)—conditionals involving costs and benefits, such as promises, persuasions (Thompson, Evans, & Handley, 2005), and so on, which are sensitive to the valence and magnitude of utility involved.

A more recent development within the new paradigm is the increasing integration of dual processing perspectives (Evans, 2008) so that probability, utility, and dual processing are now considered to be the "three pillars" of the new paradigm (Bonnefon, 2013). Dual process theories distinguish between type-1 processes—fast, intuitive, and resource-frugal—and type-2 processes—effortful, deliberative, and correlated with general ability. Dual process theories of reasoning date back to the traditional paradigm, where typically type-2 processes were thought to be identical to classical logic (Wason & Evans, 1975; Evans, Barston, & Pollard, 1983). This might explain why it took dual processing theories relatively long to be assimilated into the new paradigm. Present-day theorists, however, acknowledge this to be an oversimplification: there is no one-to-one parallel between processing and any normative system, logic or probability (Evans, 2008; Evans & Stanovich, 2013; Stanovich, 2011).

New paradigm theorists now resolve the apparent conflict by drawing on Marr's (1982) seminal distinction between levels of analysis, akin to the veteran distinction between process and product, and the Chomskyan competence-performance distinction (Chomsky, 1965). Marr's computational level of analysis, the *what* and *why* level, deals with the function computed by the system. For example, a pocket calculator computes algebra. The algorithmic level, the *how* level, is analysis of processes and representations: for example, the software of a pocket calculator. Dual processing theories provide an analysis on the algorithmic level, whereas the function computed by the system is probabilistic and utilitarian (Elqayam & Over, 2012; Oaksford & Chater, 2011, 2012, 2014). The consensus within the new paradigm, then, is that the function computed by the cognitive system—its computation—is Bayesian. The debate remains open, however, whether this computational system should also be considered normative, which is the focus of the next section.

Norm or Description?

"There is some absolute sense in which some reasoning or decision making is good, and other reasoning and decision making is bad" (Oaksford & Chater, 2007, p. 25). Some readers might be surprised by the use of such strong evaluative language in empirical science. Indeed, theories of reasoning and decision making seem to be fairly unique in cognitive psychology in their strong emphasis on issues of right and wrong, good versus bad (Evans, 2014a; although cf. Wilkinson, 2014, on normativity in theory of mind research). I use the terminology that Jonathan Evans and I coined in a recent open peer commentary treatment (Elqayam & Evans, 2011; Evans & Elqayam, 2011; collectively *E&E*) and call such approaches *normativist*. Normativism is the idea that human thinking both reflects some sort of normative standard and should conform to it (empirical and prescriptive normativism, respectively). Prominent new paradigm theories that support normativism include Stanovich and West's (Stanovich, 1999; Stanovich & West, 2000) individual differences research program and Oaksford and Chater's (1998, 2007) Bayesian rationality.

The issue with normativism is that, paradoxically, it has to draw on a non-normative type of inference. The problem starts with arbitration between normative systems. After more than half a century of contentious reasoning and decision making research, it is no secret that for any single empirical paradigm, there are at least two competing normative systems suggested as relevant in the literature. Examples are legion, and here I review only one: the (in)famous Wason selection task (Wason, 1966). In one version of this task, participants are presented with four cards, each with a letter on one side and a number of the other side, such as G, D, 4, and 8. They are also given a conditional rule such as "If there is a G on one side of the card, then there is a 4 on the other side." They are told that the rule applies to the cards and can be true or false, and they are instructed to turn over the cards—and only the cards—that can test if the rule is true or false. The task in this version is notoriously difficult: typical responses are the cards named in the task (Evans, 1972), in this case G and 4 (or sometimes just G). Only about 10% of participants provide the traditionally *normative* response (G and 8 in this case). However, this norm is based on extensional logic. In a classic contribution that shaped the new paradigm, Oaksford and Chater (1994) reanalyzed the task with Bayesian tools, proposing that participants were trying to extract the best information they could. Although not conforming to classical logic, participants were displaying *Bayesian* rationality—and being perfectly normative at that. This soon led to an intense debate on the nature of the appropriate normative system, with Oaksford and Chater basing their analyses on modal responses, and, from the other side of the debate, Stanovich and West (Stanovich, 1999; Stanovich & West, 2000) proposing that norms should be based on responses from the "best" reasoners—the cognitively able, motivated reasoners who produce the classically normative response.

What both sides of this normative debate were doing was basing their argument in favor of their preferred normative system on empirical data. This approach is all very well when one is arbitrating between *descriptive* theories: the inference is from descriptive data to descriptive theory. But if the same principle is used for arbitrating between normative theories, the result is inference from descriptions to norms, or, in perhaps more familiar terms, an "is-ought" inference. First identified by Hume (2000/1739–1740), *is-ought inference* is said to occur when an argument includes deontic terms such as *should* or *must* in the conclusion but none in the premises. For example, the notorious "survival of the fittest" of social Darwinism is based on is-ought inference, (or, as it is called in this context, the naturalistic fallacy; Moore, 1903): might ("is") is not necessarily right ("ought"). Whether inference from *is* to *ought* is invariably a fallacy is a matter for some intensive (and still ongoing) debate in the metaethics literature (Hudson, 1969; Pigden, 2010; Schurz, 1997). At best we can say that the inference is problematic and has been defended for only some well-defined special cases. Whether these cover the sort of is-ought inference committed by normativist theories is moot.

In a universe of multiple and competing theoretical norms, then, adjudicating between norms is as crucial as it is tricky. When normative theories resort to supporting their normative arguments with empirical data, they arguably commit a controversial sort of inference. The inherent paradox of normativism is that it is enabled by a non-normative inference. In E&E we therefore proposed that psychology of reasoning and decision making will be better served by a move toward a *descriptivist* paradigm, focusing on what reasoners *actually* do, rather than what they *should* do.

Bayesianism as Viewed from a Descriptivist Place

It might seem, prima facie, that a descriptivist position pulls the rug from under the new paradigm; after all, the new paradigm is firmly Bayesian. Like other formal systems in reasoning and decision making, Bayesianism doubles as a normative system and a computational-level theory; Bayesian descriptivism, then, sounds too much like an oxymoron. Giving up on Bayesian principles altogether is far too high a price: not only is it the foundation of the new paradigm, a dialogue with Bayesian principles has proved to be immensely productive in the history of reasoning and decision-making research (Hahn, 2014). Impossible as it sounds, though, we can eat the cake and keep it: maintain the indisputable advantages of Bayesian principles and still stick to a descriptivist framework. The key word is "doubles"—Bayesian formal systems can indeed be interpreted as normative as well as computational-level explanations, but these functions are separable. It helps in this context to distinguish between "soft" and "strict" Bayesian approaches (Elqayam & Evans, 2013), both of which are represented within the new paradigm. Soft Bayesianism is Bayesianism with a pinch of salt: unlike the stricter sort of Bayesianism, it adopts broad Bayesian principles such as subjectivity, uncertainty, and degrees of belief, but without committing to the normative strictures associated with systems such as subjective expected utility, the classic normative model of Bayesian decision making. Nor does soft Bayesianism postulate any strong coherence among the normative, the computational, and the algorithmic levels of analysis.

In E&E and its aftermath, we found it rather difficult to get across the distinction between computational and normative analysis (Buckwalter & Stich, 2011; Evans & Elqayam, 2011). The confusion can be traced back to a long tradition in the rationality debate, going back at least as far as Cohen (1981), which conflates normative analysis with Chomskyan competence (analogous, let us remember, to Marr's computational level of analysis). Part of the problem is Chomsky's unfortunate choice of term "competence," which can lead to misinterpretation of Chomskyan competence as synonymous with "skill" or "aptitude"; so, for example, speaker A might supposedly have better linguistic competence than speaker B. But Chomskyan competence is no more and no less than the abstract set of linguistic rules that every native speaker of a language possesses. Linguistic competence is a given in healthy adult native speakers. No one is more or less linguistically competent in a language than anyone else, unless he or she is too young to have developed it, has acquired it as a second language, or suffers from neurological damage such as aphasia.

By now it is not simple to convince reasoning and decision-making theorists that the two types of analysis can (and should!) be separated, but it is worth a try. My favorite example is that of Afro-American Vernacular English (AAVE), the version of English spoken mostly by black people in the United States (a.k.a. Black English Vernacular). For the educationalist, AAVE is a nonstandard version of English, that is to say, full of errors such as double negation. It is something to be rooted out, so that AAVE speakers in American schools can be taught "proper" "Good English" or Standard English. When a California school board announced in 1996 that it was going to recognize the status of AAVE (the language, it should be noted, spoken by most of its students), train teachers to look at it objectively, and allow its use in the classroom, a media storm erupted, ridiculing the school board for encouraging the supposedly inferior *"Ebonics"* (this is the derogatory term for AAVE; see Pullum, 1999, for an engaging and illuminating analysis).

For the linguist, however, the difference between dialects is a matter for description, not prescription. That AAVE is indisputably different from Standard English does not make it inferior, merely a different language system. Labov's (1972) classic study of AAVE (which he still calls *Black English Vernacular* or BEV) examines a range of grammatical evidence to conclude that AAVE/BEV systematically differs from other dialects of English not only in its surface structure but also in deep structure grammatical rules, most strikingly in the use of double negation. Labov points out that the double negation rule was introduced into English by prescriptive grammarians as late as the eighteenth century and that Standard English sentences such as "He doesn't know anything" already contain an implicit negation in the "anything." AAVE's "He don't know nothing" no more than does with the "nothing" explicitly what Standard English does with the "anything" implicitly. The use of double negation in AAVE, then, although socially stigmatized, cannot be attributed to any "illogicality" any more than the use of double negation in Chaucer or in "negative concord" languages such as Spanish or Hebrew (where double negation is used for emphatic purposes) is a mark of illogicality. As Pullum (1999) posits, African-American Vernacular English is not Standard English with mistakes.

More generally, a tradition going back to De Saussure (1916/1966), one of the founders of modern linguistics, firmly rejects the normative study of language in favor of descriptive analysis. Linguistics undergraduates are taught from day one that the business of the linguist is to study language, not to regulate it. As an influential textbook (Fromkin, Hyams, & Rodman, 2014, p. 290) put it: "No dialect … is more expressive, less corrupt, more logical, more complex, or more regular than any other dialect or language. … Any judgments … as to the superiority or inferiority of a particular dialect or language are social judgments, which have no linguistic or scientific basis." This clearly descriptivist stance, deeply rooted in linguistic tradition, has proved immensely productive. In the case of AAVE, for example, Labov uses its analyses to launch discussions of such major linguistic research questions as the function of syntactic transformations for the users of language. This study, and others like it, provide evidence for the heuristic value of the descriptivist approach in linguistics. The business of the teacher might be to correct, but the business of the linguist is to describe.

No more is it the business of the reasoning or the decision-making researcher to pass normative judgment. The question of Bayesianism as a competence/computational system is just as separable from its normative status as is the question of AAVE as a separate system of linguistic competence separable from the research question identifying its sociolinguistic function as low socioeconomic status marker. The normativist research questions are if reasoners should or should not conform to theorems of the probability calculus, and, when they deviate, whether they are right or wrong. This is analogous to prescriptive grammarians asking if it is "correct" to use double negation. In contrast, the descriptivist research question focuses on Bayesian rules as computational-level analysis of the rules used by reasoners. In this, what we do is analogous to what a linguist does when she constructs a theory of AAVE grammar. As a result, we can ask, for example, whether Bayesian rules provide a descriptively adequate theory of human reasoning and decision making. (As an aside, the answer we propose in Elqayam and Evans [2013] is "to some extent.") We characterize the rules governing thinking and reasoning rather than judge them.

Hahn has recently (2014) argued against the "eat your cake and have it" argument, pointing out that recent new paradigm research in reasoning has been motivated by normative concerns that cannot be rephrased as computational theories without loss of crucial scientific benefit. For example, Bayesian analysis of informal fallacies (Corner, Hahn, & Oaksford, 2011; Hahn & Oaksford, 2007; Harris, Hsu, & Madsen, 2012) examines how classic Bayesian constraints affect the way in which some arguments are perceived as better than others. Not only is this work driven by normative motivation, there are no alternative computational non-normative theories. In response I have to emphasize that it is not a matter of replacing Bayesian theory with a different computational theory. Bayesianism itself is both computational and normative at the same time. The normative "how good is the argument" research question can easily be replaced with the (no less Bayesian) descriptive "how strong is the argument" question.

Grounded Rationality

Prima facie it might seem that a descriptivist approach precludes any discussion of rationality, clearly an undesirable consequence. To clarify why this impression is unwarranted, we need to distinguish between two different types of "ought" (see E&E). In everyday speech we use terms such as *ought, should,* and *must* to express a range of meanings. A sentence such as "Jake must have made it to the train station in time" expresses an *epistemic* ought—confidence that Jake indeed made it to the train station. Such phrasing is completely innocuous: there is no value judgment involved. On the other hand, we sometimes use *evaluative* oughts, in phrases such as "This ought not to have happened," or "People should be treated equally." They are synonymous to stating "this is bad" and "treating people equally is good," respectively. The *ought* of normativism can be argued to be this sort of evaluative ought, and the "some reasoning or decision making is good, and other reasoning and decision making is bad" phrasing (Oaksford & Chater, 2007) cited above is a case in point.

There is, however, a third sort of *ought,* a *directive* or *instrumental ought;* for example, "You ought to turn left here," or "You should not switch on the air-conditioning." In this case the *ought* serves to direct action. There is nothing in the action itself that is lauded as good or condemned as bad. It just aims to achieve a goal. Perhaps you need to turn left to find the closest supermarket; perhaps it's too cold for air-conditioning, or you want to save on your electricity bill. Either way, there is nothing intrinsically good or bad about turning left or about not switching on the air-conditioner. They are not *values.* The distinction is analogous to the one proposed two decades back, between rationality$_1$ and rationality$_2$ (Evans, 1993; Evans & Over, 1996a; Evans, Over, & Manktelow, 1993). Rationality$_1$, or pragmatic or instrumental rationality, is the rationality defined by achieving one's goals. If you want to get to that supermarket, it is rational for you to turn left. Taking the right turning will get you further away. In contrast, rationality$_2$ is normative rationality—the rationality defined by conforming to a normative system. This is the rationality of normativism. A descriptivist approach, then, allows for discussions of human rationality—the instrumental, pragmatic sort of rationality.

The theory of *grounded rationality* (Elqayam, 2011, 2012) is a descriptivist theory of rationality drawing on these distinctions between rationality$_1$ and rationality$_2$, and instrumental and normative oughts, respectively. It rests on the double foundation of descriptivism and a moderate epistemic relativism. The core idea is pretty straightforward: whether a specific behavior is instrumentally rational for a specific agent (roughly, the person taking the action) would depend on the agent, her personal goals, and, importantly, the context in which she operates. For example, a curator putting together an exhibition of WWII photography need hardly avoid representative pictures—indeed, she should pursue them. In contrast, for a punter placing a bet on a horse such reliance on representativeness can lead to money loss, if it causes him to underweight the risks of his gamble. The current working definition of grounded rationality is this:

Behavior B is rational for agent A in epistemic context E if B facilitates achievement of A's goals within the constraints of E.

In this definition, "epistemic context" refers to anything that influences an agent's beliefs and desires, be it cognitive, individual, cultural, social, or any other context. The result is moderate epistemic relativism, the philosophical position that emphasizes the relativity of knowledge. For many cognitive scientists the word "relativism" is like a red rag to a bull, invoking the terrors of the relativist fallacy and a Feyerabend-style "anything goes." However, neither of these is a serious threat for *moderate* epistemic relativism of the sort advocated by Stich's (1990) seminal defense of relativist pragmatic rationality. Moderate epistemic relativism acknowledges universal constraints on cognition as well as those sensitive to epistemic context. Acknowledging a universal cognitive hard core helps moderate epistemic relativism avoid the pitfalls of radical, "anything goes" relativism.

To illustrate, let us stick with the linguistics analogy. We have seen that comparing across dialects provides a wealth of interesting linguistic data connecting to major research questions. Linguists treasure the variety of human languages. It is not by chance that they refer to extinct and endangered languages, using the same terminology that biologists use to refer to species. There is a good reason for it: comparing languages helps linguists to map out the boundaries of language universals, that is, characteristics shared by all or almost all human languages. Through this, linguists can develop better understanding of what is and what is not fundamental in human language. The question of cognitive universals versus cognitive variability is not restricted to linguistics, of course. A burgeoning literature on the topic in perception, decision making, self concepts, moral judgment, and so on (Henrich, Heine, & Norenzayan, 2010; Norenzayan & Heine, 2005) shows how deeply cognitive variability can run. As deep, for example, as the susceptibility to visual illusions: the Müller-Lyer illusion completely disappears in some populations (Segall, Campbell, & Herskovits, 1966). Nevertheless, there are still universal hard cores. One of the most well-known examples is the case of word order (Comrie, 1981; Greenberg, 1963; Hawkins, 1983): of the six possible orders of Subject (S), Verb (V), and Object (O), the vast majority of human natural languages are SVO (e.g., English) or SOV (e.g., Turkish); most of the rest are VSO and VOS, whereas OSV and OVS are extremely rare. This combination of variability alongside a universal hardcore is the basis for the sort of moderate epistemic relativism adopted by grounded rationality.

Grounded Rationality, Bounded Rationality, and the New Paradigm

Grounded rationality links to Simon's (1982) conception of bounded rationality, the proposal that the cognitive cost of computation should be taken into account as well as its benefits. Humans—or any biological computational system, for that matter— have finite resources of computation: we can pay close attention to only one thing at a time, we can only hold seven (plus/minus two) items in working memory, and we simply do not live forever—the *finitary predicament*, as Cherniak (1986) dubbed it. For Subjective Epistemic Utility (SEU), the normative model of Bayesian decision making widely adopted by economists, rationality consists of maximizing utility. That is, the rational way to act is to take into account the subjective utility (roughly, desirability or psychological value) that can be achieved by each course of action toward a goal, the probability of achieving the goal, multiplying them to calculate the SEU, then choosing whichever course of action results in the greatest overall SEU. The problem, of course, is that the mental computations needed to maximize utility can fast become intractable—too complicated to carry out. Simon therefore argues that maximization is not a practical requirement for finite creatures. If we need half an hour to decide if that strange noise coming from a dark alley is a mugger, by the time we decide to make a break for it, it might be too late. Instead, human rationality is anchored in *satisficing*— computations that are just good enough to achieve our goals, without placing unrealistic demands on finite human resources. Within the new paradigm, satisficing was incorporated into the dual-process hypothetical thinking theory (Evans, 2006, 2007), defined as the tendency of type-2 processing to latch onto the existing representation or mental model as the default unless strongly compelled to do otherwise.

What Simon essentially does is draw on an economic model of cognitive processing in which cognitive resources such as working memory and attention have cognitive utility associated with them. We can think about cognitive resources as credit in a cognitive bank account, which computations draw on to a greater or lesser degree. Within the new paradigm, Evans and Over suggested the cognate notion of epistemic utility (Evans & Over, 1996b)—the knowledge to be gained by any specific behavior carries psychological value in its own right. Mental processes, then, have both cognitive costs and cognitive gains associated with them. Once the mental cost and benefit of computations are added to the mundane (i.e., nonepistemic) cost-benefit analysis of classic economic models, the utility associated with each option can change unrecognizably, typically with the result that the satisficing option claims the greater overall utility. If Jake needs two hours of frustrated hard thinking to compute which of two online grocery sites is £1.25 cheaper for his current shopping, it might not be worth the effort—not to mention the cost associated with negative emotions (Blanchette & Caparos, 2013; Perham & Oaksford, 2005).

Where bounded rationality acknowledges universal constraints shared by all humans, grounded rationality adds relative constraints, which are taken into account in computations of epistemic utility. The cost of computations varies with the availability of cognitive resources. Simply put, not all of us have the same amount of credit in our cognitive bank account. Although for Jake the computation might not be worth the effort or the frustration, his friend Janice, who is much better with numbers, might only need five minutes to determine the same question. The same computation that is instrumentally irrational for Jake is instrumentally rational for Janice. Whether this variability is a product of cultural or individual difference is beside the point. The main thing is that cognitive effort is a cost, and this cost is relative to the agent in his or her epistemic context. Goals, too, vary as a function of epistemic context—especially epistemic (i.e., knowledge-related) goals. Effortful processing is an epistemic goal for individuals with high "thinking dispositions" score (Stanovich & West, 2000), but not for ones with low scores; people revise their epistemic goals to more modest ones as time spent on a task progresses (Ackerman, 2014); and individuals from collectivist cultures adopt more avoidance goals than individuals from individualistic cultures (Elliot, Chirkov, Kim, & Sheldon, 2001)—to name but a handful of examples.

More generally, new paradigm researchers (e.g., Evans, 2014b; Stanovich, 2013) are becoming increasingly aware of the importance of epistemic context for understating human rationality, often with a strong dual-processing element involved in the theorizing. Analytic processing, or more generally the "new mind," as Evans (2010) dubs it, allows humans a distinct kind of rationality, unshared with other animals. Subjective (Evans, 2014b) and strongly contextualized (Stanovich, 2013), this notion of rationality reflects the human condition by drawing on epistemic goals (i.e., goals of knowledge) and symbolic utility (Nozick, 1993)—the psychological value that a course of action acquires from being associated with a goal. As Stanovich (2013) memorably puts it, humans are the great social contextualizers. Both authors make a distinction between a simpler form of rationality, shared with animals ("old mind" in Evans's nomenclature, "thin rationality" in Stanovich's) and a more complex form of rationality that is distinctly human ("new mind" or "broad rationality," respectively). Epistemic rationality—the rationality of knowledge—can only be achieved by the new mind, argues Evans. In a similar vein, Stanovich points to substantial evidence that nonhuman animals often maximize utility better than humans; he then proposes to solve this apparent paradox with the notion of broad rationality as a higher-order form of rationality, with epistemic context and symbolic utility thrown in. This more complex form of rationality draws on analytic thinking in social context.

This decision-theoretic approach is much in the spirit of the new paradigm. Costs and benefits of mental processing are integrated into evaluations of instrumental rationality—a very human type of instrumental rationality that is more than just biological costs and benefits. Utility is determined within social context, another major feature of the new paradigm. Grounded rationality can be seen as an additional line of the same trend.

The notion of rationality in the new paradigm, then, is becoming increasingly richer. This is true for normative rationality as well as the more instrumental sort. As the new paradigm moves away from binary conceptions of truth, it also moves away from binary conceptions of rationality, normative rationality included. The old, "logicist" paradigm (Oaksford & Chater, 1991) was just as much binary on this front as well. Where rationality in the old paradigm consisted of conforming to validity standards of extensional logic, the notion of p-validity adopted by new paradigm researchers offers a more nuanced approach to norms of rationality. True, a binary distinction still exists: an argument is either p-valid or p-invalid. However, p-validity is now only one thread in a richer tapestry of argument characterization. Invited inferences and pragmatic argumentation, for example, are typically p-invalid, but it is possible to measure their argument strength on a continuous scale: strong pragmatic arguments are those in which the difference between the probability of the premises and the probability of the conclusion is small. For example, inference from disjunctions to conditionals, although p-invalid, is often accepted for pragmatic reasons when the probability of the disjunction is close enough to the probability of the conditional (Over, Evans, & Elqayam, 2010).

Perhaps the most striking treatment of probabilistic strength is Hahn and Oaksford's (Hahn & Oaksford, 2007) analysis of informal arguments such as circular arguments and the argument from ignorance. For example, they show that perceived strength of circular arguments is negatively associated with the probability of alternative explanations. For slippery slope arguments, they argue, utility plays a role as well. Corner et al. (2011) analyze the slippery slope argument as a *Denial of the Anetecedent* argument that can be stronger or weaker depending on its conditional probability and the utility of the consequent. (Denial of the Antecedent is a classically and probabilistically invalid conditional inference of the form "*If p, then q; not-p; therefore, not-q.*") From a descriptive point of view, there is a single psychological scale of subjective strength of arguments. I see little psychological difference between endorsing the invalid Denial of the Antecedent because it is strong and endorsing the valid *Modus Tollens* ("*If p then q; not-q; therefore, not-p*") because it is p-valid. From a normative point of view, however, the scale is broken toward the higher end by the qualitative distinction between valid and invalid (or p-valid and p-invalid) arguments. That strict Bayesians treat these with the same tools is perhaps more amenable to a descriptivist treatment than would seem at first blush.

The normative picture is equally nuanced for valid arguments. For each argument it is possible to compute a coherence probability interval with high and low boundaries (for a more detailed explanation see Over, chapter 4, this volume). Probabilities that fall outside these boundaries are incoherent in the sense that they violate the probability calculus. Note that coherence is a stricter measure than p-validity (which only supplies a minimal probability boundary), so a response can be p-valid yet incoherent relative to the Equation. Of course, it might be possible to argue that some of these incoherent responses are coherent relative to an alternative formal approach to conditionals, such as Stalnaker's (1968), which does not satisfy the Equation.[3] However, this only serves to underline the relativity of any potential norm. This new focus on coherence and p-validity opens the door to a host of descriptive, empirical questions that new paradigm researchers have only recently begun to address (Evans, Thompson, & Over, 2014). Importantly for the present context, this means that, even with dichotomous measures, normative questions no longer have a single answer. If a participant assigns an incoherent, but p-valid probability to a conclusion, is his response normative or not?

Whether the Equation itself should be considered normative—in addition to being a very good computational descriptive principle—is moot, and I am not even convinced that it is a meaningful question. Does it mean that any reasoners conforming to, say, the Stalnaker conditional are being normatively irrational? It seems more productive to cast the ensuing discussion in descriptive, computational terms.

If the discussion above makes one point salient, it is that the new paradigm is gradually moving toward a richer and more nuanced conception of rationality, one that is increasingly sensitive to epistemic utility and social context and that draws on a wider range of argument measures. It is not that the old normative ideas are discarded—it is more that they become integrated into a more complex picture. Validity still provides a significant research question, but our concepts of validity are enriched by the more descriptive—and psychologically interesting—ideas of inference strength, p-validity, and coherence. Strict conformity to SEU—Stanovich's "thin" rationality, Evans's old-mind rationality—is still of interest, but it is just one aspect of a dual approach. Note that normative rationality is by nature binary. An inference is valid or not valid. One conforms to the probability calculus or not. Any approach that allows for a more nuanced evaluation moves one step closer to descriptivism. In this way the new paradigm, in its move away from binary notions, is becoming less normativist, at least in the strict sense of the term. This makes grounded rationality very much in the spirit of the new paradigm.

Acknowledgments

I am grateful to David Over for helpful comments and discussions on a previous draft of this chapter.

Notes

1. Throughout this chapter I use the term "classical logic" as short for extensional, binary, and monotonic.

2. In this context, an argument is coherent if and only if it conforms to the probability calculus.

3. I owe this point to David Over. See Over (chapter 4, this volume) for a discussion of Stalnaker's approach.

References

Ackerman, R. A. (2014). The diminishing criterion model for metacognitive regulation of time investment. *Journal of Experimental Psychology: General, 143,* 1349–1368.

Adams, E. (1998). *A primer of probability logic.* Stanford: CLSI Publications.

Blanchette, I., & Caparos, S. (2013). When emotions improve reasoning: The possible roles of relevance and utility. *Thinking & Reasoning, 19,* 399–413.

Bonnefon, J.-F. (2009). A theory of utility conditionals: Paralogical reasoning from decision-theoretic leakage. *Psychological Review, 116,* 888–907.

Bonnefon, J.-F. (2013). New ambitions for a new paradigm: Putting the psychology of reasoning at the service of humanity. *Thinking & Reasoning, 19,* 381–398.

Buckwalter, W., & Stich, S. (2011). Competence, reflective equilibrium, and dual-system theories. *Behavioral and Brain Sciences, 34,* 251–252.

Chater, N., & Oaksford, M. (1999). The probability heuristics model of syllogistic reasoning. *Cognitive Psychology, 38,* 191–258.

Cherniak, C. (1986). *Minimal rationality.* Cambridge, MA: MIT Press.

Chomsky, N. (1965). *Aspects of the theory of syntax.* Cambridge, MA: MIT Press.

Cohen, L. J. (1981). Can human irrationality be experimentally demonstrated? *Behavioral and Brain Sciences, 4,* 317–370.

Comrie, B. (1981). *Language universals and linguistic typology: Syntax and morphology.* Oxford: Blackwell.

Corner, A., Hahn, U., & Oaksford, M. (2011). The psychological mechanism of the slippery slope argument. *Journal of Memory and Language, 64,* 133–152.

De Saussure, F. (1966). *Course in general linguistics.* New York: McGraw-Hill. (Original work published 1916.)

Douven, I., & Verbrugge, S. (2010). The Adams family. *Cognition, 117,* 302–318.

Edgington, D. (1995). On conditionals. *Mind, 104*, 235–329.

Elliot, A. J., Chirkov, V. I., Kim, Y., & Sheldon, K. M. (2001). A cross-cultural analysis of avoidance (relative to approach) personal goals. *Psychological Science, 12*, 505–510.

Elqayam, S. (2011). Grounded rationality: A relativist framework for normative rationality. In K. I. Manktelow, D. E. Over, & S. Elqayam (Eds.), *The science of reason: A Festschrift for Jonathan St. B. T. Evans* (pp. 397–420). Hove, UK: Psychology Press.

Elqayam, S. (2012). Grounded rationality: Descriptivism in epistemic context. *Synthese, 189*, 39–49.

Elqayam, S., & Evans, J. St. B. T. (2011). Subtracting "ought" from "is": Descriptivism versus normativism in the study of human thinking. *Behavioral and Brain Sciences, 34*, 233–248.

Elqayam, S., & Evans, J. S. (2013). Rationality in the new paradigm: Strict versus soft Bayesian approaches. *Thinking & Reasoning, 19*, 453–470.

Elqayam, S., & Over, D. (2012). Probabilities, beliefs, and dual processing: The paradigm shift in the psychology of reasoning. *Mind & Society, 11*, 27–40.

Elqayam, S., & Over, D. E. (2013). New paradigm psychology of reasoning: An introduction to the special issue edited by Elqayam, Bonnefon, and Over. *Thinking & Reasoning, 19,* 249–265.

Elqayam, S., Thompson, V. A., Wilkinson, M. R., Evans, J. St. B. T., & Over, D. E. (2015). Deontic introduction: A theory of inference from is to ought. *Journal of Experimental Psychology: Learning Memory and Cognition, 41*(5), 1516–1532.

Evans, J. St. B. T. (1972). Interpretation and matching bias in a reasoning task. *Quarterly Journal of Experimental Psychology, 24*, 193–199.

Evans, J. St. B. T. (1989). *Bias in human reasoning: Causes and consequences.* Brighton: Lawrence Erlbaum Associates.

Evans, J. St. B. T. (1993). Bias and rationality. In K. I. Manktelow & D. E. Over (Eds.), *Rationality: Psychological and philosophical perspectives* (pp. 6–30). London: Routledge.

Evans, J. St. B. T. (2002). Logic and human reasoning: An assessment of the deduction paradigm. *Psychological Bulletin, 128*, 978–996.

Evans, J. St. B. T. (2006). The heuristic-analytic theory of reasoning: Extension and evaluation. *Psychonomic Bulletin & Review, 13*, 378–395.

Evans, J. St. B. T. (2007). *Hypothetical thinking: Dual processes in reasoning and judgement.* Hove, UK: Psychology Press.

Evans, J. St. B. T. (2008). Dual-processing accounts of reasoning, judgment, and social cognition. *Annual Review of Psychology, 59*, 255–278.

Evans, J. St. B. T. (2010). *Thinking twice: Two minds in one brain.* Oxford: Oxford University Press.

Evans, J. St. B. T. (2012). Questions and challenges for the new psychology of reasoning. *Thinking & Reasoning, 18,* 5–31.

Evans, J. St. B. T. (2014a). Rationality and the illusion of choice. *Frontiers in Psychology, 5,* 104.

Evans, J. St. B. T. (2014b). Two minds rationality. *Thinking & Reasoning, 20,* 129–146.

Evans, J. St. B. T., Barston, J. L., & Pollard, P. (1983). On the conflict between logic and belief in syllogistic reasoning. *Memory & Cognition, 11,* 295–306.

Evans, J. St. B. T., & Elqayam, S. (2011). Towards a descriptivist psychology of reasoning and decision making. *Behavioral and Brain Sciences, 34,* 275–290.

Evans, J. St. B. T., Handley, S. H., & Over, D. E. (2003). Conditionals and conditional probability. *Journal of Experimental Psychology: Learning, Memory, and Cognition, 29,* 321–355.

Evans, J. St. B. T., Neilens, H., Handley, S. J., & Over, D. E. (2008). When can we say "if"? *Cognition, 108,* 100–116.

Evans, J. St. B. T., & Over, D. E. (1996a). *Rationality and reasoning.* Hove, UK: Psychology Press.

Evans, J. St. B. T., & Over, D. E. (1996b). Rationality in the selection task: Epistemic utility versus uncertainty reduction. *Psychological Review, 103,* 356–363.

Evans, J. St. B. T., & Over, D. E. (2004). *If.* Oxford: Oxford University Press.

Evans, J. St. B. T., Over, D. E., & Manktelow, K. I. (1993). Reasoning, decision making and rationality. *Cognition, 49,* 165–187.

Evans, J. St. B. T., & Stanovich, K. E. (2013). Dual-process theories of higher cognition: Advancing the debate. *Perspectives on Psychological Science, 8,* 223–241.

Evans, J. St. B. T., Thompson, V. A., & Over, D. E. (2014). Uncertain deduction and the new psychology of conditional inference. *Unpublished manuscript, University of Plymouth.*

Evans, J. S., & Over, D. E. (2013). Reasoning to and from belief: Deduction and induction are still distinct. *Thinking & Reasoning, 19,* 267–283.

Fromkin, V., Hyams, N. M., & Rodman, R. (2014). *An introduction to language.* Boston: Wadsworth, Cengage Learning.

Fugard, A. J. B., Pfeifer, N., Mayerhofer, B., & Kleiter, G. (2011). How people interpret conditionals: Shifts towards the conditional event. *Journal of Experimental Psychology: Learning, Memory, and Cognition, 37,* 635–648.

Gilio, A., & Over, D. E. (2012). The psychology of inferring conditionals from disjunctions: A probabilistic study. *Journal of Mathematical Psychology, 56,* 131.

Greenberg, J. H. (1963). Some universals of grammar with particular reference to the order of meaningful elements. In J. H. Greenberg (Ed.), *Universals of language* (pp. 58–90). Cambridge, MA: MIT Press.

Hahn, U. (2014). The Bayesian boom: Good thing or bad? *Frontiers in Psychology, 5* 765.

Hahn, U., & Oaksford, M. (2007). The rationality of informal argumentation: A Bayesian approach to reasoning fallacies. *Psychological Review, 114*, 704–732.

Harris, A. J. L., Hsu, A. S., & Madsen, J. K. (2012). Because Hitler did it! Quantitative tests of Bayesian argumentation using ad hominem. *Thinking & Reasoning, 18*, 311–343.

Hawkins, J. A. (1983). *Word order universals.* New York: Academic Press.

Henrich, J., Heine, S. J., & Norenzayan, A. (2010). The weirdest people in the world? *Behavioral and Brain Sciences, 33*, 61–83.

Hudson, W. D. (1969). *(Ed.) The is-ought question: A collection of papers on the central problem in moral philosophy.* London: Macmillan.

Hume, D. (2000). *A treatise on human nature.* Oxford: Clarendon Press. (Original work published 1739–1730.)

Jones, M., & Love, B. C. (2011). Bayesian fundamentalism or enlightenment? On the explanatory status and theoretical contributions of Bayesian models of cognition. *Behavioral and Brain Sciences, 34*, 169–188.

Labov, W. (1972). *Language in the inner city: Studies in the Black English Vernacular.* Philadelphia: University of Pennsylvania Press.

Manktelow, K. I., & Over, D. E. (1991). Social roles and utilities in reasoning with deontic conditionals. *Cognition, 39*, 85–105.

Manktelow, K. I., Over, D. E., & Elqayam, S. (2011). Paradigms shift: Jonathan Evans and the science of reason. In K. I. Manktelow, D. E. Over, & S. Elqayam (Eds.), *The science of reason: A Festschrift for Jonathan St B. T. Evans* (p. 397–419). Hove, UK: Psychology Press.

Marr, D. (1982). *Vision: A computational investigation into the human representation and processing of visual information.* San Francisco: W. H. Freeman.

Moore, G. E. (1903). *Principia ethica.* New York: Cambridge University Press.

Norenzayan, A., & Heine, S. J. (2005). Psychological universals: What are they and how can we know? *Psychological Bulletin, 131*, 763–784.

Nozick, R. (1993). *The nature of rationality.* Princeton, NJ: Princeton University Press.

Oaksford, M., & Chater, N. (1991). Against logicist cognitive science. *Mind & Language, 6*, 1–38.

Oaksford, M., & Chater, N. (1994). A rational analysis of the selection task as optimal data selection. *Psychological Review, 101*, 608–631.

Oaksford, M., & Chater, N. (1998). *Rationality in an uncertain world.* Hove, UK: Psychology Press.

Oaksford, M., & Chater, N. (2007). *Bayesian rationality: The probabilistic approach to human reasoning.* Oxford: Oxford University Press.

Oaksford, M., & Chater, N. (2011). Dual systems and dual processes but a single function. In K. I. Manktelow, D. E. Over, & S. Elqayam (Eds.), *The science of reason: A Festschrift for Jonathan St B. T. Evans* (pp. 339–351). Hove, UK: Psychology Press.

Oaksford, M., & Chater, N. (2012). Dual processes, probabilities, and cognitive architecture. *Mind & Society, 11*, 15–26.

Oaksford, M., & Chater, N. (2014). Probabilistic single function dual process theory and logic programming as approaches to non-monotonicity in human vs. artificial reasoning. *Thinking & Reasoning, 20*, 269–295.

Oaksford, M., Chater, N., & Larkin, J. (2000). Probabilities and polarity biases in conditional inference. *Journal of Experimental Psychology: Learning, Memory, and Cognition, 26*, 883–889.

Oberauer, K., & Wilhelm, O. (2003). The meaning(s) of conditionals: Conditional probabilities, mental models and personal utlities. *Journal of Experimental Psychology: Learning, Memory, and Cognition, 29*, 680–693.

Over, D. E. (2009). New paradigm psychology of reasoning. *Thinking & Reasoning, 15*, 431–438.

Over, D. E., Evans, J. S. T., & Elqayam, S. (2010). Conditionals and non-constructive reasoning. In M. Oaksford & N. Chater (Eds.), *Cognition and conditionals: Probability and logic in human thinking* (pp. 135–151). Oxford: Oxford University Press.

Over, D. E., Hadjichristidis, C., Evans, J. S., Handley, S. J., & Sloman, S. A. (2007). The probability of causal conditionals. *Cognitive Psychology, 54*, 62–97.

Perham, N., & Oaksford, M. (2005). Deontic reasoning with emotional content: Evolutionary psychology or decision theory? *Cognitive Science, 29*, 681–718.

Pfeifer, N., & Kleiter, G. D. (2010). Uncertain deductive reasoning. In K. I. Manktelow, D. E. Over, & S. Elqayam (Eds.), *The science of reason: A Festschrift for Jonathan St B. T. Evans* (pp. 145–166). Hove, UK: Psychology Press.

Pigden, C. R. (2010). *Hume on is and ought*. New York: Palgrave Macmillan.

Pullum, G. K. (1999). African American Vernacular English is not Standard English with mistakes. In R. S. Wheeler (Ed.), *The workings of language: From prescriptions to perspectives* (pp. 39–58). Westport, CT: Praeger.

Ramsey, F. P. (1931). *The foundations of mathematics and other logical essays*. London: Routledge and Kegan Paul.

Schurz, G. (1997). *The is-ought problem: An investigation in philosophical logic*. Dordrecht: Wolters Kluwer.

Segall, M. H., Campbell, D. T., & Herskovits, M. J. (1966). *The influence of culture on visual perception*. Oxford: Bobbs-Merrill Co.

Simon, H. A. (1982). *Models of bounded rationality*. Cambridge, MA: MIT Press.

Stalnaker, R. (1968). A theory of conditionals. *American Philosophical Quarterly Monograph Series, 2,* 98–112.

Stanovich, K. E. (1999). *Who is rational? Studies of individual differences in reasoning.* Mahwah, NJ: Lawrence Elrbaum Associates.

Stanovich, K. E. (2004). *The robot's rebellion: Finding meaning in the age of Darwin.* Chicago: University of Chicago Press.

Stanovich, K. E. (2011). *Rationality and the reflective mind.* New York: Oxford University Press.

Stanovich, K. E. (2013). Why humans are (sometimes) less rational than other animals: Cognitive complexity and the axioms of rational choice. *Thinking & Reasoning, 19,* 1–26.

Stanovich, K. E., & West, R. F. (2000). Individual differences in reasoning: Implications for the rationality debate. *Behavioral and Brain Sciences, 23,* 645–726.

Stich, S. P. (1990). *The fragmentation of reason: Preface to a pragmatic theory of cognitive evaluation.* Cambridge, MA: MIT Press.

Thompson, V. A., Evans, J. St B. T., & Handley, S. H. (2005). Persuading and dissuading by conditional argument. *Journal of Memory and Language, 53,* 238–257.

Tversky, A., & Kahneman, D. (1983). Extensional vs. intuitive reasoning: The conjunction fallacy in probability judgment. *Psychological Review, 90,* 293–315.

Wason, P. C. (1966). Reasoning. In B. M. Foss (Ed.), *New Horizons in Psychology I* (pp. 106–137). Harmandsworth, UK: Penguin.

Wason, P. C., & Evans, J. St. B. T. (1975). Dual processes in reasoning? *Cognition, 3,* 141–154.

Wilkinson, M. R. (2014). Against a normative view of folk psychology. *Frontiers in Psychology, 5,* 598.

3 The Interpretative Function and the Emergence of *Unconscious Analytic* Thought

Maria Bagassi and Laura Macchi

Consciousness is a late cerebral habit, if we consider it in terms of evolution, and of limited biological relevance, even if paranoically grandiose for the individual.

—Lamberto Maffei (2012)

The Logical Bias

Philosophy assumes the exploration of thought in its functional and structural qualities as the starting point of its investigations; more precisely, it assumes both the exploration of the instruments that contribute to the construction of truth and the exploration of the source itself that generates these instruments and guarantees them.

Thought as a living whole, creator of logical relationships but at the same time the mirror of their structure, is the subject *par excellence* of philosophical inquiry and elicits the topic of "logic" in the extended sense of the term.[1]

The aim of Aristotelian logic was to establish the truth, the ontological substance, through the study of the *argument* or the *inference* or, in other words, of what follows necessarily from certain premises, rooted in truth, as an instrument of knowledge in order to reach the truth.

After a long period of "conservatorism," starting from the beginning of the last century logic underwent a phase of intensive development, assuming a framework similar to that of the mathematical disciplines, to the extent that it became known as *mathematical logic* or *logistics*.

The aim of this renovation was to give the fundamentals of mathematics a more solid and profound base. In just a century this new mathematical logic reached a high degree of perfection and played an important role in our overall knowledge, extending far beyond the boundaries within which it had originally been created.

As a result of the work done on decontextualizing and the formalization that followed, the criteria of validity in logic gradually excluded an intuitive approach, privileging its mechanics.

It is worth noting that the development of formal logic, in fact, became a gradual process of "depsychologization" of logical language and of disambiguant simplification compared to natural language, intentionally pursued and programmatically declared by modern logic.

The psychology of thought has inherited this tradition. The purpose of the scientific, experimental approach to the study of thought was to describe human reasoning by identifying the criteria that ensure its correctness, for example, the criteria of validity of a deduction starting from a given set of premises. From an Aristotelian perspective, these criteria constitute a form of structure of foundative knowledge, general and independent of accidental and contingent particularisms, a sort of deep structure of knowledge, analogous to the deep structure of language, those linguistic universals *à la* Chomsky.

This claim to find the common element, the essence of psychological reality, underlying the multiplicity of beliefs and the multitude of situations in which thought is involved, has paradoxically transcended reality itself in a sort of metaphysics. What has, however, been lost is the complexity of reality, which was seen as an element of interference, clouding the general structure that underlies reasoning; whereas thought is of necessity contextualized, drenched in content, implicit presuppositions, beliefs, intentions and deals with reality in all its complexity.

Logic and Natural Language

Consider the case of disjunction, which has had such an impact on the fields of logic and psychology. In everyday language the use of the word *or* presupposes that the two members of the disjunction have some form of pertinent connection and that one of the two is true (although which one is true is not known).

With the aim of ensuring that its meaning would be univocal, modern logicians made disjunction independent of psychological factors and particularly of contextual knowledge. They therefore decided to extend the use of *or* (inclusive or) and also to consider the disjunction of any two sentences (members) as a combination endowed with sense even if there were no meaningful connections or links between the two disjuncts.

As for disjunction, modern logic has moved away from the use of the conditional sentence in *implication* (if … then) in everyday language acceptation. Once again, we use the *if* clause only when there is a pertinent relationship between the *if* and the *then* clauses. Modern logicians adopted the same procedure used for *or* in order to comply with the requirements of scientific languages, for which logic intended to constitute the foundation. They decided to disambiguate implication and to free it from any psychological factors, above all a pertinent relationship between antecedent and consequent. This means that the truth of the implication has been reduced to the truth or the falsity of the antecedent and of the consequent taken individually.

Therefore, modern logic uses implication in material meaning, or simply *material implication*. This is in antithesis to the use of implication in natural language, where the existence of a relationship between antecedent and consequent is a necessary condition for implication to be meaningful and true.

However, natural language is so flexible and multilayered that we can even adopt what would appear to be a totally unacceptable implication, as far as pertinence is concerned, to convey the falsity of the antecedent:

"If Peter comes this evening, [then] I'm a Dutchman."

Indeed, no word, no phrase in natural language has an univocal meaning independent of the situation on hand, of the context, and, the context being the same, depending on the intention of the speaker. Even the same individual can, of course, intend to say different things in different moments in time.

Take, for example, the stereotyped saying "war is war" or "business is business." Such tautological phrases are informative insofar as they isolate and emphasize the global meaning of the term. The subject corresponds to the general concept; the predicate highlights one particular element of the general concept to the exclusion of all others. Far from simply acknowledging the identity of the concept (war = war, business = business), these phrases provide important information: they tell us that, in times of war, the rules applied are different from those applied in times of peace, and that in business, behavior can be countenanced that would be considered unacceptable in normal, everyday relationships. Statements of this type cannot be represented as A is equal to A, but as A implies X. It is only the form of these utterances that is tautological; the message is actually informative.

Now, considering the use of negation in natural language, it can be seen that this form also can take on very different meanings, which may even be substantially opposite.[2] Imagine, for example, that you are looking for a length of cloth in a particular color, say *antique rose*, and reject the cloth that is offered as it is *powder pink* and not *antique rose*. Or that you express your opinion about a certain action saying: "That's not fair!" If we do not take into consideration the psychological characteristics, the *not antique rose* of the first example and the *not fair* of the second could both be considered as corresponding and codified as *not A*. However, with regard to their natural meaning, the *not A* of the first example corresponds to that which only *resembles* antique rose, whereas the *not A* of the second example corresponds to the *contrary* of fair.

These two negations are very different from one another in terms of the point of view of what the speaker actually meant. In the first case the negation refers to the identity within the terms of the comparison and corresponds to an acknowledgment of their similarity: the color of the cloth is *almost* antique rose, but it is not an exact match. In the second case the negation identifies with the *opposite* of what is being negated.

Then, one of the primary differences between the language of logic and natural language is that in logic negation is limited exclusively to negation for "contraposition," whereas in natural language it can also be used for "differentiation." But these are not the only differences; another very crucial difference is that in natural language the reversibility of negation by contraposition is not automatic: sometimes "not unfair" expresses fairness, but this is not always the case. An action that is defined as "not unfair" is different from an action that is described as "unfair" not because it is the opposite of unfair (i.e., definitely fair) but because it is not *completely* unfair or only unfair in some respect or just seems unfair, and so on.

When the logician borrows concepts and words from natural language, he circumscribes their meaning, eliminating any attributes that are not essential for his aims. Thus, he deviates from their significance in natural language.

Logic and natural language share a common aim, that of transmitting meaning efficaciously or, in other words, of communicating, of expressing thought. However, this objective is achieved by these two language forms in opposite ways: logic achieves a univocal communication, through simplification, eliminating any meanings that might interfere with the univocal meaning to be communicated, whereas natural language exploits the expressive richness of words, but nevertheless avoids slipping into chaos and tripping over misunderstandings, by relying on the pertinence of the meaning to the context.

In Ricoeur's words:

quand le mot accède au discours, avec sa richesse sémantique, tous nos mots etant polysémiques a quelque degré, l'univocité ou la plurivocité de notre discours n'est pas l'oeuvre des mots, mais des contextes. Dans le cas du discours univoque, c'est-à-dire du discours qui ne tolère qu'une signification, c'est la tâche du contexte d'occulter la richesse sémantique des mots, de la réduire, en établissant ce que M. Greimas appelle une isotopie, c'est-à-dire un plan de référence, une thématique, une topique identique pour tous les mots de la phrase. ... Au lieu de cribler une dimension de sens, le contexte en laisse passer plusieurs, voire en consolide plusieurs, qui courent ensemble à la manière des textes superposés d'un palimpseste ... plus d'une interprétation est alors justifiée par la structure d'un discours qui donne permission aux multiples dimensions du sens de se réaliser en même temps. Bref le langage est en fête. (Ricoeur, 1969; pp. 94, 95)

[We then understand what happens when the word reaches discourse along with its semantic richness. All our words being polysemic to some degree, the univocity or plurivocity of our discourse is not the accomplishment of words but the of contexts. In the case of univocal discourse, that is, of discourse which tolerates only one meaning, it is the task of the context to hide the semantic richness of words, to reduce it by establishing what Greimas call an isotopy, that is, a frame of reference, a theme, an identical topic for all the words of the sentence ... Instead of sifting out one dimension of meaning, the context allows several to pass, indeed, consolidates several of them, which run together in the manner of the superimposed texts of a palimpsest. ... More than one interpretation is then justified by the structure of a discourse which permits multiple dimensions of meaning to be realized at the same time. In short, language is in celebration.] (Ricoeur, 1969; Engl. transl. 1974, p. 94)

There is no hierarchical order between natural language and logical language in the sense that the former is inferior or subordinate to the latter. The two simply reflect different needs: in the first case, the need to ensure the efficacy of the communication; in the second, the need to guarantee the rigor of the inferential process. Logical discourse derives from common (or natural) discourse by a process of differentiation that, in a certain sense, establishes it as a specialist discourse. The two languages are differentiated (and in this sense different), but not in the sense of being completely separate or lacking a common basis—and even less in the sense that they represent the opposition of rationality/irrationality (Mosconi, 1990). Two constants in the history of modern logic are the tendency toward the elimination of psychological aspects and the simplification of the ambiguity of language (a justified simplification, but one that is often anti-economic from a communicative point of view). It is this presupposition that has led to only the responses falling within the logical code being considered correct and all of the others as incorrect, but we think that this prejudice is unjustified.

This was the state of the art when psychology of thought encountered logic in the 1920s. At that time logic was configured as mathematical logic, and psychology adopted it as a framework of reference. In those years, psychology was differentiating itself from philosophy and carving out a place among the sciences; to this end, to comply with requirements of "scientificity," it integrated the speculative approach adopted by philosophy with the experimental approach, basing this on logic, the discipline which constituted the foundation of the other sciences.

This then is the formal point of view that psychology of thought has used to evaluate human reasoning, judgment, decisions, thus creating the *logical-deductive* paradigm.

In fact, from the very beginning and for many decades thereafter, the privileged field of investigation was the study of syllogisms and conditional reasoning in pure logico-philosophical tradition. Since the early studies in the 1960s the predominating vision has been substantially rationalistic, from the logicistic theory of deductive reasoning to probabilistic reasoning up to Piaget's theory of epistemological genetics on cognitive development.

The Unexpected Counterpoint: The Dilemma of Error

The unexpected counterpoint that emerged was the dilemma of how to explain human error, or the deviation of human reasoning from the logical norm. This issue had already emerged (Henle, 1962), but it only became crucial in the second half of the 1960s and even more so in the 1970s as its frequency, transversality, and generality were considered to be indicative of the structural limitations of the human mind. This tendency to make mistakes translated into the concept of *bias*, indicating a systematic deviation from the norm, inherent to the human cognitive system.

From this point on experimental research on reasoning focused on the problem of human error, the incapacity to grasp the deep logical structure (for a review on this issue see Frankish & Evans, 2009). Therefore, psychology of thought had to adopt as a theoretical framework that dualistic vision—already *in nuce* from its emergence—in an attempt to explain both the human capacity for abstraction and correct reasoning, but also this tendency to error, the *bias*, the so-called cognitive illusions. As a result, from the 1970s the principal theories of thought were characterized by a dualistic vision that attributes error to thought processes that are different from those that guarantee correct reasoning.

The general denomination of these theories has brought together models that, although characterized by significant differences, share the historic dichotomy between logical and common reasoning that has always characterized the psychological theories of reasoning (Evans, Handley, Neilens, & Over, 2007; Sloman, 1996; Stanovich & West, 1998, 2000; Wason & Evans, 1975). All the dual models, over and above the various factors that characterize them, distinguish two diverse systems of reasoning: the associative (fast, automatic, and effortless—System 1) and the rule-based (slow, analytic, and effortful—System 2). The associative system reflects "similarity structure and relations of temporal contiguity" and is generally unconscious. "The other is rule based and tries to describe the world by capturing different kinds of structure, structure that is logical, hierarchical, and causal-mechanical" (Sloman, 1996, pp. 3, 6).

Philosophers and psychologists have long held and shared the view that two substantially different ways of thinking coexist in the mind. This representation is very plausible: every day we experience our capacity to reason consciously, applying a reflective process compared to the other modes that would appear to be automatic, intuitive, and that appear not to require any particular cognitive effort. Most of these automatic modes are functional, economic, the outcome of ancient, consolidated learning, and seem to elude any attempt at conscious control. As a result, thought is perceived as a superior cognitive activity, paticularly developed in humans, that identifies with awareness and constitutes a system in itself (S2). By consequence, everything that is intuitive, not conscious (S1) is perceived as being mechanical, automatic, and with little involvement of thought.

Hence, this dichotomic view of the mind focuses on one hand on the role of consciousness and on the other on the capacity for abstraction. In particular, the processing of the context in cognitive processes would be typical of System 1 (or Type 1) and was considered one of the causes of error in classical reasoning tasks, whereas the process of abstraction and de-contextualization has been considered as being emblematic of System 2 (or Type 2), ensuring that the process of reasoning is implemented correctly. Furthermore, "the fundamental computational bias in human cognition would be the tendency toward automatic contextualization of problem (System 1), … while System 2's more controlled processes would decontextualize … and would not be dominated by the goal of attributing intentionality nor the search for conversational relevance" (Stanovich & West, 2000, p. 659).[3]

Paradoxical Effects of Content in Explaining Error

A paradoxical result emerges from research regarding the effect of the content on logical abstract reasoning. In fact, if we compare the research on *belief bias* and on conditional reasoning using deontic material, what seems to be an antithetical phenomenon appears, for which the dichotomic vision has no explanation. In the first case the content provokes the bias, whereas in the latter case the deontic and familiar material activates the inferential schema for the conditional reasoning.

According to the dualistic theories of thought (Evans, 2010; Evans, Barston, & Pollard, 1983; Stupple, Ball, Evans, & Kamal-Smith, 2011), the belief bias occurs when the premises are in conflict with logical validity, as in the example "All flowers fade. All roses fade. All roses are flowers." This constitutes evidence that two systems are involved in evaluating validity and that the system of knowledge and beliefs (S1) prevails. But what type of knowledge or belief is this, if it actually defines the categorical class inclusion, implicitly activated by the terms *roses*, *flowers*, and *fade*?

It is true that from the two premises it is not possible to conclude that "roses are flowers," according to logic. However, this error from a logical point of view can be interpreted as the result of a cooperative act: concluding that roses are flowers does not add knowledge, it only allows the communicative interaction, although at the lowest level of informativeness. As in the Socratic dialogue, rhetorical questions have a common basic function from which the argument will start and will be developed. Saying that "roses are flowers" is to give its own consent on an obvious but shared knowledge. In a nonexperimental context the two premises would raise a spontaneous reply such as "We know this, so what?", the natural equivalent to the logical "nothing ensues."

In the case of the paradigmatic *selection task* (Wason, 1966),[4] on the other hand, the realistic content would activate logical inferential schemas (Johnson-Laird, Legrenzi, & Sonino Legrenzi, 1972),[5] real counterexamples (Griggs & Cox, 1982), or pragmatic reasoning schemas (Cheng & Holyoak, 1985).

At this point it is worthwhile to reflect on the strange case of the "logical aberrations" (*not p* and *q*) demonstrated, with great spirit of innovation, by Mosconi and D'Urso (1975; see Mosconi, chapter 17, this volume), using the same deontic version of the *selection* task adopted by Johnson-Laird et al. ["if a letter is sealed, then it has a

50-lire stamp on it"; selected cards: "sealed letter" (*p*) and "40-lire stamp" (*not q*)]. But this time, Mosconi and D'Urso asked the subjects to check the *inverse* rule ("if a letter is open, then it has a 40-lire stamp on it"). The cards mainly selected were again: "sealed letter" (*not p*) and "40-lire stamp" (*q*) but, this time, these cards (*not p* and *q*) were a non-sense from a logical perspective, even though they represented the crucial cases of potential violators in that context. It is worth noting that the two most selected cards were the same as those that were chosen in Johnson-Laird et al.'s original experiment, but in Mosconi and D'Urso's experiment they represented nonvalid cases.

The paradoxical effects of content that emerge from the two fields of research illustrated above cannot be explained simply by the agreement or disagreement between content and logical validity, with content always prevailing.

Actually, what this appears to mean is that independently of their logical function (*p* and *not q* vs. *not p* and *q*), the cases that are crucial for testing the rule in this specific situation ("sealed" and "40 lire") are those that are selected on the basis of the logic of pertinence and usefulness, rather than on formal logic. When we adopt the relevance criterion, we are processing information on explicit and implicit levels of knowledge, exceeding the logical layer of the conditional sentence. It is possible to understand these paradoxical results only if we consider the implications of the specific content and the relevant rule in that context. Just as content transmits implicit relations that render the material implication significant or not significant, so does the relation of inclusion in classes with categorical syllogisms, as we have seen in the belief bias.

A possible interpretation is that reasoning *always* goes beyond the logical level of meaning and so requires an intensive search for as much information as possible. Thus, the logical meaning is only one of possible interpretations, the less relevant and, therefore, the less likely to be assigned, if the "logic" intent of the task is not adequately transmitted.

In fact, natural language is ambiguous in itself, differently from specialized languages (i.e., logical and statistical ones), which presuppose a univocal, unambiguous interpretation. The understanding of what a speaker means requires a disambiguation process centered on the intention attribution.

The Emergence of the Shadow Area of Thought

Recently, certain weaknesses and criticisms of the dual theories have been discussed (Macchi, Over, & Viale, 2012). A number of compelling questions have been raised: What are the roles played by the abstraction processes, the meta-representational ability, and the working memory in distinguishing Type 1 and Type 2 processes? Is it possible to extrapolate essential factors in these processes? How do these two types of processes work and interact? In particular, what is the relation between implicit and explicit thought, between intuition and reflection, in regard to what is emerging from the shift in the paradigm of the psychology of reasoning, on which there is an ongoing debate? The new paradigm goes beyond the idealized logic of certainty, which was typical of both the old deduction paradigm and the dual process theories, toward a more realistic logic of uncertainty moving toward a general probabilistic explanation of human reasoning (Oaksford & Chater, 2006, 2013; Over, 2009). To deal with the inevitable uncertainties of reasoning in the real world, thought must of necessity process probabilities, implicit assumptions, and beliefs. This leads to a more complex view of rationality, less concerned with and less focused on the quantitative limits of conscious thought, imposed by the working memory capacity, and more centered on the ability to contextualize and exploit the complexity of what is real, moving toward a bounded-grounded form of rationality.

This has resulted in a revaluation of implicit-intuitive thought, the "shadow zone" (Macchi & Bagassi, 2012) which in the past was considered responsible for automatisms and frequently for errors in reasoning. Now the dualistic view of thought and the logical-deductive paradigm are being questioned, and this implicit-intuitive thought is acquiring a more central position in interaction with explicit thought. Intuition is able to deal rapidly with numerous items of information, interpreting them in parallel (Betsch, 2007; Betsch & Held, 2012), not limited by the working memory capacity. Contextual elements, beliefs, implicatures, and propositional attitudes are an integral part of human thought (Frankish, 2012); they contribute to conceiving the emotional response as a basis of the theory of meta-cognitive judgment, which predicts the activation of analytic processes (Thompson & Morsanyi, 2012). However, although implicit processing is losing the negative connotation that characterized it in the dual process models, it is still subordinate to explicit processing, to which some researchers continue to exclusively attribute *cognitive decoupling* and *hypothesis testing*. Although this perspective recognizes the fundamental independence of the Type 1 and Type 2 processes (also through the role of preattentive contents in reflective thought), it still attributes the ability to conceive novelty to Type 2 processes and therefore to conscious thought (Evans, 2012a, 2012b; Evans & Stanovich, 2013). Even theories that support the explicit-implicit interaction (EII—Hélie & Sun, 2010) assign the last word, the selective ability, to explicit thought; in this way the solution search process is iterated until the outcome does yield a solution. The contribution of implicit thought is almost obtained by chance, associatively, via spatial-temporal contiguity.

This phase of reflection has therefore identified the requisites for a creative process of reasoning in a conscious search, dependent on the working memory capacity or, more precisely, the capacity to manage more data and to perceive further representations of these data (take, for example, tests used to measure the working memory capacity and divergent thought). Yet again, rationality is conceived in quantitative terms and as such is measurable.

Insight Problem Solving: The Challenge of the "Mysterious Events"

The legacies from the extrapsychological disciplines such as logic, the theory of probability, and so on have conditioned the psychology of reasoning and have had a negative impact in terms of originality and adequacy in psychological research. The research on problem solving, however, was not affected by these legacies. Given that, in a manner of speaking, any situation that involves thought processes can be considered a problem, solving or attempting to solve problems is the typical and, hence, general function of thought. This then is the scenario in which the question of the relationship between explicit and implicit thought can be reconsidered in a genuine psychological perspective.

According to the consolidated theoretical tradition that started with Simon and Newell's *Human Information Processing Theory* (Simon & Newell, 1971; Newell & Simon, 1972) and has continued through to the present day (Weisberg, 2015), the labyrinth is an appropriate abstract model for human reasoning. A person faced with a problem moves within the *problem space* just as in a labyrinth; he searches for the right path, retraces his steps when he comes up against a dead end, and sometimes even returns to the starting point; he will form and apply a strategy of sorts, making a selective search in the problem space. The labyrinth model, which was devised for problems in which the difficulty lies in the calculation, the number of operations to be performed, and the quantity of data to be processed and remembered (e.g., the *Cryptoarithmetic* or the *Missionaries and Cannibals* problem), is advocated also when a change of representation is necessary, extending the selective search process to the meta-level of possible problem spaces for an alternative representation of the problem (the *Mutilated Checkerboard* problem, Kaplan & Simon, 1990).

Hence, this consolidated theoretical tradition maintains that conscious analytical thought can reorganize data if the initial representation does not work, extracting information from the failure to search for a new strategy (Fleck & Weisberg, 2013; Kaplan & Simon, 1990; Perkins, 1981; Weisberg, 2015). According to Weisberg, analytic thinking, given its dynamic nature, can produce a novel outcome; in problem solving in particular, it can generate a complex interaction between the possible solutions and the situation, such that new information constantly emerges, resulting in novelty. It could be said that, when we change strategy, we select a better route; this change is known as "restructuration without insight," which still remains on the conscious, explicit plane. Weisberg, rather optimistically, claims that, when the restructuring does not occur and the subject is at an impasse, this "may result in coming to mind of a new representation of the problem. … That new representation may bring to mind at the very least a new way of approaching the problem and, if the problem is relatively simple, may bring with it a quick and smooth solution" (2015, p. 34).

However, this concept of intelligence does not explain how we suddenly see the solution to a problem after a period of groping in the dark (the impasse), during which our conscious analytical thought has stalled and does not extract any useful information from the failure; thought is incapable of throwing light on how to reach the solution, and if the solution is hit upon, it has the characteristics of a "mysterious event."

In fact, there are situations in which the difficulty does not lie in the onerosity and complexity of the processing to be implemented, in the capacity to consciously evaluate the merits of the various possible outcomes and to settle on a suitable strategy. The issue is the difficulty in changing the representation that does not appear to involve the working memory capacity, nor the conscious retrieval from memory of solutions or crucial parts of solutions to reproduce.[6] It requires of necessity not conscious, implicit processing—incubation—and for this reason it is a challenge to the current theories of thought, regarding the central role of consciousness, given that, in these cases, restructuring (or change in representation) is not performed consciously.

In everyday life too we sometimes come up against an impasse; our analytical thought seems to fail us, and we just cannot see our way out of the dead end into which we have inadvertently strayed. We know we are in difficulty, but we have no idea what to do next. We have no strategy, and this time failure does not come to our aid with any information that could help us forge forward. In other words, we have come up against a deep impasse. These are situations that change our life radically if we do find a solution: the situation, which a moment before seemed to lead nowhere, suddenly takes on a different meaning and is transformed into something completely new. The literature discusses studies in which these situations are reproduced as experimental paradigms of insight problem solving.

Such problems look deceptively simple (Metcalfe, 1986) in that they contain little in the way of data, and the number of operations required to arrive at the solution appears limited (as for example in the *Nine-dot* problem). However, they are not simple at all. Sometimes, taken in by this apparent simplicity we are tempted by the first answer that comes to mind, but it is almost always wrong (for example, *Horse-trading* problem, Macchi & Bagassi, 2015—see appendix A), and from that point on, we are faced with an impasse. Sometimes an impasse is encountered immediately, we have no idea how to go about finding the solution, and the problem itself may initially appear impossible to solve (for example, *The Study Window* problem, Macchi & Bagassi, 2014, 2015—see appendix A).

These problems may seem bizarre or constructed to provide an intellectual divertissement, but they are a paradigmatic case of human creativity in which intelligence is at its acme. Their study provides a privileged route to understanding the processes underlying creative thought, scientific discovery and innovation, and all situations in which the mind has to face something in a new way. Insight problems have traditionally been considered tests of giftedness (see, for instance, Sternberg & Davidson, 1986), and more recently, in the debate on the dual process theories, Evans and Frankish (2009) have attributed their solution to Type 2 processes, characterized by high cognitive capacities. However, what type of intelligence is this, if it does not coincide with explicit analytical thought (Macchi, Bagassi, & Passerini, 2006)?

According to the Gestaltists, finding the solution to an insight problem is an example of creative, "productive thought." In addition to the reproductive activities of thought, there are processes that create, "produce" that which does not yet exist. This is characterized by a switch in direction that occurs together with the transformation of the problem or a change in our understanding of an essential relationship. The famous "Aha!" experience of genuine insight accompanies this change in representation, the *restructuring*. The Gestalt vision has been invoked, with different inflections, by the *special process* views of insight problem solving, investigating the qualitatively diverse processes that elude the control of consciousness through spreading activation of unconscious knowledge (Ohlsson, 2012; Ollinger, Jones, & Knoblich, 2008; Schooler, Ohlsson, & Brooks, 1993).[7] This search goes beyond the boundaries of the working memory. The process that leads to the discovery of the solution through restructuring is mainly unconscious, characterized by a period of incubation, and can only be described a posteriori (Gilhooly, Fioratou, & Henretty, 2010). The characteristic unconsciousness with which these processes are performed has led to them being defined as automatic, spontaneous associations (Ash & Wiley, 2006; Fleck, 2008; Fleck & Weisberg, 2013; Schooler et al., 1993).

On the other hand, the cognitivist approach, known as *business as usual*, requires a greater working memory capacity for explaining insight problems: intelligence fades without the "light" of consciousness and therefore identifies with conscious reflective thought.

Both of these approaches have critical aspects that reveal the complexity of the issue on hand: the first grasps the specificity of the phenomenon of discovery, which characterizes insight problems and creative thought, but is not in a position to identify the explicative processes because it does not attribute a selective quality to unconscious processes as they continue to be merely associative, automatic, and capable of producing associations that will contribute to finding the solution almost by chance. The limit of the *business as usual* approach, on the other hand, is that it levels off the specificity of genuine insight problems, lumping them together with problems that can be solved with explicit analytic thought processes. Finally, this approach makes little progress in explaining the so-called "mysterious event" in solution of insight problems, relegating them to situations of normal administration that can be dealt with by conscious analytical thought. However, when the solution is mainly the result of a covert and unconscious mind-wandering process, it cannot be attributed, even exclusively, to reflective, conscious thinking (Baird et al., 2012; Macchi & Bagassi, 2012; Smallwood & Schooler, 2006).

Restructuring as Reinterpreting: The Emergence of *Unconscious Analytic* Thought

Incubation, which remains the core issue in the renewed interest in insight problems is still a crucial question to be solved (Fleck & Weisberg, 2013; Gilhooly, Georgiou, & Devery, 2013; Gilhooly, Georgiou, Sirota, & Paphiti-Galeano, 2015; Macchi & Bagassi, 2012; Sio & Ormerod, 2009).

A heterogeneous complex of unresolved critical issues underlies the research on this subject (for a review, see Sio & Ormerod, 2009) and still revolves around the controversy of the relationship between the conscious and unconscious layers of thought in the solution of insight problems.

However, the various mechanisms that have been proposed only describe the characteristics of the solution but do not explain the processes of reasoning that have made the solution possible (these include, for example, eliciting new information, selective forgetting, strategy switching, and relaxing self-imposed inappropriate constraints).[8]

Moreover, a general characteristic that is common to the literature on insight problems in general, and in particular on the incubation-solution relationship, is the total absence of an analysis of the types of difficulty found in individual insight problems. In other words, what makes a difficult insight problem difficult? What kinds of difficulty are we facing? If it were possible to lay them out as a continuum in an ascending order of difficulty, we would see that, in fact, the difficulty correlates with the incubation, and this, in turn, with the possible discovery of the solution, thus allowing the restructuring process to occur. Therefore, incubation may offer a measure of the degree and type of difficulty of the problem, as it may vary in length, depending on the degree of gravity of the state of impasse (Macchi & Bagassi, 2012; Segal, 2004).[9]

At this point, the question is what kind of unconscious intelligent thought operates during the incubation to solve these special problems. Through experiments in brain imaging, it is now possible to identify certain regions of the brain that contribute both to unconscious intuition and to the processing that follows. Jung-Beeman, Bowden, Haberman, Frymiare, Arambel-Liu, Greenblatt, et al. (2004) found that creative intuition is the culmination of a series of transitional cerebral states that operate in different sites such as the anterior cingulate of the prefrontal cortex and the temporal cortex of both hemispheres and for different lengths of time. According to these authors, creative intuition is a delicate mental balancing act that requires periods of concentration from the brain but also moments in which the mind wanders and retraces its steps, in particular during the incubation period or when it comes up against a dead end.

Incubation is actually the necessary but not sufficient condition to reach the solution. It allows the process but does not guarantee success; however, if it is inhibited, for example by compelling participants to verbalize, the solution process will be impeded.

The study of the verbalization effect, indeed, offers a promising line of research[10] to study the thought processes underlying the solution. In a recent study (Macchi &

Bagassi, 2012), the "verbalization" procedure was adopted as an indirect method of investigating the kind of reasoning involved in two classical insight problems, the *Square and Parallelogram* (Wertheimer, 1925) and the *Pigs in a Pen* (Schooler et al., 1993); the investigation focused on whether concurrent serial verbalization would disrupt insight problem solving.[11] The hypothesis was that correct solutions would be impaired if a serial verbalization procedure were to be adopted, as it would interfere with unconscious processing (incubation). We found that the percentage of participants who successfully solved insight problems when verbalizing the process used to reach the solution was inferior to that of the control subjects who were not instructed to do so. In the *Square and Parallelogram* problem there were 80% correct responses in the no-verbalization condition versus only 37% correct responses in the verbalization condition. The difference increased in the *Pigs in a Pen* problem: the percentage of correct responses was 12% in the verbalization condition and 87% in the no-verbalization condition. Our hypothesis was further confirmed by a study on the *Mutilated Checkerboard* problem (Bagassi, Franchella, & Macchi, 2015), in which the no-verbalization condition significantly increased the number of solutions with respect to the control condition of verbalization, this latter being in accordance with the procedure adopted by Kaplan and Simon (1990).

Schooler, Ohlsson, and Brooks (1993) also investigated the effect of verbalization on insight, suggesting that language can impair solution and therefore thinking. They claim "that insight involves processes that are distinct from language" (p. 180), given the nonverbal characteristic of perceptual processes. This view follows the traditional dichotomous theory, according to which language, considered extraneous to the higher-order cognitive processes involved in solution, impairs thinking.

We take a different view than Schooler et al. (1993), although their study was extremely stimulating and innovative; in our opinion, language too has a nonreportable side in its implicit, unconscious dimension, which belongs to the common experience and domain. In fact language as a communicative device is realized by a constant activity of disambiguation, by covert, implicit, unconscious processes, nonreportable in *serial* verbalization. When the participants in these studies were still exploring ways of discovering a new problem representation, they were not able to express consciously and therefore to verbalize their attempts to find the solution. Indeed, our data showed that serial "on-line" verbalization, compelling participants to "restless" verbalization, impairs reasoning in insight problem solving; this provides support to the hypothesis of the necessity of an incubation period during which the thinking processes involved are mainly unconscious. During this stage of wide-range searching, the solution still has to be found, and verbalization acts as a constraint, continuously forcing thought back to a conscious, explicit level and maintaining it in the impasse of the default representation. Conscious explicit reasoning elicited by verbalization clings to the default interpretation, thus impeding the search process, which is mainly unconscious and unreportable.[12]

Hence, we speculate that the creative act of restructuring implies high-level implicit thought, a sort of *unconscious analytic thought,* informed by relevance, where *analytic thought* is not to be understood in the sense of a gradual, step-by-step simplification of the difficulties in the given problem, but as the act of grasping the crucial characteristics of its structure.

The same data are seen in a different light and new relations are found by exploring different interpretations, neither by exhaustive searches nor by abstractions but by involving a relationship between the data that is most pertinent to the object of the problem. In this way, each individual stimulus takes on a different meaning with respect to the other elements and to the whole, contributing to a new representation of the problem, to the understanding of which it holistically concurs. Indeed, the original representation of the data changes when a new relation is discovered, giving rise to a gestalt, a different vision of the whole, which has a new meaning.

In other words, solving an insight problem—restructuring—means discovering a new perspective, a different sense to the existing relations. The interrelations between the elements of the default interpretation have to be loosened in order to perceive new possibilities, to grasp among many salient cues what is the most pertinent with the aim of the task, or in other words, to reach the understanding. This type of process cuts across both conscious and unconscious thought.

Recent experimental studies and theoretical models are now starting to consider the link between explicit and implicit information processing, in that continuous fluctuation of thought between the focus on the task, the data, and the explicit request, and then the withdrawal into an internal dimension, tempering the processing of the external stimuli to slip into an *internal train of thought* (stimulus-independent thought—*SIT*)[13] so as to allow goals other than those that caused the impasse to be considered. This internal processing that takes place during incubation has a neural correlate in the "default mode network" (*DMN;* Raichle, MacLeod, Snyder, Powers, Gusnard, & Shulman, 2001), that is, in a coordinate system of neural activities that continue in the absence of an external stimulus.[14] Moreover, "the sets of major brain networks, and their decompositions into subnetworks, show close correspondence between the independent analyses of resting and activation brain dynamics" (Smith, Fox, Miller, Glahn, Fox, Mackay, et al., 2009, p. 13040). It is interesting to note the close correspondence, as neural networks activated, between the brain activity when focused on a task and that recorded during the resting state. The brain seems to work dynamically in different ways on the same task, at a conscious and not-conscious layer, as the substrate of a restless mind (Smallwood & Schooler, 2006; Smallwood, McSpadden, Luus, & Schooler, 2008).

During incubation, when an overall spreading activation of implicit, unconscious knowledge is under way, in the absence of any form of conscious control, *relevance constraint* allows multilayered thinking to discover the solution, as a result of the restless *mind wandering* between the implicit and the explicit levels in search of the relationship of the data that would finally offer an exit from the impasse.

Rather than abstracting from contextual and more specific elements, as in accordance with the logical approach, we exploit these elements, grasping the gist that provides the maximum of information in view of the aim; we can indeed abstract, if this is the acknowledged purpose. Giving sense to something, *understanding*, does not derive from a summation of semantic units, each with a univocal, conventional meaning. These are simply inputs for that activity of thought, which is crucial to dynamically attributing the most relevant relationship to the recognized aim, in an inferential game that can result in interpretations that can be quite different from those originally intended. Depending on which (new) aim of the task is assumed, the relationship between the items of information changes; when the relationship is changed, the meaning that each element takes on changed too. Incredibly, the result is not cognitive chaos, but a sharper understanding.

The Interpretative Function of Thought

The search for meaning, in view of an objective, characterizes every activity of the human cognitive system at all levels, from perception to language and to reasoning.

For instance, syntax in itself does not give direct access to meaning, only on the basis of the rules that discipline it. Similarly, perceiving external reality is not just a question of registering stimuli; sensorial data have to be interpreted and organized, and relations created.

When analyzing sensory experience it is important to realize that our conscious sensations differ *qualitatively* from the physical properties of stimuli because, as Kant and the idealist predicted, the nervous system extracts only certain pieces of information from each stimulus while ignoring others. It then *interprets* this information within the constraints of the brain's intrinsic structure and previous experience. Thus, ... sounds, words, music, ... color, tones ... are mental creations constructed by the brain out of sensory experience. They do not exist as such outside the brain. (Gardner & Johnson, 2013, p. 455)

Even when we use an object, the interpretative function comes into play, as Rizzolatti and Strick (2013) suggest:

... objects usually have more than an affordance and may be grasped in several ways. How does the brain determine which is optimal? Behavioral analysis of grasping, for instance, reveals that nonvisual factors determine the choice of affordance and thus how an object will actually be grasped. These factors relate both to what the object is for and the individual's intent at that moment. ... Thus, when an unconventional use of an object is intended, the prefrontal input could override the selection of *standard* affordances and select those affordances that are congruent with the individual's intention. (pp. 421–422)

Routine activities such as recognizing a face, admiring a landscape, or looking at a painting, activities that seem to be nothing more than input from our visual capacity, are actually "analytical triumphs" made possible through our ability to grasp sensorial information holistically and attribute a meaning to it, so that in perceiving a scene, for example, we respond to a whole in which the relationship we establish between the individual parts is crucial.[15]

Our view, therefore, is that this *interpretative function* is a characteristic inherent to all reasoning processes and is an adaptive characteristic of the human cognitive system in general. Rather than abstracting from contextual elements, this function exploits their potential informativeness (Levinson, 1995, 2013; Mercier & Sperber, 2011; Sperber & Wilson, 1986/1995; Tomasello, 2009). It guarantees cognitive economy when meanings and relations are familiar, permitting recognition in a "blink of an eye."

This same process becomes much more arduous when meanings and relations are unfamiliar, obliging us to face the novel. When this happens, we have to come to terms with the fact that the usual, default interpretation will not work, and this is a necessary condition for exploring other ways of interpreting the situation. A restless, conscious and unconscious search for other possible relations between the parts and the whole ensues until everything falls into place and nothing is left unexplained, with an interpretative heuristic-type process.

In our view, the way the human cognitive system functions is recognizable, and it continuously occurs in the process involved in the understanding of an utterance. In fact, language and thought share a unitary cognitive activity, addressed by the interpretative function and grounded on common, universal interactive abilities: "the ability to make models of the other, to 'read' the intentions behind actions, to make rapid interactional moves in an ongoing sequence of actions structured at many levels" (Levinson, 1995, pp. 225–226).

This meta-representational ability always implies an analytic, multilayered reasoning, which works on a conscious and unconscious layer by processing explicit and implicit contents, presuppositions, and beliefs, but is always informed by relevance.[16] Tracing communicative behavior back to its meaning is impossible to compute logically, given that what is said presupposes many plausible, possible premises; nevertheless, the absolute inferential miracle of communication continues to take place.

Sometimes, however, our initial interpretation does not support understanding, and misunderstanding is inevitable; as a result, sooner or later we come up against an impasse. We are able to get out of this impasse by neglecting the default interpretation and looking for another one that is more pertinent to the situation and which helps us grasp the meaning that matches both the context and the speaker's intention; this requires continuous adjustments until all makes sense.

For instance, grasping a witty comment and understanding rhetorical figures of speech such as irony and metaphor are the result of an interpretative process adhering to or detracting from the explicit meaning as the case requires, often all in the twinkling of an eye.

This analysis allows us to consider insight problems as paradigmatic examples of *misunderstanding*, in that they arise from a *qui pro quo*, a glitch in communications. When insight problems are used in research, it could be said that the researcher sets a trap, more or less intentionally, inducing an interpretation that appears to be pertinent to the data and to the text; this interpretation is adopted more or less automatically because it has been validated by use. But in point of fact the solution presupposes a different interpretation that is only theoretically possible, just as when someone makes a witty comment, playing on an unusual interpretation of a sentence.

Indeed, the solution—*restructuring*—is a *re*-interpretation of the relationship between the data and the aim of the task, a search for the appropriate meaning carried out at a deeper level, not by automaticity. If this is true, then a disambiguant reformulation of the problem that eliminates the trap into which the subject has fallen, should produce restructuring and the way to the solution.

The Retrieved Psychological Wholeness in Creative Intelligence

Efficacious reformulation is a sort of "litmus paper," indirect evidence that the reasoning applied by those who found the solution during the incubation period consists of a highly pertinent disambiguation process performed by *unconscious analytical thought*. Therefore, another mode of processing emerges in the activity of unconscious thought, compared to that traditionally characterized as associative-automatic; this is similar to the mode proposed for the conscious analytic thought that operates the function of decoupling of hypothetical thinking (reflective mind) compared to the executive function of sustaining decoupled representations (algorithmic mind) (Stanovich, 2009).

The hypothesis regarding the effect of reformulation has been confirmed in classical insight problems such as the *Square and the Parallelogram*, the *Pigs in a Pen,* the *Bat & Ball* and the *Study Window* in recent studies (Macchi & Bagassi, 2012, 2015) which showed a dramatic increase in the number of solutions (appendix B).

In their original version these problems are true brain teasers, and the majority of participants in these studies needed them to be reformulated in order to reach the solution; it is likely that the few who were able to "see" the solution in the original formulation could be endowed with a high level of cognitive ability of the type we call *interpretative talent*. It could be argued that they possess a particular skill in perceiving the real aim of the task, *re*-interpreting all the informative elements of the situation and recognizing the specific intent behind the problem, thus *re*-contextualizing the task rather than *de*-contextualizing it. This skill would activate an implicit, increasingly focused selective search exploiting unconscious knowledge by covert mind wandering that eludes the control of consciousness.

At this point, it could be speculated that the high-level cognitive abilities required for insight problem solving are the same as those needed to solve classic reasoning tasks. As has been previously discussed, there is a consolidated line of research that claims that the ability to abstract (cognitive decoupling) is a *sine qua non* condition for correct reasoning and for hypotheses testing (Stanovich & West, 1998, 1999, 2000). However, a number of interesting studies in the past ten years have shown that beliefs affect both intuitive and reflective reasoning (Verschueren & Schaeken, 2005) and that many cognitive biases are independent of cognitive abilities (Evans et al., 2007; Stanovich & West, 2008). The question of the negative effect of the content (beliefs, context) on abstract reasoning is once more a subject of debate.

In spite of the significance of these findings, which show the almost ineradicable impact of beliefs on reasoning, in an updated version of dual-process theories Evans and Stanovich (2013) consider reasoning abilities of the higher order to be still identified with hypothetical thought, which is supported by cognitive decoupling and highly correlated to working memory capacity.

This issue can be explored by revisiting the classic studies conducted on probabilistic reasoning with people trained in logic and mathematics, inviting further speculation about the type of cognitive capabilities of those gifted people who solved the same tasks without such training.

A first critical question emerges from an analysis of the results obtained by Tversky and Kahneman (1983) working with *informed* and even *sophisticated* subjects in the field of statistics. Paradoxically, when attempting to solve the well-known *Linda* problem, the majority of these subjects committed the so-called *conjunction fallacy*. In our view, the fact that people trained in decoupling failed to recognize the inclusion rule, which is a relatively elementary logical rule, raises questions that need to be addressed.

With a view to further investigating this topic, we also conducted a study with statistically sophisticated subjects, the majority of whom solved the *Medical Diagnosis* (MD) problem (66% of 35 subjects) but were not able to solve the less computationally complex *Linda* problem (only 14% of the subjects did not commit the conjunction fallacy) (Macchi, Bagassi, & Ciociola, unpublished manuscript). These results suggest that the skills of these subjects allowed them to grasp the informativeness of the data and the aim of the task in Bayesian problems (MD) but did not help them in focusing on the logical relationships of the items in the *Linda* problem. In this case, the misleading contextualization of the task, which focused on an irrelevant sketch of personality, hindered even statistically sophisticated subjects from grasping the intention of the experimenter concerning the inclusion-class rule.

Analogously, in a recent experiment involving the belief bias and the Wason *Selection Task,* we found that the performance of subjects trained in logic did not differ significantly to that of naive participants who had not received such training. The Wason *Selection Task* was used with abstract material: only 13% of the informed subjects gave the correct response, a very similar result to that obtained by the naive group (5%). Paradoxically, a number of subjects were able to execute the material implication truth table correctly, but were not able to identify the only case that was crucial for falsifying the rule. The same paradoxical result occurred with the *Invalid-Believable* syllogism, showing that the influence of beliefs on reasoning is analogous in the two groups (correct responses, respectively, 10% and 4%; appendix C).

How can it be that subjects trained in logic or statistics had such difficulty with these tasks? One would have expected that their training would have facilitated them in decontextualizing and in abstracting, but this was not the case. In the absence of specific indications they reasoned in the same manner as the naive group and used the conversational rules of natural language to interpret. However, when the instructions expressly required an evaluation of the validity of the syllogism according to the criteria of logic, both the statistically trained subjects and the logically naive group "limited" their evaluation simply to the logical validity, reducing the belief bias (respectively, 66% and 41%). Even in the paradigmatic Wason *Selection Task,* once the unidirectional sense of the material implication had been clarified simply by adding the phrase "but not vice-versa" to the rule, the percentage of correct responses increased dramatically to over 50% in both groups.

In tasks of this kind, even though the syllogism is the prototype of deductive reasoning and the selection task is an example of the material implication, subjects are able to grasp their specific meaning, particular to a specialized language, only if adequately transmitted.

In order for the problem to be solved correctly, the interpretation of the task requires a "particularized" formulation (see Grice, 1975; Levinson, 1995, 2000). In the absence of an appropriate translation from the logical to the natural code or of an adequate signaling of the code used, it is only to be expected that the understanding of the syllogism as a discourse proposed by the researcher (and consequently the response to it) is based on discoursive-conversational rules that do not necessarily correspond to the logical code to which the researcher is referring (even taking into account that the particular knowledge of the informed subjects may compensate for the lack of indication or "translation").

We could further speculate that the few who gave the correct answer to the *Linda* problem, the Wason *Selection Task,* or the syllogisms had a particular aptitude for grasping the intention of the researcher and the aim of the task rather than the ability to decouple from contents and contexts. In doing so, they revealed a form of "interactional intelligence."

We suggest that cognitive decoupling could be seen as a form of sophisticated interpretative ability that permits the subject to decouple, but to decouple from the default interpretation and identify that which is most pertinent to the task on hand. A particularly interesting question is how the working memory capacity can support hypothetical thinking. The capacity refers only to the quantitative dimension of conscious information processing the working memory can deal with (as required in working memory capacity tests). It can manage various tasks concurrently but with a limited quantity of information at a time. When our usual thought processes fail, we decouple from the default representation and try new paths until we finally find the solution. When this is not possible at the explicit level, the limits of the working memory are exceeded, activating a preconscious or even unconscious processing that is unconstrained by the attentional focus of the working memory.

Cognitive decoupling assumes continuous "fluctuations in focus between external stimuli ['on task' thoughts] and stimulus-independent thought" (Schooler et al., 2011), which is quite different from a process of distraction. It would appear, therefore, that the crucial element in decoupling is the qualitative dimension. What distinguishes a quirky representation from a representation that, even if it is abstract, still provides an explanation of the phenomenon? In our opinion, it is pertinence, as pertinence exercises a constraint, in this case, a qualitative constraint, on our thought processing, so that an abstraction becomes an explanation.

This view would appear to converge with the *Global Workspace* theory proposed by Baars (1989) and its neuronal correlates of consciousness (Dehaene & Changeux, 2011; Kandel, 2012) as well as with the concept of the restless mind, now known as DMN, which supports the idea that a coordinated system of neural processes continues to process the external stimulus while decoupling from it. Moreover, these forms of decoupling "afford an opportunity for dishabituation," also because there is no evidence that the default interpretation has been abandoned, increasing the chances of starting a new train of thought (Schooler et al., 2011, p. 324).

At this point in the debate, what has changed in our concept of human rationality? It could be said that, in spite of the quantitative constraints that characterize our thought processes, we are subtly adaptive to novelty because we use default strategies and construct new ones as required, always applying heuristic procedures to grapple with problems. Although heuristics are judged to be *weak methods* and do not guarantee success in problem solving, heuristics are the only means available as exhaustive searching does not characterize human reasoning.

What happens when the difficulty does not lie in the computational onerosity of the problem but in the need for a new representation? In this case our mind, free from the quantitative limits of the consciousness, fluctuates between the conscious and unconscious levels in a wide-range analytical search.

Paradoxically, it is due to these constraints that analytical thought can achieve scientific discovery and create poetry. As psychoanalysis predicted, unconscious processes appear to underlie almost every aspect of our conscious existence, promoting a creative and imaginative dimension capable of *re*-describing reality, revealing unedited dimensions.

Notes

1. We quote here the reflections of the philosopher and psychologist Paolo Bozzi (1930–2003) of the school of Gaetano Kanizsa and Vittorio Benussi.

2. The following considerations on the negation are derived from numerous conversations on this issue with Giuseppe Mosconi, always a source of inspiration and intellectual divertissement.

3. In the last decades there has been a growing interest in the dual process theories of reasoning from different perspectives, which however still share the dichotomic view (Evans, 2012; Evans & Frankish, 2009; Frederick, 2005; Kahneman, 2011; Manktelow, Over, & Elqayam, 2011; Stanovich & Toplak, 2012).

4. In the Wason *selection task* (1966), the subjects were presented with the following sentence, "*if there is a vowel on one side of the card, then there is an even number on the other side*" (*p→q*), together with four cards, each of which had a letter on one side and a number on the other side ("A," "B," "2," "3"), and either would be face upward. The task was to select just those cards that would have to be turned over in order to discover whether the rule was true or false. The results of this study showed that only a minority select "3" (*not q*) (bias toward verification).

5. In order to determine whether the use of realistic materials would improve performance in this task, a well-known experiment was performed by Johnson-Laird, Legrenzi, and Sonino Legrenzi (1972). The *Post-office Worker* task involved selecting from a set of envelopes those that, if they were turned over, could violate a given rule. The rule concerned the realistic relation, "*if a letter is sealed, then it has a 50 lire stamp on it.*" Almost all of the subjects made the correct answer with the realistic material ("sealed letter" and "40 lire stamp"—*p* and *not q*).

6. With regard to this issue, it is interesting to take a closer look at the *nine-dot* problem as investigated by Weisberg and Alba (1981). The authors gave the subjects—experiment 1, fourth group—the hint "line 1 + line 2," which is the crucial part of the solution. In fact "their latter 10 attempts were carried out on forms of the problem that contained the first two lines of the solution, and the subjects were so informed" (p. 173). Our claim is that, in this case, the solution was not obtained by restructuring after impasse but by retrieving and using knowledge available in memory. What still has to be explained is the process of thinking of the very few, who do actually solve the problem of the nine dots, without hints. This indeed can be considered and constitutes a "mysterious event."

7. Jung-Beeman et al. (2004) and Bowden et al. (2005) allege the same processes as an explanation for CRA problems, which they define as "mini-insight" problems, with the support of neuroimaging methods and electrophysiological recordings. The transferability of their results to the

processes involved in real insight problems is still a debatable issue (Masoni, Praticò, Bagassi, & Macchi, 2015).

8. The most debated topics, which are also the subject of experimental research, include the duration of the periods of preparation and incubation; the difficulty of the assigned tasks, demanding versus undemanding during the incubation period; and types of insight problem, visual versus linguistic.

9. The existence of multiple types of incubation effect (Sio & Ormerod, 2009) would, in our view, indicate that the impasse has different levels of difficulty. For example, compare the *Candle Problem* (Duncker, 1945) to the *Square and the Parallelogram Problem* (Wertheimer, 1925).

10. A recent contribution by Fleck and Weisberg (2013) takes the opposite view: according to these authors, verbalization does not have a nullifying effect, and therefore they propose that the restructuring that takes place after a period of incubation and accompanied by insight is only one of the possible ways of solving the problem.

11. We adopted two reformulated versions of the two problems (experiment 1; Macchi & Bagassi, 2012). The results obtained using these versions were significantly different from the results obtained with the control versions.

12. Dehaene and Changeux (2011), with their model Global Neuronal Workspace (GNW), provide indirect evidence of these processes through converging neuroimaging and neurophysiological data. Verbalization would prevent the long-distance axons of GNW neurons to broadcast specific neural representations to many other processors brain-wide. Global broadcasting allows information to be more efficiently processed (because it is no longer confined to a subset of nonconscious circuits but can be flexibly shared by many cortical processors) and eventually to be verbally reported (because these processors include those involved in formulating verbal messages).

13. Stmulus-independent thought occupies almost of 50% waking thought (Killingsworth & Gilbert, 2010, p. 932; Schooler et al., 2011).

14. "In cognitive neuroscience, researchers have examined neural processes that occur in the absence of an explicit task (such as in the resting state) in part due to an observation of a coordinated system (including the posterior parietal cingulate, the medial prefrontal cortex (medial PFC) and the medial temporal lobes),–now known as the 'default mode network' (DMN) that exhibits neural activity that often continues in the absence of an external task" (Schooler et al., 2011, p. 319).

15. "… human neuroimaging methods and electrophysiological recordings during conscious access, under a broad variety of paradigms, consistently reveal a late amplification of relevant sensory activity, long-distance cortico-cortical synchronization at beta and gamma frequencies, and 'ignition' of a large-scale pre-fronto-parietal network" (Dehaene & Changeux, 2011, p. 209).

16. The hypothesis that interpretation uses the conscious-unconscious level of thought is supported by the studies on selective spatial attention in cluttered displays of Dehaene and Changeux (2011). According to these authors the gate to *the conscious access* (working memory) is

always the outcome of a *nonconscious selection*, "... separation of relevant versus irrelevant information, based on its saliency or relevance to current goals ... that regulates which information reaches conscious processing" (pp. 201, 202).

References

Ash, I. K., & Wiley, J. (2006). The nature of restructuring in insight: An individual-differences approach. *Psychonomic Bulletin & Review, 13*, 66–73.

Baars, B. J. (1989). *A cognitive theory of consciousness.* Cambridge, MA: Cambridge University Press.

Bagassi, M., Franchella, M., & Macchi, L. (2015). High cognitive abilities or interactional intelligence in insight problem solving? Manuscript under review.

Baird, B., Smallwood, J., Mrazek, M. D., Kam, J. W. Y., Franklin, M. S., & Schooler, J. W. (2012). Inspired by distraction: Mind wandering facilitates creative incubation. *Psychological Science, 23*, 1117–1122.

Betsch, T. (2007). The nature of intuition and its neglect in research on judgment and decision making. In H. Plessner, C. Betsch, & T. Betsch (Eds.), *Intuition in judgment and decision making* (pp. 3–22). Mahwah, NJ: Lawrence Erlbaum Associates.

Betsch, T., & Held, C. (2012). Run and jump modes of analysis. *Mind and Society, 11*(1), Special issue on: Dual process theories of human thought: The debate, 69–80.

Bowden, E. M., Jung-Beeman, M., Fleck, J., & Kounios, J. (2005). New approaches to demystifying insight. *Trends in Cognitive Sciences, 9*, 322–328.

Cheng, P., & Holyoak, K. (1985). Pragmatic reasoning schemas. *Cognitive Psychology, 17*, 391–416.

Dehaene, S., & Changeux, J.-P. (2011). Experimental and theoretical approaches to conscious processing. *Neuron, 70*, 200–227.

Duncker, K. (1945). *On problem solving. Psychological monographs,* Vol. 58 (270). Berlin: Springer. (Original work in German, *Psychologie des produktiven Denkens,* published 1935.)

Evans, J. S. B. T. (2010). *Thinking twice. Two minds in one brain.* Oxford: Oxford University Press.

Evans, J. S. B. T. (2012a). Dual-process theories of reasoning: facts and fallacies. In K. J. Holyoak & R. G. Morrison (Eds.), *The Oxford handbook of thinking and reasoning.* (pp. , 115-133). Oxford: Oxford University Press.

Evans, J. S. B. T. (2012b). Spot the difference: distinguishing between two kinds of processing. *Mind & Society,* Special issue on: Dual process theories of human thought: The debate, *11*(1), 121–131.

Evans, J. S. B. T., Barston, J. L., & Pollard, P. (1983). On the conflict between logic and belief in syllogistic reasoning. *Memory & Cognition, 11*(3), 285–306.

Evans, J. S. B. T., & Frankish, K. (Eds.). (2009). *In two minds*. Oxford: Oxford University Press.

Evans, J. S. B. T., Handley, S., Neilens, H., & Over, D. E. (2007). Thinking about conditionals: A study of individual differences. *Memory & Cognition, 35*, 1772–1784.

Evans, J. S. B. T., & Stanovich, K. E. (2013). Dual-process theories of higher cognition: Advancing the debate. *Perspectives on Psychological Science, 8*, 223–241, 263–271.

Fleck, J. I. (2008). Working memory demands in insight versus analytic problem solving. *European Journal of Cognitive Psychology, 20*, 139–176.

Fleck, J. I., & Weisberg, R. W. (2013). Insight versus analysis: Evidence for diverse methods in problem solving. *Journal of Cognitive Psychology 25*(4), 436–463.

Frankish, K. (2012). Dual systems and dual attitudes. *Mind and Society, 11*(1), Special issue on: Dual process theories of human thought: The debate,41–51.

Frankish, K., & Evans, J. S. B. T. (2009). The duality of mind: An historical perspective. In J. S. B. T. Evans & K. Frankish (Eds.), *In two minds* (pp. 2–29). Oxford: Oxford University Press.

Frederick, S. (2005). Cognitive reflection and decision making. *Journal of Economic Perspectives, 19*(4), 25–42.

Gardner, E. P., & Johnson, K. O. (2013). Sensory coding. In E. R. Kandel, J. Schwartz, T. Jessell, S. Siegelbaum, & A. J. Hudspeth (Eds.), *Principles of neural science* (pp. 449–474). New York: Mc-Graw-Hill.

Gilhooly, K. J., Fioratou, E., & Henretty, N. (2010). Verbalization and problem solving: Insight and spatial factors. *British Journal of Psychology, 101*, 81–93.

Gilhooly, K. J., Georgiou, G. J., & Devery, U. (2013). Incubation and creativity: Do something different. *Thinking & Reasoning, 19*, 137–149.

Gilhooly, K. J., Georgiou, G. J., Sirota, M., & Paphiti-Galeano, A. (2015). Incubation and suppression processes in creative problem solving. *Thinking & Reasoning, 21*(1), 130–146.

Grice, H. P. (1975). Logic and conversation. In P. Cole & J. Morgan (Eds.), *Syntax and semantics* (Vol. 3). *Speech acts* (pp. 41–58). New York: Academic Press.

Griggs, A., & Cox, J. M. (1982). The elusive thematic-materials effect in Wason's selection task. *British Journal of Psychology, 73*(2), 407–420.

Hélie, S., & Sun, R. (2010). Incubation, insight, and creative problem solving: A unified theory and a connectionist model. *Psychological Review, 117*, 994–1024.

Henle, M. (1962). On the relation between logic and thinking. *Psychological Review, 69*, 366–378.

Johnson-Laird, P. N., Legrenzi, P., & Sonino Legrenzi, M. (1972). Reasoning and a sense of reality. *British Journal of Psychology, 63*(3), 395–400.

Jung-Beeman, M., Bowden, E. M., Haberman, J., Frymiare, J. L., Arambel-Liu, S., Greenblatt, R., et al. (2004). Neural activity when people solve verbal problems with insight. *PLoS Biology, 2*(4), 1–11.

Kahneman, D. (2003). A perspective on judgment and choice: mapping bounded rationality. *American Psychologist, 58*(9), 697–720.

Kahneman, D. (2011). *Thinking: Fast and slow*. New York: Farrar, Straus and Giroux

Kandel, E. R. (2012). *The age of insight*. New York: Random House.

Kaplan, C. A., & Simon, H. A. (1990). In search of insight. *Cognitive Psychology, 22*, 374–419.

Killingsworth, M. A., & Gilbert, D. T. (2010). A wandering mind is an unhappy mind. *Science, 330*, 932.

Levinson, S. C. (1995). Interactional biases in human thinking. In E. N. Goody (Ed.), *Social intelligence and interaction* (pp. 221–261). Cambridge: Cambridge University Press.

Levinson, S. C. (2000). *Presumptive meanings: The theory of generalized conversational implicature*. Cambridge, MA: MIT Press.

Levinson, S. C. (2013). Cross-cultural universals and communication structures. In M. A. Arbib (Ed.), *Language, music, and the brain: A mysterious relationship* (pp. 67–80). Cambridge, MA: MIT Press.

Macchi, L., & Bagassi, M. (2012). Intuitive and analytical processes in insight problem solving: A psycho-rhetorical approach to the study of reasoning. *Mind & Society*, Special issue on: Dual process theories of human thought: The debate, *11*(1), 53–67.

Macchi, L., & Bagassi, M. (2014). The interpretative heuristic in insight problem solving. *Mind & Society, 13*(1), 97–108.

Macchi, L., & Bagassi, M. (2015). When analytic thought is challenged by a misunderstanding. *Thinking & Reasoning, 21*(1), 147–164.

Macchi, L., Bagassi, M., & Ciociola, P. B. Normative performances in classical reasoning tasks: high cognitive abilities or interactional intelligence? Unpublished manuscript.

Macchi, L., Bagassi, M., & Passerini, G. (2006). *Biased communication and misleading intuition of probability*. International Workshop on "Intuition and affect in risk perception and decision making," Bergen (Norway), November 3-4, oral presentation.

Macchi, L., Over, D., & Viale, R. (Eds.). (2012). Special issue: Dual process theories of human thought: The debate. *Mind & Society, 11*, 1.

Maffei, L. (2012). *La libertà di essere diversi*. Bologna: Il Mulino.

Manktelow, K. I., Over, D. E., & Elqayam, S. (2011). Paradigms shift: Jonathan Evans and the science of reason. In K. I. Manktelow, D. E. Over, & S. Elqayam (Eds.), *The science of reason: A Festschrift in Honour of Jonathan St. B.T. Evans* (p. XX). Hove, U.K.: Psychology Press.

Masoni, L., Praticò, P. Bagassi, M., & Macchi, L. (2015). Insight and CRA problems: A methodological issue. Manuscript under review.

Mercier, H., & Sperber, D. (2011). Why do humans reason? Arguments for an argumentative theory. *Behavioral and Brain Sciences, 34*, 57–111.

Metcalfe, J. (1986). Feeling of knowing in memory and problem solving. *Journal of Experimental Psychology: Learning, Memory, and Cognition, 12*(2), 288–294.

Mosconi, G. (1990). *Discorso e pensiero*. Bologna: Il Mulino.

Mosconi, G., & D'Urso, V. (1974). *Il farsi e il disfarsi del problema*. Firenze: Giunti-Barbera.

Mosconi, G., & D'Urso, V. (1975). *The selection task from the standpoint of the theory of double code*. In International Conference on "Selection Task," Trento.

Newell, A., & Simon, H. A. (1972). *Human problem solving*. Englewood Cliffs, NJ: Prentice-Hall.

Oaksford, M., & Chater, N. (2006). *Bayesian Rationality*. Oxford: Oxford University Press.

Oaksford, M., & Chater, N. (2013). Dynamic inference and everyday conditional reasoning in the new paradigm. *Thinking & Reasoning, 19*, 346–379.

Ohlsson, S. (1984). Restructuring revisited I: Summary and critique of the Gestalt theory of problem solving. *Scandinavian Journal of Psychology, 25*, 65–78.

Ohlsson, S. (2012). The problems with problem solving: Reflections on the rise, current status, and possible future of a cognitive research paradigm. *Journal of Problem Solving, 5*(1), 101–128.

Ollinger, M., Jones, G., & Knoblich, G. K. (2008). Investigating the effect of mental set on insight problem solving. *Experimental Psychology, 55*(4), 270–282.

Over, D. E. (2009). New paradigm psychology of reasoning: Review of "Bayesian Rationality." *Thinking & Reasoning, 15*, 431–438.

Perkins, D. (1981). *The mind's best work*. Cambridge, MA: Harvard University Press.

Raichle, M. E., MacLeod, A. M., Snyder, A. Z., Powers, W. J., Gusnard, D. A., & Shulman, G. L. (2001). A default mode of brain function. *Proceedings of the National Academy of Sciences of the United States of America, 98*(2), 676–682.

Ricoeur, P. (1969). *Le conflit des interprétations: Essais d'herméneutique*. Paris: Éditions du Seuil; translated into English as *Conflict of Interpretation* (Ed. Don Ihde), Northwestern University Press: Evanston, Illinois, 1974.

Rizzolatti, G., & Strick, P. L. (2013). Cognitive functions of the premotor systems. In E. R. Kandel, J. Schwartz, T. Jessell, S. Siegelbaum, & A. J. Hudspeth (Eds.), *Principles of neural science* (pp. 412–425). New York: McGraw-Hill.

Schooler, J. W., Ohlsson, S., & Brooks, K. (1993). Thoughts beyond words: When language overshadows insight. *Journal of Experimental Psychology. General, 122*(2), 166–183.

Schooler, J. W., Smallwood, J., Christoff, K., Handy, T. C., Reichle, E. D., & Sayette, M. A. (2011). Meta-awareness, perceptual decoupling and the wandering mind. *Trends in Cognitive Sciences, 15,* 319–326.

Segal, E. (2004). Incubation in insight problem solving. *Creativity Research Journal, 16,* 141–148.

Simon, H. A., & Newell, A. (1971). Human problem solving: The state of theory. *American Psychologist, 21*(2), 145–159.

Sio, U. N., & Ormerod, T. C. (2009). Does incubation enhance problem solving? A meta-analytic review. *Psychological Bulletin, 135*(1), 94–120.

Sloman, S. A. (1996). The empirical case for two systems of reasoning. *Psychological Bulletin, 119,* 3–22.

Smallwood, J., & Schooler, J. W. (2006). The restless mind. *Psychological Bulletin, 132,* 946–958.

Smallwood, J., McSpadden, M., Luus, B., & Schooler, J. W. (2008). Segmenting the stream of consciousness: The psychological correlates of temporal structures in the time series data of a continuous performance task. *Brain and Cognition, 66,* 50–56.

Smith, S. M., Fox, P. T., Miller, K. L., Glahn, D. C., Fox, P. M., Mackay, C. E., Filippini, N., Watkins K. E., Toro, R., Laird, A. R., & Beckmann, C. F. (2009). Correspondence of the brain's functional architecture during activation and rest. *Proceedings of the National Academy of Sciences of the United States of America, 106*(31), 13040–13045.

Sperber, D., & Wilson, D. (1995). *Relevance: Communication and cognition.* Oxford: Blackwell. (Original work published 1986.)

Stanovich, K. E. (2009). Distinguishing the reflective, algorithmic and autonomous minds: Is it time for a three-process theory? In J. S. B. T. Evans & K. Frankish (Eds.), *In two minds* (pp. 55–87). Oxford: Oxford University Press.

Stanovich, K. E., & Toplak, M. E. (2012). Defining features versus incidental correlates of Type 1 and Type 2 processing. *Mind & Society,* Special issue on: Dual process theories of human thought: The debate, *11*(1), 3–13.

Stanovich, K. E., & West, R. F. (1998). Individual differences in rational thought. *Journal of Experimental Psychology. General, 127,* 161–188.

Stanovich, K. E., & West, R. F. (1999). Discrepancies between normative and descriptive models of decision making and the understanding/acceptance principle. *Cognitive Psychology, 38,* 349–385.

Stanovich, K. E., & West, R. E. (2000). Individual differences in reasoning: Implications for the rationality debate? *Behavioral and Brain Sciences, 23,* 645–726.

Stanovich, K. E., & West, R. F. (2008). On the relative independence of thinking biases and cognitive ability. *Journal of Personality and Social Psychology, 94,* 672–695.

Sternberg, R. J., & Davidson, J. E. (Eds.). (1986). *Conceptions of giftedness*. New York: Cambridge University Press.

Stupple, E. J. N., Ball, L. J., Evans, J. S. B. T., & Kamal-Smith, E. (2011). When logic and belief collide: Individual differences in reasoning times support a selective processing model. *Journal of Cognitive Psychology*, *23*(8), 931–941.

Thompson, V., & Morsanyi, K. (2012). Analytic thinking: Do you feel like it? *Mind & Society*, *11*(1), Special issue: Dual process theories of human thought: The debate, 1–13.

Tomasello, M. (2009). *Why we cooperate*. Cambridge, MA: MIT Press.

Tversky, A., & Kahneman, D. (1983). Extension versus intuitive reasoning: The conjunction fallacy in probability judgment. *Psychological Review*, *90*(4), 293–315.

Verschueren, N., & Schaeken, W. (2005). A dual-process specification of causal conditional reasoning. *Thinking & Reasoning*, *11*(3), 239–278.

Wason, P. C. (1966). Reasoning. In B. M. Foss (Ed.), *New horizons in psychology 1* (pp. 135–151). Harmondsworth: Penguin.

Wason, P. C., & Evans, J. S. B. T. (1975). Dual processes in reasoning? *Cognition*, *3*, 141–154.

Weisberg, R. W. (2015). Toward an integrated theory of insight in problem solving. *Thinking & Reasoning*, *21*(1), 5–39.

Weisberg, R. W., & Alba, J. W. (1981). An examination of the alleged role of "fixation" in the solution of several "insight" problems. *Journal of Experimental Psychology. General*, *110*(2), 169–192.

Wertheimer, M. (1925). *Drei Abhandlungen zur Gestalttheorie*. Erlangen: Verlag der Philosophischen Akademie.

Appendix A

Horse-Trading Problem (Macchi & Bagassi, 2015)

A man bought a horse for £70 and sold it for £80, then he bought it back for £90 and sold it for £100. How much did he make?

The majority of problem solvers come up with one of two answers:

• £20 (41%), the correct answer (the dealer makes £10 on the first transaction, 70–80, and £10 on the second, 90–100).
• £10 (49%) (the dealer makes £10 on the first transaction, but loses this gain when he buys the horse back for £90. He then makes £10 on the last transaction, when he sells the horse again for £100).

Table 3.1

Results obtained with the *Horse-Trading* problem (*n* = 90)

"20"	"10"	"0"	"30"	other
37 (41%)	44 (49 %)	6 (7%)	1 (1%)	2 (2%)

Study Window Problem (Mosconi & D'Urso, 1974)

The study window measures 1 m in height and 1 m wide. The owner decides to enlarge it and calls in a workman. He instructs the man to double the area of the window without changing its shape so that it still measures 1 m by 1 m. The workman carried out the commission. How did he do it?

The solution is to be found in a square (geometric form) that "rests" on one of its angles, thus becoming a rhombus (phenomenic form). Now the dimensions given are those of the two diagonals of the represented rhombus (ABCD).

All the participants to the study (30) declared the problem impossible to solve (Macchi & Bagassi, 2014).

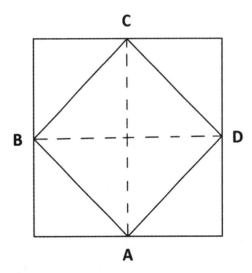

Figure 3.1
The *Study Window* problem solution.

Appendix B

Square and Parallelogram Problem (Wertheimer, 1925)

Given that AB = a and AG = b, find the sum of the areas of *square ABCD* and *parallelogram EBGD*.

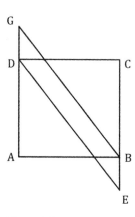

Figure 3.2
The Square and Parallelogram problem

Experimental Version Given that AB = a and AG = b, find the sum of the areas of the two *partially overlapping figures*.

Pigs in a Pen Problem (Schooler et al., 1993)

Nine pigs are kept in a square pen. Build two more square *enclosures* that would put each pig *in a pen* by itself.

Experimental Version Nine pigs are kept in a square pen. Build two more squares that would put each pig by itself.

A

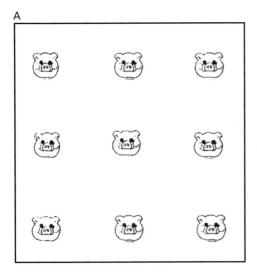

B

Figure 3.3
The *Pigs in a Pen* problem (A); solution (B).

Bat and Ball Problem (Frederick, 2005; Kahneman, 2003)

A bat and a ball cost $1.10 in total. The bat costs $ 1.00 more than the ball. How much does the ball cost?

Experimental Version A bat and a ball cost $1.10 in total. The bat costs $ 1.00 more than the ball. Find the cost *of the bat and* of the ball.

Study Window Problem

The study window measures 1 m in height and 1 m wide. The owner decides to enlarge it and calls in a workman. He instructs the man to double the area of the window without changing its shape and so that it still measures one meter by one meter. The workman carried out the commission. How did he do it?

Experimental Version

The study window measures 1 m in height and 1 m wide. The owner decides to enlarge it and calls in a workman. He instructs the man to double the area of the window: the workman *can change the orientation* of the window but not its shape and so that it still measures one meter by one meter. The workman carried out the commission. How did he do it?

[The underlining and the italicization (not included in the texts submitted to the subjects) stress the crucial points manipulated in the experimental versions.]

Results

Outcomes of these studies are summarized in table 3.2.

Table 3.2

Percentages of correct solutions with reformulated experimental versions

Problems	Control Version	Experimental Version
Square and parallelogram	9 (19%)	28 (80%)
	$n = 47$	$n = 35$
Pigs in a pen	8 (38%)	20 (87%)
	$n = 25$	$n = 23$
Bat and ball	2 (10%)	28 (90%)
	$n = 20$	$n = 31$
Study window	0 (0%)	21 (66%)
	$n = 30$	$n = 30$

Appendix C

Syllogisms

Valid-believable syllogisms: "All animals are mortal. All humans are animals. All humans are mortal."
Invalid-believable syllogisms: "All flowers fade. All roses fade. All roses are flowers."

Table 3.3
Percentages of correct solutions with *Syllogisms*

	Control		Experimental Version	
	Naive	Informed	Naive	Informed
Valid-believable (humans)	47 (94%)	77 (96%)	48(94%)	27 (93%)
	$n = 50$	$n = 80$	$n = 51$	$n = 29$
Invalid-believable (roses)	2 (4%)	8 (10%)	21 (41%)	20 (66%)
	$n = 50$	$n = 80$	$n = 51$	$n = 30$

Wason Selection Task

Control: "If there is a vowel on one side of the card, then there is an even number on the other side."
Experimental version: "If there is a vowel on one side of the card, then there is an even number on the other side, _but not vice-versa_."

Table 3.4
Percentages of correct solutions with *Wason Selection Task*

Control		Experimental Version	
Naive	Informed	Naive	Informed
2 (5%)	12 (13%)	29 (55%)	18 (60%)
$n = 53$	$n = 90$	$n = 52$	$n = 30$

II The Paradigm Shift: The Debate

4   The Paradigm Shift in the Psychology of Reasoning

David Over

Probabilistic, and specifically Bayesian, accounts of cognition are having a profound impact on cognitive psychology (Chater & Oaksford, 2008), and the psychology of reasoning in particular has been deeply affected, resulting in what can only be called a paradigm shift (Elqayam & Over, 2013; Manktelow, Over, & Elqayam, 2011; Oaksford & Chater, 2007; Over, 2009). Claims about paradigm changes are common in academic subjects, but in this instance, the recent advances apply a new normative standard for logical reasoning, and if a change in fundamental logic is not a paradigm shift, what could be?

The traditional paradigm in the psychology of reasoning concentrated primarily, and even exclusively in many works, on binary distinctions: between truth and falsity, consistency and inconsistency, possibility and impossibility, and what does or does not necessarily follow from *assumed* premises (Johnson-Laird & Byrne, 1991, is an influential example of this paradigm). It required participants in its experiments to assume that arbitrary premises given to them were true. Inferences from actual *beliefs* were held to be biases or fallacies, and in some leading old paradigm theories, departures from extensional logic were supposed fallacies (see again Johnson-Laird & Byrne, 1991). Even a disinclination to draw inferences from totally unbelievable assumptions—such as that all dogs are cats, and all cats are rats—was labeled "belief bias." There was no recognition that most useful everyday and scientific reasoning, whether deductive or inductive, is from beliefs that are held with more or less confidence and that the degree of confidence in the beliefs affects confidence in the conclusions (Elqayam & Over, 2013; Evans & Over, 2013; Gilio & Over, 2012).

The new paradigm in the psychology of reasoning is Bayesian in a wide sense, focusing on subjective probability and utility. It is concerned with inferences from *degrees of belief* and also with reasoning that can be affected by utility judgments (Oaksford & Chater, 2007; Over, 2009). The traditional, old binary paradigm could not integrate the psychology of reasoning with the study of subjective probability and utility in judgment and decision making. Suppose that we are trying to decide whether to borrow some money, or support a political party in an election, or look for a certain symptom of a disease. We could theoretically just assume that, if we take out the loan—vote for the political party—find the symptom—then we will be able to afford the repayments—the party will fulfill its promises—the disease will be present. But clearly, for effective decision making, we must assess how probable these conditionals are—how likely it is that, if we take out the loan, then we will be able to afford the repayments, and equally for the other conditionals.

Conditionals and Probability

The best way to explain more fully the difference between the old and new paradigms in the psychology of reasoning is by using conditionals. Indicative, counterfactual, deontic, and other conditionals are found everywhere in human reasoning and are fundamental to it. A good exemplar of an old paradigm analysis of conditionals can be found in perhaps the most influential book in this paradigm, Johnson-Laird and Byrne (1991). Their example of an indicative conditional is:

(1) If Arthur is in Edinburgh, then Carol is in Glasgow.

Johnson-Laird and Byrne note that (1) is true when Arthur is in Edinburgh and Carol is in Glasgow, and that (1) is false when Arthur is in Edinburgh and Carol is not in Glasgow. They then ask whether (1) is true or false when Arthur is not in Edinburgh. They reply, "It can hardly be false, and so, since the proposition calculus allows only for truth or falsity, it must be true" (p. 7, and see also p. 74). The unquestioned presuppositions of the old paradigm are crystal clear here. Indicative conditionals in natural language are only to be classified as true or false, and the binary extensional logic of the propositional calculus is uncritically accepted as the normative standard for what this classification should be.

The propositional calculus is the logic for the *material conditional* of elementary extensional logic. If (1) is such a conditional, then it is logically equivalent to the disjunction:

(2) Either Arthur is not in Edinburgh or Carol is in Glasgow.

Claiming that natural language conditionals such as (1) are true when their antecedents are false immediately implies one of the "paradoxes of the material conditional": that it is logically valid to infer these conditionals from the negations of their antecedents (see further Johnson-Laird & Byrne, 2002, on the supposed logical "validity" of the paradoxes). There are long-standing arguments in philosophical logic and more recently in the psychology of reasoning against identifying natural language conditionals with material conditionals because of the resulting paradoxes (Bennett, 2003; Edgington, 1995; Evans & Over, 2004).

Consider just one paradoxical result of holding that (1) is true simply because it is false that Arthur is in Edinburgh, as Johnson-Laird and Byrne contend. In that case, (1) will become more and more probable as it becomes more and more probable that Arthur will not go to Edinburgh, even if Carol has absolutely no reason to go to Glasgow and every intention of staying home in Melbourne to look after her young daughter, who would otherwise be left alone. Byrne and Johnson-Laird (2010) have made the general claim that the "correct" probability of an indicative conditional *if p then q*, P(*if p then q*), is the probability of *not-p or q*, P(*not-p or q*). But to return to one of our examples, suppose that we are becoming more and more unlikely to take out the loan because of increasingly well-founded fears that we will be unable to afford the repayments. Claiming P(*if p then q*) = P(*not-p or q*) in general would force us to conclude that it is becoming more and more likely that, if we do take out the loan, then we will be able to afford the repayments. Clearly again, this account of conditionals cannot be integrated with the study of decision making.

There are analyses and logics for the natural language indicative conditional that do not make it equivalent to the material conditional of propositional calculus. The account of Stalnaker (1968, 1975) has stimulated many others. In his logical analysis, (1) is false when the closest possibility in which Arthur does go to Edinburgh is one in which Carol stays in Melbourne. This analysis has intuitive appeal given the information we have added to the example. With Arthur not going to Edinburgh, we might only have to imagine a relatively slight change to envisage a possibility in which he does go to Edinburgh, and yet Carol continues to stay in Melbourne to look after her daughter. The paradoxes of the material conditional are not logically valid for Stalnaker's conditional.

Stalnaker proposed an extension of the *Ramsey test* as a psychological process for thinking about the closest possibility and assessing the probability of a conditional (Ramsey, 1929/1990b; and see also Edgington, 1995). In this test as Stalnaker formulates it, people would make a judgment about (1) by supposing that Arthur goes to Edinburgh, making whatever changes might be necessary to preserve consistency, and then judging how far that makes it believable that Carol goes to Glasgow. The result of a Ramsey test would be that the probability of (1) is equated with the conditional probability that Carol goes to Glasgow given that Arthur goes to Edinburgh. This judgment could be made in a number of ways. It might most simply result from recalling that Arthur has gone to Edinburgh many times in the past, but Carol has never, under any circumstances, gone to Glasgow. This recollection would make (1) highly improbable by the Ramsey test: (1) would not be highly probable, or "true," merely because Arthur was unable to go to Edinburgh on this occasion.

Consider an example that has a more scientific justification:

(3) If Arthur has chronic hyperglycemia, then he has diabetes.

We can reinforce all our points using this example. For a diagnosis, we cannot just assume (3) as a major premise and its antecedent as a minor premise, in order to infer by modus ponens (MP) that Arthur definitely has diabetes. For a rational MP inference, leading to rational medical decision making, we must assess the probability of (3) and also how likely Arthur is to have chronic hyperglycemia. Scientific research can determine the conditional probability of diabetes given chronic hyperglycemia. Applying a Ramsey test to the major premise, medical doctors could use their knowledge of the scientific findings and of background facts about Arthur to judge how likely he was to have diabetes given that he had chronic hyperglycemia. To determine how likely the minor premise was, that Arthur has chronic hyperglycemia, the doctors could conduct medical tests using blood or urine samples. The tests might not be completely decisive because of possible false positives, but these can be taken into account in Bayesian reasoning. When the probabilities of both the major and minor premises turn out to be quite high, the doctors could justifiably infer, from a Bayesian point of view, that it is quite probable that Arthur has diabetes. Even if ordinary people cannot themselves fully engage in this medical reasoning, they can learn to trust the probability judgments of doctors, and in general, people do have relatively more confidence in the conclusions of MP inferences from assertions by relatively high-level experts, such as senior doctors compared to medical students (Stevenson & Over, 2001).

Most logicians and philosophers would argue that there is a close relation of some type between the subjective probability of conditionals such as (1) and (3) and the corresponding conditional probability (see Bennett, 2003, for a survey). Stating that this relation is identity between the two probabilities is so important that it has sometimes been called *the Equation* (Edgington, 1995):

$P(if\ p\ then\ q) = P(q|p)$

A Bayesian account of conditional reasoning, in terms of conditional probability, follows directly from the Equation (Oaksford & Chater, 2007; Pfeifer & Kleiter, 2010), which has been tested in experiments as the descriptive conditional probability hypothesis (about singular conditionals; see Cruz & Oberauer, 2014, on general conditionals). There has been very strong experimental support for this hypothesis and so of the Equation as descriptive of people's probability judgments about conditionals (Douven & Verbrugge, 2010; Evans, Handley, Neilens, & Over, 2007; Evans, Handley, & Over, 2003; Fugard, Pfeifer, Mayerhofer, & Kleiter, 2011; Oberauer & Wilhelm, 2003; Over, Hadjichristidis, Evans, Handley, & Sloman, 2007; Politzer, Over, & Baratgin, 2010; Singmann, Klauer, & Over, 2014). Attempts to counter this evidence from a mental model point of view (Byrne & Johnson-Laird, 2010; Girotto & Johnson-Laird, 2010) are vitiated by modal fallacies (Milne, 2012; Over, Douven, & Verbrugge, 2013; Politzer et al., 2010). A conditional that satisfies the Equation has been called an *Adams conditional*, a *conditional event*, a *probability conditional*, or a *suppositional conditional* (Adams, 1998; de Finetti, 1995; Evans & Over, 2004; Fugard et al., 2011). We use the term "probability conditional" here for such a conditional.

The material conditional and Stalnaker's conditional are always made true or false by a state of affairs, whether actual or possible, but Lewis (1976) proved that the probability of such a conditional cannot logically be the conditional probability (see Douven & Dietz, 2011, for important points). Stalnaker was simply mistaken in claiming that the probability of his conditional is the conditional probability. His conditional is not a probability conditional, as we have just defined it. It may be that the semantics of the natural language conditional is like Stalnaker's but that people use some rough heuristic to assess its probability as the conditional probability or that these two probabilities are too close to each other to separate in experiments that are easy for psychologists to design (Evans & Over, 2004). However, there are other well-known theories of the conditional in logic and philosophy that are not threatened by Lewis's proof.

According to Adams (1975), an indicative conditional is not made true or false by any state of affairs, actual or possible, and only expresses the subjective conditional probability. This could be called the *no-truth-value* view. It might be thought that this proposal could not be taken descriptively as a psychological account because ordinary people will often say that conditionals are "true" or "false," but it cannot be dismissed quite so easily. The terms "true" and "false" have a range of uses in natural language, and people are happy at times to call totally subjective judgments "true" or "false." For example, people may say that it is "true" that champagne should be chilled before drinking and "false" that the finest claret should be, but they are arguably doing no more than subjectively endorsing or rejecting these judgments and not taking them to be made true or false by some objective state of affairs. One way indeed to look at the no-truth-value view of indicative conditionals is that it makes them closely comparable to deontic conditionals about subjective matters (see Manktelow & Over, 1991, and Oaksford & Chater, 2007, on deontic conditionals). Nevertheless, there are psychological results that go against this no-truth-value theory.

Table Truth Tasks

Psychologists have long studied truth table tasks (Evans & Over, 2004, pp. 34–35). In these tasks participants are given a conditional, *if p then q*, and are asked whether the four Boolean states represented in a truth table—the *p & q, p & not-q, not-p & q*, and *not-p & not-q* possibilities—make the conditional true or false. Participants tend to respond that *p & q* makes the conditional true, *p & not-q* makes it false, but that the *not-p* states are irrelevant to its truth and falsity. When given the option, they say that the *not-p* cases make the conditional neither true nor false (Politzer et al., 2010). In light of these results people would say that (1) is true when Arthur is in Edinburgh and Carol is in Glasgow, false when Arthur is in Edinburgh and Carol is not in Glasgow, but that Arthur's not being in Edinburgh is irrelevant to the truth or falsity of (1) and does not make it either true or false. This response pattern was traditionally called *the defective truth table*.

Consider actually observing a *not-p* state for (1), e.g., looking at Arthur and Carol together in Melbourne, which is an instance of *not-p & not-q* for (1). The indicative (1) is, at least in some sense, "void" or "empty" in this case. No one would look directly at these two in Melbourne while saying, "If this man is in Edinburgh then this woman is in Glasgow," and there is certainly no reason to conclude that this natural language conditional must be "true" because the propositional calculus exists. What might be used in this kind of context would be a counterfactual, "If this man were in Edinburgh then this woman would be in Glasgow," and the probability of this counterfactual could be the conditional probability (Gilio & Over, 2012; Politzer et al., 2010; and see Over et al., 2007, on the probability of counterfactuals).

The no-truth-value view cannot apparently explain why the *p & q* state of affairs makes *if p then q* true, and the *p & not-q* state makes *if p then q* false, but the "defective" truth table results support another theory of the conditional that it is not subject to Lewis's proof. This is the *three-value* account, which is one way to interpret Adams (1998). In it, the indicative conditional of natural language can be made true, false, or "void" by states of affairs (Edgington, 1995; Evans & Over, 2004; Pfeifer & Kleiter, 2010). The indicative conditional, *if p then q*, is made true by a *p & q* state of affairs, made false by a *p & not-q* state of affairs, but is "void" and neither true nor false in a *not-p* state of affairs. This three-value account view can be traced back to de Finetti (1936/1995), and the "defective" truth table should be called the *de Finetti table* after de Finetti (Politzer et al., 2010), who first proposed it for what he called the conditional event and which we are terming the *probability conditional* (see table 4.1). The use of "defective" might suggest that the participants are "defective" in the negative sense for producing it, although it is well justified on logical and psychological grounds (see also Baratgin, Over, & Politzer, 2013).

The indicative conditional in the three-value theory is closely comparable to a conditional bet. Politzer et al. (2010) compared the following in an experiment:

(4) If the chip is square (*s*), then it will be black (*b*).

Table 4.1
The "defective" 2×2 de Finetti table for *if p then q*

p \ q	1	0
1	1	0
0	I	I

1 = true, 0 = false, and I = irrelevant for truth or falsity.

(5) I bet you 1 Euro that if the chip is square (*s*), then it will be black (*b*).

The participants were shown a distribution of chips that were square or round and black or white and were told that a chip was going to be randomly drawn from this distribution. The indicative conditional (4) was about this random chip, and Mary makes the conditional bet (5) about it with Peter. According to the three-value view, there is this close relation between (4) and (5) when the random chip is selected. The *s* & *b* outcome makes (4) true, and Mary wins her bet (5), the *s* & *not-b* outcome makes (4) false and Mary loses her bet (5), and the *not-s* outcomes make (4) a void assertion and (5) a void bet. To be more precise about the last case, when a round chip is selected, there is no indicative fact to make (4) true or false (although a counterfactual could still be used), and Mary neither wins nor loses her bet (5), with no money changing hands between her and Peter. Politzer et al. (2010) confirmed this three-value account by finding a close relation between (4) and (5) in how participants classified (4) as true, false, and neither true nor false, and (5) as won, lost, and neither won nor lost.

Consider questions about the probability of the truth of (4) and the probability of winning the bet (5). Politzer et al. (2010) argued that these questions presuppose that there is nonvoid indicative assertion and a nonvoid bet. The first question is then about the probability of the *s* & *b* outcome given that there is a nonvoid indicative assertion, which is $P(s \ \& \ b|s) = P(b|s)$, and the second question is about probability of the *s* & *b* outcome given that there is a nonvoid bet, which is again $P(s \ \& \ b|s) = P(b|s)$. Thus, the answers to these questions should conform to the Equation, and Politzer et al. (2010) confirmed that this is so, with a minority conjunctive response, $P(s \ \& \ b)$, which is found in certain materials (see also Evans et al., 2007, on this response and how it is found in participants of relatively low cognitive ability).

The three-value theory aims to describe when an actual state of affairs makes a conditional true. It does not apply to other uses of "true" in natural language (Politzer et al., 2010). People will no doubt say that *if p then p* is "true," or "certainly true," without implying that *p* is true. These phrases can be used simply because $P(if \ p \ then \ p) = P(p|p) = 1$. More generally, to say *It is true that p*, for any *p* whether a conditional or not, can simply be strong way of asserting or emphasizing *p* in what has been called a *pleonastic* or *redundant* use of "true" (Edgington, 2003; Ramsey, 1927/1990a). The same points made above about the no-truth-value view and pragmatic uses of "true" also apply to the three-value position. People can apply "true" to *if p then q* to endorse it strongly, or to indicate pragmatically that their confidence in it is above a contextually presupposed threshold, and when $P(if \ p \ then \ q) = P(q|p) = 1$, it is the highest possible threshold. Table 4.1 does not account for all uses of "true" and "false" and has to be supplemented by the Ramsey test in theories of the conditional. Some uses of "true" are the result of a Ramsey test on *if p then q* yielding an especially high conditional probability, with

Table 4.2

The 3×3 de Finetti table for *if p then q*

p \ q	1	U	0
1	1	U	0
U	U	U	U
0	U	U	U

1 = true, 0 = false, and U = uncertain.

P(q|p) at 1 or close to it. Some uses of "false" are the result of an especially low conditional probability from the Ramsey test, with *P(q|p)* at 0 or close to it.

Politzer et al. (2010) followed the traditional practice in truth table studies of investigating a 2×2 table, in which the four possible states could only be classified as true or false and not as themselves uncertain. Table 4.1 is a 2×2 de Finetti table: *p* and *q* are true or false in it and never have the third value of being uncertain. But clearly this is special case. We would usually be uncertain whether Arthur had chronic hyperglycemia and diabetes when we heard (3) and started to gather evidence about it. We might also catch only a glimpse of a chip or see it only in bad light and be uncertain whether it was square, or black. Baratgin et al. (2013) point out that de Finetti (1936/1995) considered this more general case and presented table 4.2 for it. They extended the material of Politzer et al. (2010) to this case by obscuring the chips shown to participants with a "filter" and confirmed that the participants' responses in the extended task tended to correspond the 3×3 de Finetti table shown as table 4.2.

There is the still more general possibility in which the "uncertain" classification for *p* and *q* is broken down into different degrees of subjective probability. For example, as we gathered more and more evidence about (3), we would judge it more and more likely, or unlikely, that Arthur had chronic hyperglycemia and diabetes, and confidence in the other possible Boolean states would change as well, leading to more or less confidence in *if p then q*. There is a *many-value* extension of the three-value table in the normative literature, in which when a *not-p* state holds, the value of *if p then q* in the resulting table becomes the subjective conditional probability itself, *P(q|p)*. This table could be called the *Jeffrey table*, Jeffrey (1991a) having been the first to propose it (see also Edgington, 1995 and Stalnaker & Jeffrey, 1994). In the Jeffrey table the conditional is "void" in a *not-p* case as an indicative statement, in that the actual state of affairs does not make it true or false. However, it does have a subjective probability, *P(q|p)*, as represented in table 4.3, which is not of course a single third value but could be any degree of belief. Baratgin et al. (2013) suggest that their experimental technique could be extended to explore this normative proposal experimentally.

Table 4.3
The Jeffrey table for *if p then q*

p $\quad q$	1	0
1	1	0
0	$P(q|p)$	$P(q|p)$

1 = true, 0 = false, and $P(q|p)$ = the subjective conditional probability of q given p.

Probabilistic Validity and Coherence

The psychology of reasoning has to adopt "new" definitions of validity and consistency to study inferences from degrees of belief rather than arbitrary assumptions: probabilistic validity and probabilistic consistency. These definitions are relatively new to the psychology of reasoning, but they can be introduced by referring to a classic article in judgment and decision making by Tversky and Kahneman (1983).

Tversky and Kahneman found that people sometimes commit the *conjunction fallacy*, which is to be inconsistent with probability theory by judging that the probability of a conjunction, $P(p \& q)$, is strictly greater than the probability of one of its conjuncts, $P(p)$. In their most famous example participants were given a description of a young woman: "Linda is 31 years old, single, outspoken and very bright. She majored in philosophy. As a student, she was deeply concerned with issues of discrimination and social justice and also participated in antinuclear demonstrations."

Tversky and Kahneman predicted that participants would make probability judgments about Linda by reference to a stereotype activated by this description. In participant, they would judge the probability of

(6) Linda is a bank teller and is active in the feminist movement

to be greater than the probability of

(7) Linda is a bank teller.

Participants did tend to make this judgment, and because (6) is of the form $p \& q$ and (7) p, they committed the conjunction fallacy. Such inconsistency with probability theory could be called probabilistic inconsistency (and it sometimes is), but the standard term for it is *incoherence*. Tversky and Kahneman noted that to be *coherent* is "… to satisfy the constraints of probability theory," and that "… the normative theory of judgment under uncertainty has treated the coherence of belief as the touchstone of human rationality." The underlying justification for this normative standard is the *Dutch Book* theorem (de Finetti, 1974; Howson & Urbach, 2006), which proves that to be incoherent is to be exposed to a Dutch book, a series of bets that one can only lose ("coherence" and "incoherence" are the standard translations of terms used by de Finetti). Studies of whether people are coherent in their probability judgments form a substantial part of the field of judgment and decision making, and yet the old paradigm in the psychology of reasoning failed to make use of coherence as a normative standard. This failure severely limited the extent to which the psychology of reasoning could be relevant to the study of judgment and decision making, but this limitation is now being overcome in the new paradigm in the psychology of reasoning.

Tversky and Kahneman (1983) themselves made a connection between coherence and logical validity. It is logically valid to infer p from $p \& q$, an inference that is called *&-elimination*, and therefore, $P(p \& q)$ should not, normatively, be greater than $P(p)$. Violating this relation leads directly to a Dutch book (as illustrated in Gilio & Over, 2012). Much more generally, there is a necessary connection between logical validity and coherent probability judgments, and we can use this relation to define logical validity itself, giving us *probabilistic validity* or *p-validity*. A single premise inference is *p*-valid if and only if the probability of the premise cannot be coherently greater than the probability of the conclusion. More informally, a *p*-valid inference cannot take us from a high-probability premise to a conclusion of lower probability: it preserves our degree of confidence in the premise (as long as our probability judgments are coherent).

The paradoxes of the material conditional are *p*-invalid inferences for the probability conditional. For example, it is *p*-invalid to infer *if p then q* from *not-p*, since $P(not\text{-}p)$ can be coherently greater than $P(q|p)$. It is *p*-invalid to infer (1) from the belief that Arthur is not in Edinburgh. We can have a high degree of belief that Arthur is not in Edinburgh but a low degree of belief that, if he is there, Carol is in Glasgow. We can coherently have a high degree of belief that Carol is going to be in Melbourne to look after her daughter no matter where Arthur is.

Another example of a single-premise inference that is valid for the material conditional but *p*-invalid for the probability conditional is inferring *if not-p then q* from *p or q*. Suppose a game show presenter puts a prize behind door A before the show and then tells a contestant during the show that the prize is behind door A or door B. The contestant might infer with confidence that, if the prize is not behind A, then it is behind B. However, the game show presenter could more justifiably disbelieve that conditional and infer instead with confidence that, if the prize is not behind A, then it has been stolen (perhaps there have been recent thefts in the studio). In general, it can be justifiable to have high confidence in $P(p\ or\ q)$ and low confidence in $P(if\ not\text{-}p\ then\ q) = P(q|not\text{-}p)$, and indeed people do not always have high confidence in this inference (Cruz, Baratgin, Oaksford, & Over, 2015; Over, Evans, & Elqayam, 2010; also see Gilio & Over, 2012, for a full probabilistic analysis of this inference).

Tversky and Kahneman did not investigate whether people commit the conjunction fallacy when they actually infer p from $p \& q$ in this *p*-valid inference. What Tversky and Kahneman found was that people commit the fallacy when they make probability judgments about p and $p \& q$ outside the context of an inference and separated by other probability judgments, or at least without an explicit inference form. But an important question for psychologists of reasoning is precisely that: whether people commit the conjunction fallacy by assigning a higher probability to $p \& q$ than to p when they infer the latter as a conclusion from the former as a premise (see Cruz et al., 2015, on this question).

It could not be more important for psychologists of reasoning to study how far people are coherent and conform to p-validity generally in their inferences. For inferences of more than one premise, the definition of p-validity is extended in the following way (Adams, 1998). Let the subjective *uncertainty of an assertion* be defined as 1 minus its probability. For example, the uncertainty of p is $1 - P(p)$. Then an inference is *p-valid* if and only if the uncertainty of the conclusion cannot coherently exceed the sum of the uncertainties of the premises (see Gilio, 2002, for an alternative but equivalent way to define p-validity, called there p-entailment, and a study of it in depth). Informally and intuitively, a p-valid inference cannot take us from relatively low uncertainty in the premises to relatively high uncertainty in the conclusion. To extend the example, consider the inference of *&-introduction*: inferring $p \& q$ from the two separate premises p and q. This is a p-valid inference because $1 - P(p \& q)$ is never coherently greater than $1 - P(p) + 1 - P(q)$. To make the example more specific, suppose some people have degrees of belief in p and q such that $P(p) = P(q) = 0.6$, and they infer $p \& q$ from these uncertain beliefs. Then the uncertainty of the premises for them is $1 - 0.6 + 1 - 0.6 = 0.8$, and to conform to p-validity, they cannot assign a probability to $P(p \& q)$ that is less than 0.2. The uncertainty for them should not be greater than 0.8 for this conclusion.

We can be even more precise formally by using probability theory to derive a coherent probability interval for $P(p \& q)$ given $P(p)$ and $P(q)$:

$$\max[0, P(p) + P(q) - 1)] \leq P(p \& q) \leq \min[P(p), P(q)]$$

The above interval incorporates Tversky and Kahneman's (1983) point: $P(p \& q)$ should not be greater than the minimum of $P(p)$ and $P(q)$. But as can also be seen in the above, $P(p \& q)$ can also fallaciously be too low, given specific $P(p)$ and $P(q)$. We do not commit the conjunction fallacy, as it is usually defined, by judging $P(p \& q) = 0.1$ when $P(p) = P(q) = 0.6$, but we are incoherent because 0.1 is less than the minimum value, 0.2, given by the interval.

Probability intervals have been derived for a wide range of inferences, including conditional inferences (Coletti & Scozzafava, 2002; Pfeifer & Kleiter, 2009, 2010). Consider MP as an example. As an inference from beliefs, people use MP to infer a degree of belief in q from degrees of belief in the major premise, *if p then q*, and the minor premise, *& p*. Suppose the premise beliefs are such that $P(if\ p\ then\ q) = 0.7$ and $P(p) = 0.6$. For example, Arthur might tell his friends that (3) is well established scientifically and that he has chronic hyperglycemia, but they might not fully believe him (perhaps Arthur has misunderstood his doctor in the past). From these beliefs as premises, the friends could use MP to infer that he has diabetes. What should their degree of confidence be in this conclusion? We can answer this question, assuming that $P(if\ p\ then\ q) = P(q|p)$, by applying the total probability theorem from probability theory:

$$P(q) = P(p)P(q|p) + P(not\text{-}p)P(q|not\text{-}p)$$

Putting the relevant degrees of belief in the above results in:

$$P(q) = (0.6)(0.7) + (0.4)P(q|not\text{-}p)$$

Arthur's friends may not be able to make a judgment about $P(q|not\text{-}p)$, the probability that he has diabetes given that he does not have chronic hyperglycemia. But $P(q|not\text{-}p)$ has a minimum value of 0 and a maximum value of 1, and that implies that the degree of belief in the MP conclusion q, that Arthur has diabetes, should have (0.6) $(0.7) = 0.42$ as its minimum value and $(0.6)(0.7) + 0.4 = 0.82$ as its maximum value. Thus, to be coherent following their MP inference, Arthur's friends should have degrees of belief that he has diabetes between this lower value of 0.42 and this upper value of 0.82. With degrees of belief in this interval for the conclusion of MP, they will clearly conform to p-validity as well: the uncertainty of their conclusion will not be greater than the sum of the uncertainties of their premises. But note that they will be incoherent by being overconfident if their degree of belief is greater than 0.82 that Arthur has diabetes. The old paradigm could not even express the significant notion that people can be overconfident in the conclusion of MP, which was endorsed at ceiling in traditional experiments on conditional inference.

There are unending studies of a relatively small number of conditional inferences in the old paradigm: of MP and also of modus tollens (MT), *Affirmation of the consequent* (AC), and *Denial of the antecedent* (DA). Parallel with extensional logic, MP and MT are both p-valid, and AC and DA are p-invalid, and there are also coherent probability intervals for MT, AC, and DA. However, it is only very recently that psychologists have started to investigate whether people's probability judgments for the conclusions of these inferences conform to p-validity (which is only logically required for MP and MT) and are in the coherent probability intervals for them (Evans, Thompson, & Over, 2015; Pfeifer & Kleiter, 2010; Singmann et al., 2014). A significant result at this early stage is the probabilistic equivalent of the endorsement at ceiling of MP in traditional experiments. People tend to comply with p-validity for MP and to be in its coherent probability interval. One reason this finding is of great interest is that a form of MP is of central importance in the Bayesian account of belief updating over time. Belief updating over time is yet another topic that was almost totally ignored in the old paradigm. It studied only "synchronic" or "static" inference from assumptions treated as if timelessly true. But clearly people's degrees of beliefs change over time as they get more evidence for or against them, or for other reasons, and this belief updating should be right at the center of the new probabilistic approach to the psychology of reasoning.

Dynamic Reasoning and Belief Updating

People tend to perform inferences from their beliefs, and not from arbitrary assumptions, and to change over time by updating or revising their beliefs and what they infer from them. Consider an example used by Hadjichristidis, Sloman, and Over (2014). Suppose that Linda has some reason to believe that she is pregnant. Not being sure that this is so, she buys a pregnancy test one afternoon to get better grounds for a higher degree of belief. The test is one of the best on the market, and although a false positive is always possible, she is very highly confident that, if the test is positive, then she is pregnant. She takes the test in the evening, and it is positive. She now updates her belief and becomes almost certain that she is pregnant, and she will make an appointment with her doctor to get even better evidence for this belief.

Linda is using *dynamic* reasoning, and *dynamic MP*, to update her beliefs over time, making her more and more confident that she is pregnant. Assuming $P(if\ p\ then\ q)$ = $P(q|p)$, dynamic MP is equivalent to *conditionalization* in Bayesianism. In Bayesian accounts of scientific inference (Howson & Urbach, 2006), scientists begin with some reason to suppose a hypothesis h holds, giving them some reasonably high degree of "prior" belief in it, say, $P_1(h)$. They plan to run an experiment that might give them some relevant evidence e. They make "likelihood" judgments about how probable e is given h, $P_1(e|h)$, compared to how probable e is given *not-h*, $P_1(e|not\text{-}h)$. They can then use Bayes's theorem to infer a relatively high degree of belief in h given e, $P_1(h|e) > P_1(h)$. After they actually run the experiment and find, as they hope, that e definitely holds, $P_2(e) = 1$, they use Bayesian *strict*, or *simple*, conditionalization to infer a *posterior* degree of belief in h, $P_2(h) = P_1(h|e)$, that is higher than their prior degree of belief, $P_1(h)$. In the new psychology of reasoning Oaksford and Chater (2013) have recast their earlier Bayesian theory of conditional inference (Oaksford & Chater, 2007) as an account of dynamic reasoning.

Strict conditionalization and its equivalent of dynamic MP are justified if and only if $P_1(h|e) = P_2(h|e)$. Conditional probabilities that do not change over time are said to be *rigid* or *invariant* (Hadjichristidis et al., 2014 Oaksford & Chater, 2013). Linda, for example, has to remain confident in the pregnancy test from the afternoon when she buys it to the evening when she takes it. If she were to notice before she took the test that she had been sold one that was beyond its "use-by" date, she would lose confidence in it. She would no longer be highly confident that she was pregnant if the test were positive: the conditional probability—that she was pregnant given a positive result—would have failed to be rigid. The invariance, or otherwise, of people's probability judgments about conditionals over time is a new topic to be explored in the new paradigm.

In many ordinary and scientific contexts it can be uncertain whether a test result is positive or some other evidence holds. Linda may look at the result of the pregnancy test when she has had too much to drink, or has lost her glasses, and she may be somewhat unsure whether she is looking at a positive result or not. Similarly, scientists can be uncertain, due to observational or other problems, about evidence e, $P_2(e) < 1$. However, they could use *Jeffrey conditionalization*, or *Jeffrey's rule*, to update their belief in hypothesis h (Jeffrey, 1991b):

$$P_2(h) = P_2(e)P_1(h|e) + P_2(not\text{-}e)P_1(h|not\text{-}e)$$

The justification of Jeffrey conditionalization also depends on the rigidity of the conditional probabilities: $P_2(h|e) = P_1(h|e)$ and $P_2(h|not\text{-}e) = P_1(h|not\text{-}e)$. Clearly, if rigidity holds, then Jeffrey's rule reduces to the theorem of total probability, which is used, as noted above, to derive the coherent probability interval for "static" MP. Oaksford and Chater (2007, 2013) have referred to Jeffrey's rule as a way of extending their account of dynamic reasoning to fully uncertain contexts. People might not fully conform to it (Hadjichristidis et al., 2014 and compare, Zhao & Osherson, 2010, and Murphy, Chen, & Ross, 2012), but investigations of it and dynamic reasoning in general are at a very early stage (see Baratgin & Politzer, 2010, on different notions of conditionalization).

Conclusion

Little can be learned about everyday or scientific reasoning by studying only inferences from fixed assumptions. The results are paradoxical when psychologists take the propositional calculus, not only as the correct normative standard for the indicative conditional of natural language but as the inspiration for their psychological theory of this conditional. With these limitations it is no wonder that the traditional psychology of conditional reasoning was not integrated with the study of judgment and decision making. The psychology of conditionals should be about degrees of beliefs (and also desire in the case of deontic conditionals), just as judgment and decision making is.

Arbitrary assumptions do not help people make rational decisions, and absolute certainty cannot usually be found, assumed, or presupposed in the real world. To be rational, or even irrational in way that is not completely absurd, people must make inferences from their beliefs and judge how confident they should be in the conclusions that they infer from their mostly uncertain beliefs. They cannot assume fixed epistemic states of mind but must update their degrees of beliefs over time. The psychology of reasoning should provide a theory of how people perform inferences from their beliefs and change their beliefs in a dynamic process of reasoning.

Tversky and Kahneman (1983) made a connection between judgment and decision making and the psychology of reasoning by pointing out that $P(p \& q)$ should not be greater than $P(p)$ because there is a valid inference from $p \& q$ to p. But the old, traditional paradigm in the psychology of reasoning did not respond by developing the point. It ignored the general relation between probability and valid inference, and this also greatly limited the extent to which it could be related to judgment and decision making. It did not even try to find out whether the conjunction fallacy tends to be committed when p is actually inferred from p & q. In contrast, the new probabilistic psychology of reasoning is much better placed to discover whether bringing beliefs together in explicit inferences increases conformity with p-validity and improves coherence (Cruz et al., 2015; Evans et al., 2015).

The probabilistic psychology of conditional reasoning avoids the paradoxes of the material conditional and other unjustified inferences, such as inferring, in general, a high probability for *if not-p then q* from a high probability for *p or q*. It validates a more limited range of inferences (Adams, 1998), and this fundamental change in logic supports the statement that there has been a paradigm shift in the psychology of reasoning to a more general Bayesian and probabilistic approach. The traditional distinctions do not completely disappear. The concept of p-validity itself still implies a difference between deduction and induction. Deduction has p-valid inferences that do not increase uncertainty; induction has inferences that can be probabilistically strong but do increase uncertainty (Evans & Over, 2013; and see Singmann & Klauer, 2011; Singmann et al., 2014 on the psychological reality of the distinction).

There is an inductive, heuristic inference from Linda's description to a degree of belief in the conjunction that she is a bank teller and a feminist. This belief might be a useful expectation about Linda, but the inference greatly increases uncertainty (for most people) from high confidence in the descriptive premises. If deduction from the uncertain conjunction is then used to infer a degree of belief that Linda is a bank teller without increasing uncertainty, all is well. But if the heuristic is applied to infer a lower degree of belief that Linda is a bank teller than that she is a bank teller and a feminist, the conjunction fallacy will be committed. The deductive-inductive distinction is still there, but it is interpreted in a new way. The two types of inference can fail in practice to work well together, resulting in biases and fallacies. There are logically "transparent" extensional contexts in which the conjunction and other fallacies tend not to be committed (Barbey & Sloman, 2007; Over, 2007; Tversky & Kahneman, 1983), but more research is needed on how deduction and induction work together in uncertain reasoning and why they sometimes fail to do so. The ultimate goal should be to integrate reasoning research much more closely with the study of judgment and decision making.

References

Adams, E. (1975). *The logic of conditionals: An application of probability to deductive logic*. Dordrecht: Reidel.

Adams, E. (1998). *A primer of probability logic*. Stanford: CLSI Publications.

Baratgin, J., Over, D. E., & Politzer, G. (2013). Uncertainty and de Finetti tables. *Thinking & Reasoning, 19*, 308–328.

Baratgin, J., & Politzer, G. (2010). Updating: A psychological basic situation of probability revision. *Thinking & Reasoning, 16*, 245–287.

Barbey, A. K., & Sloman, S. A. (2007). Base-rate respect: From ecological rationality to dual processes. *Behavioral and Brain Sciences, 30*, 241–254.

Bennett, J. (2003). *A philosophical guide to conditionals*. Oxford: Oxford University Press.

Byrne, R. M. J., & Johnson-Laird, P. N. (2010). Conditionals and possibilities. In M. Oaksford & N. Chater (Eds.), *Cognition and conditionals: Probability and logic in human thought* (pp. 55–68). Oxford: Oxford University Press.

Chater, N., & Oaksford, M. (Eds.). (2008). *The probabilistic mind: Prospects for Bayesian cognitive science*. Oxford: Oxford University Press.

Coletti, G., & Scozzafava, R. (2002). *Probabilistic logic in a coherent setting*. Dordrecht: Kluwer Academic Publishers.

Cruz, N., Baratgin, J., Oaksford, M., & Over, D. E. (2015). Bayesian reasoning with ifs and ands and ors. *Frontiers in Psychology, 6*, 192.

Cruz, N., & Oberauer, K. (2014). Comparing the meanings of "if " and "all." *Memory & Cognition, 42*, 1345–1356.

de Finetti, B. (1974). *Theory of probability*. Chichester: Wiley.

de Finetti, B. (1995). The logic of probability. Translated in R. B. Angell (Ed.), The logic of probability. *Philosophical Studies, 77*, 181–190. (Original work published 1936.)

Douven, I., & Dietz, R. (2011). A puzzle about Stalnaker's hypothesis. *Topoi, 30*, 31–37.

Douven, I., & Verbrugge, S. (2010). The Adams family. *Cognition, 117*, 302–318.

Dummett, M. (2000). *Elements of intuitionism*. Oxford: Clarendon.

Edgington, D. (1995). On conditionals. *Mind, 104*, 235–329.

Edgington, D. (2003). What if? Questions about conditionals. *Mind & Language, 18*, 380–401.

Elqayam, S., & Over, D. E. (2013). New paradigm psychology of reasoning: An introduction to the special issue edited by Elqayam, Bonnefon, & Over. *Thinking & Reasoning, 19*, 249–265.

Evans, J. St. B. T., Handley, S., Neilens, H., & Over, D. E. (2007). Thinking about conditionals: A study of individual differences. *Memory & Cognition, 35*, 1772–1784.

Evans, J. St. B. T., Handley, S., & Over, D. E. (2003). Conditional and conditional probability. *Journal of Experimental Psychology: Learning, Memory, and Cognition, 29*, 321–335.

Evans, J. St. B. T., & Over, D. E. (2004). *If*. Oxford: Oxford University Press.

Evans, J. St. B. T., & Over, D. E. (2013). Reasoning to and from belief: Deduction and induction are still distinct. *Thinking & Reasoning, 19*, 268–283.

Evans, J. St. B. T., Thompson, V. A., & Over, D. E. (2015). *Uncertain deduction and conditional reasoning. Frontiers in Psychology, 6*, 398.

Fugard, J. B., Pfeifer, N., Mayerhofer, B., & Kleiter, G. D. (2011). How people interpret conditionals: Shifts toward conditional event. *Journal of Experimental Psychology: Learning, Memory, and Cognition, 37*, 635–648.

Gilio, A. (2002). Probabilistic reasoning under coherence in System P. *Annals of Mathematics and Artificial Intelligence, 34*, 5–34.

Gilio, A., & Over, D. E. (2012). The psychology of inferring conditionals from disjunctions: A probabilistic study. *Journal of Mathematical Psychology, 56*, 118–131.

Girotto, V., & Johnson-Laird, P. N. (2010). Conditionals and probability. In M. Oaksford & N. Chater (Eds.), *Cognition and conditionals: Probability and logic in human thought* (pp. 103–115). Oxford: Oxford University Press.

Hadjichristidis, C., Sloman, S. A., & Over, D. E. (2014). Categorical induction from uncertain premises: Jeffrey's doesn't completely rule. *Thinking & Reasoning, 20*, 405–431.

Howson, C., & Urbach, P. (2006). *Scientific reasoning: The Bayesian approach* (3rd ed.). Chicago: Open Court.

Jeffrey, R. C. (1991a). Matter of fact conditionals. *Aristotelian Society Supplementary Volume, 65*, 161–183.

Jeffrey, R. C. (1991b). *The logic of decision* (2nd ed.). Chicago: University of Chicago Press.

Johnson-Laird, P. N., & Byrne, R. M. J. (1991). *Deduction.* Hove, London: Erlbaum.

Johnson-Laird, P. N., & Byrne, R. M. J. (2002). Conditionals: A theory of meaning, pragmatics and inference. *Psychological Review, 109*, 646–678.

Lewis, D. K. (1976). Probabilities of conditionals and conditional probabilities. *Philosophical Review, 85*, 297–315.

Manktelow, K. I., & Over, D. E. (1991). Social roles and utilities in reasoning with deontic conditionals. *Cognition, 39*, 85–105.

Manktelow, K. I., Over, D. E., & Elqayam, S. (2011). *Paradigm shift: Jonathan Evans and the science of reason*. In K. I. Manktelow, D. E. Over, & S. Elqayam (Eds.), The science of reason: A Festschrift for Jonathan St B T Evans (pp. 1–16). Hove, UK: Psychology Press.

Milne, P. (2012). Indicative conditionals, conditional probabilities, and the "defective truth-table": A request for more experiments. *Thinking & Reasoning, 18*, 196–224.

Murphy, G. L., Chen, S. Y., & Ross, B. H. (2012). Reasoning with uncertain categories. *Thinking & Reasoning, 18*, 81–117.

Oaksford, M., & Chater, N. (2007). *Bayesian rationality: The probabilistic approach to human reasoning*. Oxford: Oxford University Press.

Oaksford, M., & Chater, N. (2013). Dynamic inference and everyday conditional reasoning in the new paradigm. *Thinking & Reasoning, 19*, 346–379.

Oberauer, K., & Wilhelm, O. (2003). The meaning(s) of conditionals: Conditional probabilities, mental models and personal utilities. *Journal of Experimental Psychology: Learning, Memory, and Cognition, 29*, 680–693.

Over, D. E. (2007). The logic of natural sampling. *Behavioral and Brain Sciences, 30*, 277.

Over, D. E. (2009). New paradigm psychology of reasoning. *Thinking & Reasoning, 15*, 431–438.

Over, D. E., Douven, I., & Verbrugge, S. (2013). Scope ambiguities and conditionals. *Thinking & Reasoning, 19*, 284–307.

Over, D. E., Evans, J. St. B. T., & Elqayam, S. (2010). Conditionals and non-constructive reasoning. In M. Oaksford & N. Chater (Eds.), Cognition and conditionals: Probability and logic in human thinking (pp. 135–151). Oxford: Oxford University Press.

Over, D. E., Hadjichristidis, C., Evans, J. S. B. T., Handley, S. J., & Sloman, S. A. (2007). The probability of causal conditionals. *Cognitive Psychology, 54*, 62–97.

Pfeifer, N., & Kleiter, G. D. (2009). Framing human inference by coherence based probability logic. *Journal of Applied Logic, 7*, 206–217.

Pfeifer, N., & Kleiter, G. D. (2010). The conditional in mental probability logic. In M. Oaksford & N. Chater (Eds.), Cognition and conditionals: Probability and logic in human thinking (pp. 153–173). Oxford: Oxford University Press.

Politzer, G., Over, D. E., & Baratgin, J. (2010). Betting on conditionals. *Thinking & Reasoning, 16*, 172–197.

Ramsey, F. P. (1990a). Facts and propositions. In D. H. Mellor (Ed.), *Philosophical papers* (pp. 34–51). Cambridge: Cambridge University Press. (Original work published 1927.)

Ramsey, F. P. (1990b). General propositions and causality. In D. H. Mellor (Ed.), *Philosophical papers* (pp. 145–163). Cambridge: Cambridge University Press. (Original work published 1927.)

Singmann, H., & Klauer, K. C. (2011). Deductive and inductive conditional inferences: Two modes of reasoning. *Thinking & Reasoning, 17*, 247–281.

Singmann, H., Klauer, K. C., & Over, D. E. (2014). New normative standards of conditional reasoning and the dual-source model. *Frontiers in Psychology, 5*, 316.

Stalnaker, R. (1968). A theory of conditionals. In N. Rescher (Ed.), Studies in logical theory (pp. 98–112). Oxford: Blackwell.

Stalnaker, R. (1975). Indicative conditionals. *Philosophia, 5*, 269–286.

Stalnaker, R., & Jeffrey, R. (1994). Conditionals as random variables. In E. Eells & B. Skyrms (Eds.), Probability and conditionals (pp. 31–46). Cambridge: Cambridge University Press.

Stevenson, R. J., & Over, D. E. (2001). Reasoning from uncertain premises: Effects of expertise and conversational context. *Thinking & Reasoning, 7*, 367–390.

Tversky, A., & Kahneman, D. (1983). Extensional vs. intuitive reasoning: The conjunction fallacy in probability judgment. *Psychological Review, 90*, 293–315.

Zhao, J., & Osherson, D. (2010). Updating beliefs in light of uncertain evidence: Descriptive assessment of Jeffrey's rule. *Thinking & Reasoning, 16*, 288–307.

5 Knowing Enough to Achieve Your Goals: Bayesian Models and Practical and Theoretical Rationality in Conscious and Unconscious Inference

Mike Oaksford

Theoretical or epistemic rationality is concerned with the relationships among an agent's beliefs; that is, if you believe x what else should you believe? The standard view is that these transitions between beliefs are rational if they conform to the corresponding logical relations between propositions (Smith, 2004). So, to use conditional inference as an example, if S believes *if p then q* and p are true, then S should believe q to be true by an application of the logical law of *modus ponens* (MP). Recent Bayesian approaches in the psychology of reasoning have questioned this standard binary truth functional approach, arguing that beliefs are graded and probabilistic (Oaksford & Chater, 1994, 2007, 2009). They also adopt "the Equation" (Edgington, 1995), that the probability of a conditional *if p then q* equals the conditional probability, $\Pr(q|p)$. In this view if S believes *if p then q* to degree a and comes to believe p to degree 1, that is, $\Pr(p) = 1$, then S should now believe q to the same degree as the conditional probability, $\Pr(q) = \Pr(q|p) = a$, by Bayesian conditionalization. The move to probabilities directly raises issues of practical rationality. Practical or instrumental rationality provides a suitable means for achieving one's goals regardless of the nature of those goals (in particular whether they are morally or ethically "correct"). The laws of probability, to which rational transitions between graded beliefs are now expected to conform, are traditionally justified in terms of their practical rationality (Joyce, 2004). So the standard Dutch book arguments propose that conforming to the rules of probability is rational because not doing so could lead one to take actions that are self-defeating, that is, placing bets that one is bound to lose.[1] Here, "practically self-defeating" means failing to maximize expected utility.

The new paradigm in the psychology of reasoning (Over, 2009; Manktelow, 2012; and a special issue of *Thinking and Reasoning:* Elqayam & Over, 2013) has embraced the fact that utilities and probabilities jointly influence human reasoning (Bonnefon, 2013). For example, there has been research on utility conditionals in reasoning (Bonnefon, 2009; Evans, Neilens, Handley, & Over, 2008) and in Bayesian argumentation (Corner, Hahn, & Oaksford, 2011). These conditionals are of the form *if you take action A it will lead to a consequence C*, where C has a clearly defined utility. Reasoning with such rules frequently violates standard logical laws (Bonnefon & Hilton, 2004) but in many cases can be shown to conform to the principle of maximizing expected utility (Corner et al., 2011) or to sensible heuristic approximations to it (Bonnefon, 2009).

In this chapter we argue that in everyday reasoning in the real world there is invariably a trade-off between achieving our goals in a timely way and the search for truth; that is, there is a trade-off between our epistemic and practical goals. In other words our everyday reasoning, although practically rational, may not always fulfill the demands of theoretical rationality. We first rehearse the Dutch book arguments for the rationality of the laws of probability to show how practical concerns of maximizing expected utility work in these proofs. We then show that accounts of unconscious inference (Helmholtz, 1866/2000) in perception and action proposed by the Bayesian brain hypothesis directly embody the trade-off between epistemic and practical goals (Clark, 2013; Dayan & Hinton, 1996; Friston, 2005). We then show that recent Bayesian approaches to explicit conscious reasoning are directly related to these approaches to unconscious inference and that they also rely on constructing a generative model of the world (Ali, Chater, & Oaksford, 2011; Fernbach & Erb, 2013; Sloman, 2005; Sloman, Barbey, & Hotaling, 2009; Sloman & Lagnado, 2005). Moreover, we argue that in contexts where explicit verbal reasoning is involved the trade-off between theoretical and practical rationality recurs in the need to balance the costs of reasoning against the likelihood of achieving one's goals. We propose that reasoners must take into account cognitive efficiency (Sperber & Wilson, 1986) and the social distribution of knowledge. We show how these considerations generalize to argumentation and how they may address some outstanding problems for Bayesian approaches to the psychology of reasoning. We conclude that human reasoning is rational but, like the laws of probability, only with respect to practical rationality.

The Dutch Book Argument

The Dutch book argument for the laws of probability theory aims to show that violating these laws is practically irrational (Joyce, 2004). The argument begins with some relatively self-evident assumptions, the most important of which is the *expected utility thesis*. This thesis states that an act best satisfies a rational agent's desires if and only if it is the one that maximizes his or her subjective expected utility. The expected utility thesis shows that with a few further assumptions an agent will reveal the strength of her beliefs in her betting behavior. These assumptions are: "(a) the agent desires only money; (b) her desire for money does not vary in changes in her fortune; and (c) she is not averse to risk or uncertainty" (from Joyce, 2004, p. 136). Beginning with a confidence measure defining the agent's degree of belief in a proposition x, $c(x)$, the aim is to show that the measure c must respect the laws of probability.

Borrowing heavily from Joyce (2004, p. 136), consider a wager W such that one receives £110 if Arsenal wins and £10 if Arsenal loses. Clearly whether Arsenal wins or not is not dependent on the wager, and it is certain that if Arsenal wins, a person taking the bet will win £110, and if Arsenal loses, she will win £10. The fair price that a person will pay for this wager, £f, will be the sum at which he or she is indifferent between £f and the wager W. Assuming that that person's degree of belief that Arsenal will win is $c(x)$, the expected utility thesis dictates that £f will be W's expected payoff, that is, £f = $E(W)$ = c(Arsenal wins) × £110 + [1 − c(Arsenal wins)] × £10. We can then immediately see that this person's confidence that Arsenal will win, c(Arsenal wins) = $(f − £10)/(£110 − £10)$. Assuming the person is indifferent between the wager and £60, this means that his degree of belief in Arsenal winning is 0.6. Consequently we can infer people's degrees of belief from their betting behavior.

So far we still do not know whether $c(x)$ respects the laws of probability. This is what the Dutch book theorems establish. By the expected utility thesis and (a) to (c), people should be willing to trade any set of wagers for their fair price. Moreover, they should be willing to trade any set of fair prices for the corresponding wagers. Take the following sets of wagers and fair prices:

W_A = (£100 if Arsenal wins, £0 else), £f_A = £25

W_C = (£100 if Chelsea wins, £0 else), £f_C = £25

$W_{A\ or\ C}$ = (£100 if Arsenal OR Chelsea wins, £0 else), £$f_{A\ or\ C}$ = £60

Arsenal and Chelsea are playing each other, and all bets are off if there is a draw. In this example someone should be indifferent between the first two wagers plus £60 (W_A, W_C, £60) and the last wager plus £50 ($W_{A\ or\ C}$, £50). However, W_A and W_C are equivalent, in terms of payoffs, to $W_{A\ or\ C}$. Consequently, the extra £10 paid for the first bundle of wagers and fair prices yields no potential gain; that is, the set of fair prices is irrational. So if you found someone to accept these wagers from you at these "fair" prices, you would be guaranteed to make £10 whatever the outcome, and this person would then become a money pump. It is therefore clear that one should be indifferent between the first two wagers, W_A and W_C, and $W_{A\ or\ C}$ in these circumstances, and therefore £$f_{A\ or\ C}$ should equal £f_A + £f_C. Consequently, $c(x)$ should conform to the additivity axiom of probability theory, that is, $Pr(x\ or\ y) = Pr(x) + Pr(y)$, when x and y are logically incompatible (both teams cannot win). Similar Dutch book arguments can be made for all the other laws of probability theory.

This argument shows that "it is practically irrational to hold beliefs that violate the laws of probability" (Joyce, 2004, p. 136). If our degrees of belief do not conform to the laws of probability, then we are bound to take actions, that is, place bets, from which we are guaranteed to incur a loss. That is, we will fail to maximize expected utility. The Dutch book is probably the best-known practical justification for the laws of probability. However, to move beyond (a)–(c) above, that is, to lift restrictions on what people desire beyond money, requires moving to Savage's (1954) axioms that established the coherence of expected utility and probabilistic consistency at the same time. This approach still offers a practical justification for the laws of probability. Both the Dutch book argument and Savage's axioms have been accused of offering a purely pragmatic justification for the rationality of beliefs. This approach is seen as inadequate because it does not account for the way beliefs are sensitive to evidence and how they play important roles in theoretical reasoning. This has led to attempts to "depragmatize," as Joyce (2004) puts it, arguments for probabilistic consistency. Recent approaches such as that of Joyce (1998) argue that probabilistic consistency can be justified in purely epistemic terms by relating it to the accuracy of beliefs.

The fact that such justifications can be found is primarily of philosophical concern. Bayesian approaches have been the most successful attempts to articulate the nature of scientific inference (Earman, 1992; Howson & Urbach, 1989). Scientific is our best model of a purely rational epistemic enterprise, where the goal is truth. If scientific inference only had a pragmatic justification, it would not appear to be able to improve the accuracy of our beliefs. However, from a psychological point of view, we are concerned with everyday human reasoning in the real world. So, although psychological scientists share these epistemic goals, the empirical question is what real human reasoners do in their everyday lives. What we suggest is that in everyday reasoning in the real world there is invariably a trade-off between achieving our goals in a timely way and the search for truth; that is, there is a trade-off between our epistemic and practical goals. People only want to know enough to achieve their practical goals.

Practical Rationality in Everyday Human Reasoning

In explicating the trade-off between practical and epistemic goals and how it is captured by Bayesian models, we begin with a simple example. We then show how it is dealt with in unconscious inference in the Bayesian brain theory. We then introduce current Bayesian accounts of conditional reasoning—reasoning using what in English is rendered as *if … then*—and show that similar issues recur.

Suppose that on Bodmin Moor someone saw what could be a domestic cat or a panther. She may believe quite strongly that what she is seeing is a panther because of the potential consequences of disbelieving it, although it is difficult to know quite what the correct description of her attitude is in this situation. She may just be hedging her bets: the costs are so great if it is a panther, she will act as if it were. However, if she is taken aback and feels fear as a result of what she sees, her belief that it is a panther must be quite strong. At the very least she would be manifesting the behavior of someone who believes this strongly. In this section we argue that examples like this are readily dealt with by the theory of unconscious inferences (Helmholtz, 1866/2000) involved in perception and action emerging in the Bayesian brain hypothesis (Clark, 2013; Dayan & Hinton, 1996; Friston, 2005), which suggests that practical considerations are always active in implicit, unconscious inference.

The Bayesian Brain

Our current best bet about how the brain constructs reality from perturbations of its sensory surfaces in perception is based on Bayesian inferential processes. In the Bayesian brain theory (Clark, 2013; Dayan & Hinton, 1996; Doya, Ishii, Pouget, & Rao, 2006; Friston, 2005), hypotheses about the hidden causes of those perturbations, that is, the objects and events in the world (the priors for which are derived from the data using empirical Bayesian methods: Carlin & Louis, 2000), generate predictions for the states of an organism's sensory surfaces in a hierarchical Bayesian model. The difference between the predictions and the actual states of the organism's sensory surfaces (the data) creates a prediction error, which is fed back up the hierarchy. The idea is to select the hypothesis that minimizes the prediction error. Minimizing prediction error is the same as minimizing uncertainty or entropy (Shannon & Weaver, 1949) or maximizing information gain (Lindley, 1956). In the psychology of reasoning Oaksford and Chater (1994, 1996) used maximizing information gain to account for data selection behavior in Wason's (1968) selection task.

In the Bayesian brain theory there are two ways to minimize entropy (average surprise), either by optimizing internal predictions in perception or by acting so that sensory data better match internal predictions. So an agent's beliefs, or *generative model*, not only relates hidden causes (the world) to sensory events but also includes future states and actions and their consequences. The agent then uses this model to generate policies to act on that minimize surprise about the outcomes of her actions. Friston, Schwartenbeck, FitzGerald, Moutoussis, Behrens, and Dolan (2013; see also Schwartenbeck, FitzGerald, Dolan, & Friston, 2013) argue that the value of a policy, π, for example, to approach or avoid the object in the beast of Bodmin example, in a certain state, s, is equal to the difference between two probability distributions. The first distribution is over outcome states given a policy and the current state, $Pr(S|s, \pi)$. The second distribution is the probability of outcome states based solely on prior beliefs, that is, the agents generative model, m, which represents goal states that the agent desires, for example, $Pr(S|m)$. The difference between these distributions is given by the Kullback–Leibler distance (which equals expected information gain; see Oaksford & Chater, 1996, appendix), so the value of a policy given the current state is $V(\pi|s) = -D[Pr(S|s, \pi)||Pr(S|m)]$.[2] Friston et al. (2013) and Schwartenbeck et al. (2013) observe that this distance can be decomposed into two terms:

$$-D[Pr(S|s, \pi)||Pr(S|m)] = H[Pr(S|s, \pi)] + \sum_{S} Pr(S|s, \pi).u(S\,|\,m) \qquad (5.1)$$

The first term is the entropy over the outcomes or goal states. The second term is the expected utility over outcomes. It important to bear in mind that, in this set up, utility ($u()$) is conceived of in terms of surprise. That is, an agent values unsurprising events. We further discuss these two terms after first seeing how equation 5.1 applies to our example.

The predictions for the value of policies are generated by the same generative model used to perceive the current state. Perception and action in this theory and in other accounts (Gibson, 1950, 1979) are inseparable and mutually informing. Consequently, hypotheses about states, s, also receive support from the consequences of possible policies for action based on those states. Assume there are two policies, approach or avoid. The approach policy has the potential to produce a very surprising outcome, being mauled, assuming the current state is one of observing a panther. Consequently, surprise will be minimized by avoidance, which makes sense only if the agent interprets what it is seeing as a panther. This line of argument suggests that what we believe at the relatively low perceptual level depends on the practical issues concerning the possible outcomes of our actions and how we value those outcomes. Indeed, the reaction of being taken aback and fear at this unconscious interpretation may well be a reaction to the anticipation of a surprising event (Joffily & Coricelli, 2013; Moore & Oaksford, 2000).[3]

The two components of the decomposition in equation 1 correspond to two different strategies. If surprising outcomes with high utilities are not predicted, then agents may *explore* outcomes by minimizing entropy, the first term. On the other hand, when some outcomes have high utilities, then the second term will be weighted more highly, and people will attempt to *exploit* the situation to maximize expected utility. These two different strategies correspond to disinterested (maximizing information gain) and interested (maximizing expected utility) inquiry discussed in the reasoning literature (Chater, Crocker, & Pickering, 1998; Chater & Oaksford, 1999; Klauer, 1999; Oaksford, Chater, & Grainger, 1999; Sperber, Cara, & Girotto, 1995). In the context of explicit, conscious inference in verbal reasoning in Wason's selection task, Oaksford et al. (1999) argued that there is evidence for both strategies when appropriate. In the context of implicit, unconscious inference (Helmholtz, 1866/1962) in perception and action, Friston et al. (2013) and Schwartenbeck et al. (2013) argue that both strategies are always in evidence but differentially weighted. That is, in unconscious inference there is always a trade-off between our epistemic and our practical goals, between exploration to reduce ignorance and exploitation to get what we want.

The Bayesian brain approach to unconscious inference suggests that there are two rational inferential strategies determining behavior in perception and action. This approach suggests that at this level, practical concerns are invariably in play. We now suggest that in human reasoning at the conscious verbal level similar constraints are also operative.

Explicit Verbal Reasoning

So far we have looked at the unconscious inferences involved in perception and action. We have considered how these unconscious inferences can be considered as products of rational Bayesian inference. Moreover, this has shown that maximizing utility, that is, practical concerns, figures in the inferences about what is out there and how to act. We now argue that frequently, especially when things go wrong, explicit verbal reasoning interposes between perception and action. We suggest that exactly the same trade-offs occur in the actual contexts in which conscious verbal reasoning normally occurs.

We first observe that the need for a theory of conscious verbal reasoning to address our practical concerns has been pointed out in the philosophical literature by Harman (1968; following Sellars, 1963). They argued that understanding psychological states or beliefs required an account of how they figure in the *evidence-inference-action* game. We need to know what perceptual evidence would lead us to acquire a belief (the object is a panther), how these beliefs lead to other beliefs (inference: *if panther then low utility of an encounter*), and how these beliefs lead to action (avoidance). Of course, the inferences involved may be unconscious: as in the case of the beast of Bodmin, we may only become aware of the fact that the inference has been drawn by our emotional reaction. We now argue that considering how conscious verbal reasoning arises in the evidence-inference-action game may inform our current accounts of conditional reasoning.

Conditional inference is explicit verbal reasoning with *if … then*, such as in the logical inference rule, *modus ponens*, for example, *if the key is turned* (p), then *the car starts* (q), *the key is turned*, therefore, *the car starts*, an example that we will use throughout. As in this example, most current research in this area has focused on such causal conditionals (Cummins, 1995; Evans, Handley, & Bacon, 2009; Over, Hadjichristidis, Evans, Handley, & Sloman, 2007), where the antecedent (p) of a conditional describes a cause of the effect described in the consequent (q). Three other inferences other than MP have been empirically investigated.[4] These are denying the antecedent (DA, *if p then q, not-p; therefore, not-q*); affirming the consequent (AC, *if p then q, q, therefore p*), and *modus tollens* (MT, *if p then q, not-q, therefore, not-p*). MP and MT are logically valid inferences, and DA and AC are logical fallacies. Typically people do not show the logical patterns of performance with purely alphanumeric stimuli. With causal materials inference patterns that are completely nonlogical, according to standard binary logic, predictably occurs.

In particular it seems that, when provided with a causal conditional, people automatically retrieve information about *enabling conditions* and *alternative causes,* and effects are observed like those found when this information is explicitly provided, as in Byrne (1989). Take the following examples of explicit conditional premises:

If you turn the key the car starts. (2)

If battery is dead the car does not start. (2′)

If you turn the key the car starts.   (3)

If you hot-wire it the car starts.   (3')

   Adding (2'), which describes a disabling condition (having a dead battery), leads people to draw fewer MP and MT inferences to the conclusion that the car starts (MP) or that the key was not turned (MT) when told about the categorical premises, "the key is turned" (MP) or "the card did not start" (MT), respectively. Similar effects are observed for materials pretested for possible disabling conditions even when the second premise was not explicitly presented (Cummins, 1995; Cummins et al., 1991). This pattern of reasoning, where the addition of the information in (2') and (3') loses conclusions that were previously available, is *nonmonotonic* (Oaksford & Chater, 1991, 1994, 2014; Stenning & van Lambalgen, 2005).

   Recently it has been proposed that (2) and (2') and (3) and (3') are represented cognitively in causal Bayes nets (Ali, Chater, & Oaksford, 2011; Fernbach & Erb, 2013). Figure 5.1 shows the representations proposed. These are both cases of a collider structure, and the arrows indicate the causal direction from cause to effect. In (3) and (3') the parents of the effect node are alternative causes. In (2) and (2') one of the parents of the effect node is a cause (2); the other a disabling condition (2').[5] These causal models each represent probability distributions over the three discrete random variables in each diagram. The parents of the effect node (q) are conditionally independent given the effect. The probabilities assigned simply indicate that causes are probabilistically sufficient for their effects and that the disabler is probabilistically sufficient for the nonoccurrence of the effect. The CBNs in figure 5.1 are very simple generative models of the worlds described in the conditional premises. Consequently, on this account, there is some theoretical continuity between unconscious implicit inference in perception and action and conscious explicit verbal reasoning. Both can be seen as Bayesian inference. Conditional inference is implemented in these models by Bayesian conditionalization.

   To our knowledge conditional inferences of this type have rarely ever been placed in an appropriate context where they may actually occur to a reasoner in everyday life such as, for example, when leaving home to go to work and consequently needing to

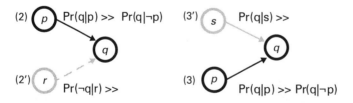

Figure 5.1
Causal Bayes net representations of (2) and (2') with a cause (p) and a disabler (r) and (3) and (3') with a cause (p) and an alternative cause (s).

start the car. Recently, Oaksford and Chater (2014) briefly considered the consequences of contextualizing reasoning in this way. This discussion was in the context of the problems raised by the nonmonotonic nature of the inferences observed with materials such as (2) and (2′) and (3) and (3′). Nonmonotonic reasoning raises profound problems for cognitive theories of reasoning, primarily in connection with the frame problem (Fodor, 1983, 2001; Oaksford & Chater, 1991, 1995). Oaksford and Chater (2014, in press-a,-b) argued that the context-bound and goal-directed nature of most reasoning in practical situations may significantly delimit the scope of these problems. We now argue that the issues raised in that discussion can be viewed from a rational perspective as a trade-off between exploration and exploitation just as in unconscious inference. But first we work through Oaksford and Chater's (2014) example to highlight the issues relevant to our current discussion.

Suppose Jane is at home with the goal of getting to work. This will involve her using her car. Figure 5.2 shows the sequence of generative models created in working memory as Jane reasons about how best to achieve this goal. Her subgoal is to

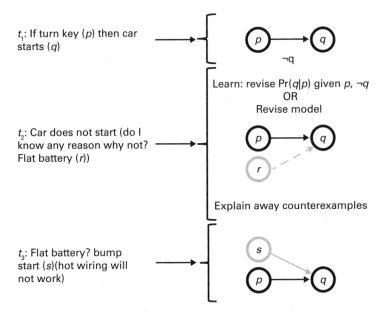

Figure 5.2
The sequence of causal models created in response to the failure of the car to start given the key was turned (t_1); a defeater is accessed at t_2 (flat battery); an alternative cause that will start the car even though the battery is dead is accessed at t_3 (jump starting). Hot wiring requires a charged battery (this is how criminals steal cars), although jump starting does not (it just requires jump leads to connect to another car's battery).

start the car, which triggers the dependency expressed in the rule *if she turns the key the car starts*. So at t_1 she therefore turns the key, predicting that the car will start. She will do this without any consideration of possible defeaters or alternative causes. This is because this is the normal way of starting a car. The inference to the need to turn the key and the prediction that the car will start at t_1 is most likely unconscious. However, suppose at t_2 this action fails to produce the desired effect—the car does not start. At this point fully explicit conscious reasoning processes must be engaged by this surprising event. She then has to consider possible defeaters. The defeaters she knows about will be prepotent in memory triggered by the context she is now in. Perhaps she considers that the battery is dead. This defeater explains away the apparent counterexample to the original dependency and triggers an alternative cause that will overcome the defeater such as jump starting. The defeater must be recruited first because it determines the alternative cause: hot wiring will not work with a dead battery. If jump starting does not work at t_3, then this is probably as far as Jane can go. She has reached the limit of her knowledge of the way cars work. The issues raised by the frame problem suggest that Jane should stand around attempting to access distant knowledge that might resolve the problem. However, it is obvious what she would do in this situation: call the AAA (Automobile Association) or a similar roadside assistance organization. Her goal is to get to work within a reasonable time frame. Her overarching goal is not to discover the truth about why the car did not start. Moreover, she is aided in achieving her goal by the fact that knowledge is socially distributed. AAA men have the knowledge that she lacks. So her goal is much more easily achieved by contacting the AAA. Doing so saves taxing her cognitive resources any further. Thus, the social distribution of knowledge allows people to avoid such unnecessary cognitive effort.

This example raises many issues. For example, the way that deictic context provides a set of rich cues to the information we are likely to need in order to achieve our goals. Oaksford and Chater (2014) discuss some of these issues with respect to the problems Fodor (1983, 2001) observed for computational accounts of everyday inference. The particular issue that we pick up on here concerns the fact that this passage of reasoning has an obvious stopping point. That is, there are clear criteria for when a practical decision needs to be made to stop reasoning and pursue a different strategy. Within modern technological societies, knowledge is massively distributed across society. There are experts on every topic. Consequently, explicit reasoning processes at the individual level (note that all the inferences at t_1 could be unconscious), triggered by the problems we encounter in the real world, often have obvious alternative strategies. The social distribution of knowledge means that most of us possess only very shallow knowledge of the world in which we need to act. There are always cases when we must stop reasoning and "phone a friend."

Bayesian Brains Again   The issues raised in this example relate directly to the decomposition of the Kullback–Leibler distance (equation 1), which we introduced in discussing unconscious inference. The explicit reasoning triggered by the unexpected event of the car not starting is aimed at making that event predictable by constructing a new model of the situation that minimizes entropy. That is, explicit verbal reasoning in this context is an exploratory strategy, not of external states but of possible states that render the unexpected event predictable. Actions must then be selected that achieve the goal of changing the external state—getting the car to start—given this new initial state. At each point in the reasoning process the goal is to minimize entropy. That is, conscious individual verbal reasoning should be regarded as a prototypically human exploratory strategy to reduce uncertainty about possible future events in prediction and about possible past events in explaining away the unexpected (i.e., supposing the battery was dead explains away the predictive failure).

The distribution of knowledge in human societies and the ability to communicate about complex matters using language massively expands the range of exploratory information-gathering strategies to reduce ignorance and uncertainty that are available to humans. Consequently, the "phone a friend" strategy is simply an extension of exploratory, entropy-minimizing strategies that transcend the individual to exploit the social distribution of knowledge. This can involve reading a book or talking to friends who know more about cars than Jane does. However, Jane also has a practical goal: she wants to get to work in good time.

In this practical context, exploration or information gathering (Pirolli & Card, 1999) has associated costs: reasoning or phoning a friend takes time. Each exploratory action has an expected benefit and an associated time cost that must be traded against the increasing expected cost over time of not achieving one's goal. At some point the increasing expected costs associated with failing to start the car must exceed the expected benefits of continued information search (Fu & Gray, 2006). The situation is slightly more complex because other than starting the car Jane has a deadline on her overarching goal of getting to work. This suggests that there is a point in time at which continued effort to start the car could not achieve the goal of getting to work on time. Consequently a time utility function (Horvitz & Rutledge, 1991) needs to be introduced to reflect this deadline. This will also have to factor in the time to instigate alternative strategies. These too may involve exploiting the social distribution of knowledge (call the AAA) but may also involve other strategies with associated costs (call a cab). In sum, the second component of the decomposition in equation 1, maximizing expected utility, is always in play in most passages of reasoning in the everyday world that are aimed at resolving practical problems. At some point, continued reasoning is not rational because it could no longer achieve one's goals. Consequently, in conscious everyday verbal reasoning in the real world, just as in the unconscious inference in perception and action, there is a trade-off between our epistemic and our practical goals. We now argue that these considerations generalize directly to argumentation.

Reasoning and Argumentation   Recently, it has been suggested that reasoning usually has an argumentative goal (Hahn & Oaksford, 2007; Mercier & Sperber, 2011). For example, Hahn and Oaksford (2007, p. 705) argued that:

Typically, reasoning takes place in the service of argumentation, that is, in the attempt to persuade yourself or others of a particular position. Argumentation is the overarching human activity that studies of deductive reasoning, inductive reasoning, judgment, and decision making are really required to explain. So one might attempt to convince someone else to accept a controversial standpoint p by trying to persuade them that p is actually a logical consequence of their prior beliefs or current commitments; or that p has strong inductive support; or, when p is an action, that p will help to achieve their current goals.

Fulfilling argumentative goals brings many practical issues to the fore. For example, assume person A is trying to persuade person B to give up smoking. The argument goes as follows:

A: You should give up smoking because if you smoke you will get lung cancer.

B: Many people smoke and don't get lung cancer.

A: Maybe, but it will negatively affect your health.

B: What is your evidence, and anyway whatever happens is already predetermined in my genes.

One would hope that the negative health consequences of smoking were common knowledge and that little argument would be required. However, logically B's refutation of A's opening gambit is correct. A's retreat to the more general claim is probably sound. But B's response, a demand for more evidence and the introduction of a further point, means that practically the argument has probably run its course. A may be able to drag up some further evidence of negative health effects, but is it worth her time and effort? B's final point suggests it might not be. Neither B nor A can resolve whether B is genetically predisposed to smoking-related diseases. Moreover, B's invocation of this point suggests a fatalistic attitude against which further argument on A's behalf is probably futile.

This example suggests that there are epistemic restrictions on how far an argument can go. If no more evidence immediately comes to A's mind, then the argument is practically at an end (although it may be rejoined at a later date, after A has phoned a friend). Moreover, A needs to make a practical judgment about how likely further argument is to succeed against the background of B's fatalism. A also needs to weigh these against her other priorities (perhaps she is trying to sell B nicotine patches) and the time she has available to try and convince B to give up. These practical judgments about when the argument needs to stop are determined by ongoing assessments of whether further arguments are available to be put forward and the likelihood of success. In other words in practical domains such as argumentation, there are stopping rules and a range of concerns with well-defined utilities. That is, in argumentation as well as in explicit reasoning there is a trade-off between practical and epistemic goals.

Modus Ponens **and Group Reasoning** We have argued (a) that conscious verbal reasoning often occurs as a result of the failure of a spontaneous unconscious inference, and (b) that these are typically practical contexts where an obvious strategy to reduce uncertainty is to confer with others with more knowledge of the situation. Here we argue that these two factors may lead to an understanding of two phenomena that seem initially problematic for the Bayesian approach. The first is why in conditional reasoning, especially with abstract material, the MP inference is endorsed so strongly (Schroyens & Schaeken, 2003). This has typically been argued to provide strong evidence for logical insight. From a probabilistic perspective there is no reason to expect this result. MP is drawn with a probability of about 0.97. In current Bayesian theory (Oaksford & Chater, 2007, 2013), this high value is not consistent with the much lower endorsements of MT or with the patterns of inference observed with DA and AC. Oaksford and Chater (2013) show how starting with such a high value for MP as an estimate of the conditional probability, $Pr(q|p)$, sensible predictions can be arrived at for the remaining inferences. However, this still does not fully explain its initial high value.

The conditional is usually used to describe the dependencies in the world that can be used to predict what will happen next in order to guide our actions. In practical contexts such as starting the car, one has no choice but to turn the key predicting the car will start. After all this is the choice that should minimize surprise, as it is the action most likely to produce this outcome. As we argued in the above example, such inferences are unconscious, and we only become conscious of the reasoning process involved when it goes wrong—the car does not start. If, as it has been argued, conditionals describe reliable dependencies of this form (Ali et al., 2011; Sloman, 2005; Sloman & Lagnado, 2005; Sloman et al., 2009), then allowing them to guide our expectations in attempting to fulfill our goals is more like a reflex action, a consequence of unconscious inference, rather than conscious, verbal reasoning. Consequently, from a practical perspective the very high endorsements of MP should come as no surprise.

Group reasoning is another area that probabilistic approaches to human reasoning have not addressed. However, as we have just seen, like Mercier and Sperber (2011; Sperber & Mercier, 2012), Bayesian theorists have suggested that reasoning most often has argumentative ends (Hahn & Oaksford, 2007). That is, reasoning usually occurs in a social context. Moreover, as we have just argued, the shallow nature of much individual human knowledge suggests that much reasoning rapidly transcends the individual when predictions fail. The advantage of reasoning in a group is that goals will be more clearly defined by the group, who will be simultaneously engaged in evaluating the reasoning of others. Reasoning in a group also has the advantage of saving time because of the social distribution of knowledge. Although one member may be incapable of seeing anything wrong in his or her own reasoning, others can point this out. For example, Jane's attempt to jump start the car may have failed because she misdiagnosed the fault. Because this is the only diagnosis she has available, she cannot personally move beyond this. But the moment she phones a friend and is asked whether the engine turned over, it will be clear that there are other alternatives. People are aware not only of the social distribution of knowledge but of the time savings that can accrue by exploiting this fact in reasoning. It makes practical sense to collaborate.

Conclusions: Rationality

The psychology of reasoning has always been taken to address the question of whether people are rational. That is, the experimental results of reasoning experiments where the results are predictable from logic or probability theory bear on whether we are rational or not. These comparisons to normative theories only address the issue of human rationality given the justifications for the laws they prescribe. As we have seen in the Dutch book argument, the justification for the rationality of the laws of probability is practical, based on maximizing expected utility. We suggested that in the real world in which people must act, issues of practical rationality will always be a concern. We saw in unconscious inference in perception and action that what we believe— how we interpret sensory evidence—is influenced both by prediction errors and by the utilities of possible actions that we might take consequent on that interpretation. As we saw, we may become aware of the belief we have just formed only as a result of the consequent emotional reaction to anticipated surprise. At this unconscious level in Bayesian brain theory, reducing uncertainty (minimizing entropy) and maximizing expected utility always work together, although they are differentially weighted. The processes are rationally captured by Bayesian inference, and clearly practical concerns are in evidence at this unconscious level.

Explicit conscious verbal reasoning, which must occasionally intervene between perception and action, is a much more time-consuming and knowledge-limited process. As we argued, it most often occurs in practical contexts when things go wrong in attempting to achieve our goals. As in the Dutch book theorems, it is not rational to act in a way that does not achieve our goals. This simple fact makes time costs relevant, as many of the goals we want to achieve have deadlines, such as the need to get to work by a certain time. Time can be built into utility functions to express whether further reasoning is likely to achieve the goal. However, in practical contexts, where knowledge is socially distributed, the phone-a-friend strategy is always available. Again, a decision needs to be made about one's likelihood of accessing, by continued reasoning, the solution to one's problems so that one's goals can be achieved. This all makes rational sense, but only from the perspective of practical rationality.

The social distribution of knowledge also suggests that in most practical contexts the issues raised by the frame problem will not arise. The problem is that for nonmonotonic reasoning with conditional knowledge such as *if I turn the key the car starts*, potentially (McDermott, 1987) invokes the whole of world knowledge in order to determine whether one can actually predict that the car will start. The analogue in practical action would be checking every possible cause of the car not starting before turning the key. As we argued above, this is practically infeasible, and indeed, the act of turning the key predicting that the car will start is an unconscious almost reflexive inference. Moreover, the social distribution of knowledge suggests that most of us possess only a limited amount of this information, which is hence not available to be accessed to serve inference. It is not irrational to act based on what you know. It is irrational not to consult others who may know more than you to achieve your goals.

As we mentioned in discussing the Dutch book argument, such purely practical justifications for the laws of probability may not satisfy all the requirements of theoretical rationality. Presumably, what we should believe about the world is the truth, and the truth should be independent of our goals. Consequently, it is a good thing that there are now purely epistemic justifications for the laws of probability that can underpin Bayesian analysis of scientific inference. However, in the practical contexts in which reasoning processes are most often invoked for everyday reasoners, practical concerns will always be relevant to achieving their goals. So in conscious everyday verbal reasoning in the real world, just as in the unconscious inference in perception and action, there is a trade-off between our epistemic and our practical goals. Consequently, human reasoning is rational, but it is limited to practical rationality in most everyday contexts.

Acknowledgments

This work was supported by a grant from the ANR Chorus 2011 (project BTAFDOC).

Notes

1. We should note that the principle of conditionalization, which involves belief change, does not have a standard Dutch book justification. These justifications simply specify that the agent is rational at any particular point in time if his or her beliefs at that time conform to the laws of probability. They do not license any particular rule, such as Bayesian conditionalization, for belief change. We would have to move to diachronic Dutch book justifications, which have their problems (e.g., van Fraassen, 1981).

2. For discrete random variables, the Kullback–Leibler distance from P_0 to P_1,

$$D(\mathrm{Pr_0}||\mathrm{Pr_1}) = \sum_x \mathrm{Pr_0}(x)ln\left(\frac{\mathrm{Pr_0}(x)}{\mathrm{Pr_1}(x)}\right)$$

As this is the log of a ratio, it is 0 if P_0 and P_1 assign the same probability to each event x in the event space over which they assign probabilities.

3. Both Moore and Oaksford (2000) and Joffily and Coricelli (2013) have suggested, as Moore and Oaksford (2000) put it, "what you feel is what you don't know." These authors also discuss how negative emotions triggered by increased uncertainty increase learning rates (Joffily & Coricelli, 2013; Moore & Oaksford, 1999, 2000, 2002) and how positive emotions associated with reduced uncertainty decrease learning rates (Joffily & Coricelli, 2013).

4. Note that there are also many more inference rules that attach to the conditional that could be investigated.

5. We leave out the complexity that for (3) and (3′) the integration rule is given by a noisy-OR gate, whereas for (2) and (2′) the integration rule is given by the noisy-AND-NOT gate (see, Oaksford & Chater, in press).

References

Ali, N., Chater, N., & Oaksford, M. (2011). The mental representation of causal conditional inference: Causal models or mental models. *Cognition, 119*, 403–418.

Bonnefon, J. F. (2009). A theory of utility conditionals: Paralogical reasoning from decision theoretic leakage. *Psychological Review, 116*, 888–907.

Bonnefon, J.-F. (2013). New ambitions for a new paradigm: Putting the psychology of reasoning at the service of humanity. *Thinking & Reasoning, 19*, 381–398.

Bonnefon, J. F., & Hilton, D. J. (2004). Consequential conditionals: Invited and suppressed inferences from valued outcomes. *Journal of Experimental Psychology: Learning, Memory, and Cognition, 30*, 28–39.

Byrne, R. M. J. (1989). Suppressing valid inferences with conditionals. *Cognition, 31*, 61–83.

Carlin, B. P., & Louis, T. A. (2000). *Bayes and empirical Bayes methods for data analysis* (2nd ed.). London: Chapman & Hall/CRC.

Chater, N., Crocker, M., & Pickering, M. (1998). The rational analysis of inquiry: The case of parsing. In M. Oaksford & N. Chater (Eds.), *Rational models of cognition* (pp. 441–468). Oxford: Oxford University Press.

Chater, N., & Oaksford, M. (1999). Information gain and decision-theoretic approaches to data selection. *Psychological Review, 106*, 223–227.

Clark, A. (2013). Whatever next? Predictive brains, situated agents, and the future of cognitive science. *Behavioral and Brain Sciences, 36*, 181–253.

Corner, A., Hahn, U., & Oaksford, M. (2011). The psychological mechanism of the slippery slope argument. *Journal of Memory and Language, 64*, 133–152.

Cummins, D. D. (1995). Naïve theories and causal deduction. *Memory & Cognition, 23*, 646–658.

Cummins, D. D., Lubart, T., Alksnis, O., & Rist, R. (1991). Conditional reasoning and causation. *Memory & Cognition, 19*, 274–282.

Dayan, P., & Hinton, G. (1996). Varieties of Helmholtz machine. *Neural Networks, 9*, 1385–1403.

Doya, K., Ishii, S., Pouget, A., & Rao, P. N. (2006). *The Bayesian brain*. Cambridge, MA: MIT Press.

Earman, J. (1992). *Bayes or bust?* Cambridge, MA: MIT Press.

Edgington, D. (1995). On conditionals. *Mind, 104*, 235–329.

Elqayam, S., & Over, D. E. (2013). New paradigm psychology of reasoning: An introduction to the special issue edited by Elqayam, Bonnefon, and Over. *Thinking & Reasoning, 19*, 249–266.

Evans, J. S. B. T., Handley, S., & Bacon, A. (2009). Reasoning under time pressure: A study of causal conditional inference. *Experimental Psychology, 56*, 77–83.

Evans, J. S. B. T., Neilens, H., Handley, S., & Over, D. (2008). When can we say "if"? *Cognition, 108*, 100–116.

Fernbach, P. M., & Erb, C. D. (2013). A quantitative causal model theory of conditional reasoning. *Journal of Experimental Psychology: Learning, Memory, and Cognition, 39*, 1327–1343. doi:10.1037/a0031851.

Fodor, J. A. (1983). *The modularity of mind*. Cambridge, MA: MIT Press.

Fodor, J. A. (2001). *The mind doesn't work that way*. Cambridge, MA: MIT Press.

Friston, K. (2005). A theory of cortical responses. *Philosophical Transactions of the Royal Society of London. Series B, Biological Sciences, 360*, 815–836.

Friston, K., Schwartenbeck, P., FitzGerald, T., Moutoussis, M., Behrens, T., & Dolan, R. J. (2013). The anatomy of choice: Active inference and agency. *Frontiers in Human Neuroscience, 7*, 598. doi:10.3389/fnhum.2013.00598.

Fu, W., & Gray, W. D. (2006). Suboptimal tradeoffs in information seeking. *Cognitive Psychology, 52*(3), 195–242. doi:10.1016/j.cogpsych.2005.08.002.

Gibson, J. J. (1950). *The perception of the visual world*. Boston: Houghton Mifflin.

Gibson, J. J. (1979). *The ecological approach to visual perception*. Boston: Houghton Mifflin.

Hahn, U., & Oaksford, M. (2007). The rationality of informal argumentation: A Bayesian approach to reasoning fallacies. *Psychological Review, 114*(3), 704–732. doi:10.1037/0033-295X.114.3.704.

Harman, G. (1968). Three levels of meaning. *Journal of Philosophy, 65*, 590–602.

Helmholtz, H. (2000). Concerning the perceptions in general. In S. Yantis (Ed.), *Visual perception: Essential readings* (pp. 24–44). New York: Psychology Press. (Original work published 1866.)

Horvitz, E., & Rutledge, G. (1991). Time-dependent utility and action under uncertainty. In *Proceedings of the Seventh Conference on Uncertainty in Artificial Intelligence* (pp. 151–158). San Mateo, CA: Morgan Kaufmann.

Howson, C., & Urbach, P. (1989). *Scientific reasoning: The Bayesian approach.* Chicago and La Salle, IL: Open Court.

Joffily, M., & Coricelli, G. (2013). Emotional valence and the free-energy principle. *PLoS Computational Biology, 9*(6), e1003094. doi:10.1371/journal.pcbi.1003094.

Joyce, J. M. (1998). A nonpragmatic vindication of probabilism. *Philosophy of Science, 65,* 575–603.

Joyce, J. M. (2004). Bayesianism. In A. R. Miele & P. Rawling (Eds.), *The Oxford handbook of rationality* (pp. 132–155). Oxford: Oxford University Press.

Klauer, K. C. (1999). On the normative justification for information gain in Wason's selection task. *Psychological Review, 106,* 215–222.

Lindley, D. V. (1956). On a measure of the information provided by an experiment. *Annals of Mathematical Statistics, 27,* 986–1005.

Manktelow, K. I. (2012). *Thinking and reasoning.* Hove, East Sussex: Psychology Press.

McDermott, D. (1987). A critique of pure reason. *Computational Intelligence, 3,* 151–160. doi:10.1111/j.1467-8640.1987.tb00183.x.

Mercier, H., & Sperber, D. (2011). Why do humans reason? Arguments for an argumentative theory. *Behavioral and Brain Sciences, 34*(2), 57–74. doi:10.1017/S0140525X10000968.

Moore, S. C., & Oaksford, M. (1999). Feeling low but learning faster: The long term effects of emotion on human cognition. In *Proceedings of the 21st Annual Conference of the Cognitive Science Society* (pp. 411–415). Hillsdale, NJ: Lawrence Erlbaum Associates.

Moore, S. C., & Oaksford, M. (2000). Is what you feel what you don't know? *Behavioral and Brain Sciences, 23,* 211–212.

Moore, S. C., & Oaksford, M. (2002). Some long-term effects of emotion on cognition. *British Journal of Psychology, 93,* 383–395.

Oaksford, M., & Chater, N. (1991). Against logicist cognitive science. *Mind & Language, 6,* 1–38. doi:10.1111/j.1468-0017.1991.tb00173.x.

Oaksford, M., & Chater, N. (1994). A rational analysis of the selection task as optimal data selection. *Psychological Review, 101,* 608–631.

Oaksford, M., & Chater, N. (1995). Theories of reasoning and the computational explanation of everyday inference. *Thinking & Reasoning, 1,* 121–152.

Oaksford, M., & Chater, N. (1996). Rational explanation of the selection task. *Psychological Review*, *103*, 381–391.

Oaksford, M., & Chater, N. (2007). *Bayesian rationality*. Oxford: Oxford University Press.

Oaksford, M., & Chater, N. (2009). Precis of "Bayesian rationality: The probabilistic approach to human reasoning." *Behavioral and Brain Sciences*, *32*, 69–84.

Oaksford, M., & Chater, N. (2013). Dynamic inference and everyday conditional reasoning in the new paradigm. *Thinking & Reasoning*, *19*, 346–379. doi:10.1080/13546783.2013.808163.

Oaksford, M., & Chater, N. (2014). Probabilistic single function dual process theory and logic programming as approaches to non-monotonicity in human vs. artificial reasoning. *Thinking & Reasoning*, *20*, 269–295. doi:10.1080/13546783.2013.877401.

Oaksford, M., & Chater, N. (in press-a). Probabilities, causation, and logic programming in conditional reasoning: A reply to Stenning and van Lambalgen. *Thinking and Reasoning*.

Oaksford, M., & Chater, N. (in press-b). Causal models and conditional reasoning. In M. Waldmann (Ed.), *Oxford handbook of causal cognition*. Oxford: Oxford University Press.

Oaksford, M., Chater, N., & Grainger, B. (1999). Probabilistic effects in data selection. *Thinking & Reasoning*, *5*, 193–243.

Over, D. E. (2009). New paradigm psychology of reasoning: Review of "Bayesian Rationality." *Thinking & Reasoning*, *15*, 431–438. doi:10.1080/13546780903266188.

Over, D., Hadjichristidis, C., Evans, J. S. B. T., Handley, S., & Sloman, S. (2007). The probability of causal conditionals. *Cognitive Psychology*, *54*, 62–97.

Pirolli, P., & Card, S. (1999). Information foraging. *Psychological Review*, *106*(4), 643–675. doi:10.1037/0033-295X.106.4.643.

Savage, L. (1954). *The foundations of statistics* (2nd ed.). New York: Dover.

Schroyens, W., & Schaeken, W. (2003). A critique of Oaksford, Chater, and Larkin's (2000) conditional probability model of conditional reasoning. *Journal of Experimental Psychology: Learning, Memory & Cognition*, *29*, 140–149.

Schwartenbeck, P., FitzGerald, T., Dolan, R. J., & Friston, K. (2013). Exploration, novelty, surprise, and free energy minimization. *Frontiers in Psychology*, *4*, 710. doi:10.3389/fpsyg.2013.00710.

Sellars, W. (1963). Abstract entities. *Review of Metaphysics*, *16*, 627–671.

Shannon, C. E., & Weaver, W. (1949). *The mathematical theory of communication*. Urbana: University of Illinois Press.

Sloman, S. A. (2005). *Causal models: How people think about the world and its alternatives*. New York: Oxford University Press.

Sloman, S., Barbey, A., & Hotaling, J. (2009). A causal model theory of the meaning of cause, enable, and prevent. *Cognitive Science*, *33*, 21–50.

Sloman, S. A., & Lagnado, D. A. (2005). Do we do? *Cognitive Science, 29*, 5–39. doi:10.1207/s15516709cog2901_2.

Smith, M. (2004). Humean rationality. In A. R. Miele & P. Rawling (Eds.), *The Oxford handbook of rationality* (pp. 75–92). Oxford: Oxford University Press.

Sperber, D., Cara, F., & Girotto, V. (1995). Relevance theory explains the selection task. *Cognition, 57*, 31–95.

Sperber, D., & Mercier, J. (2012). Reasoning as social competence. In H. Landemore & J. Elster (Eds.), *Collective wisdom: Principles and mechanisms* (pp. 368–392). Cambridge: Cambridge University Press.

Sperber, D., & Wilson, D. (1986). *Relevance: Communication and cognition.* Oxford: Blackwell.

Stenning, K., & van Lambalgen, M. (2005). Semantic interpretation as computation in nonmonotonic logic: The real meaning of the suppression task. *Cognitive Science, 29*, 919–960. doi:10.1207/s15516709cog0000_36.

van Fraassen, B. C. (1981). A problem for relative information minimizers in probability kinematics. *British Journal for the Philosophy of Science, 32*, 375–79.

Wason, P. C. (1968). Reasoning about a rule. *Quarterly Journal of Experimental Psychology, 20*, 273–281.

6 Logic, Probability, and Inference: A Methodology for a New Paradigm

Jean Baratgin and Guy Politzer

The New Reasoning Paradigm

Psychologists who investigate human reasoning define human rationality by using formal models that predict individuals' optimal behavior with respect to their objectives. In particular, some formal models have generally been considered as norms of reference for rational inference. Traditionally, psychologists have assumed that different psychological processes correspond to the different formal domains of reasoning: decision making, probability judgment, inductive reasoning, and deductive reasoning. Three main formal models have been used as references to study these domains: subjective expected utility for decision making, the Bayesian model for probability judgment and induction, and binary classical logic for deduction. It must be kept in mind that the choice of a specific model, which is guided by theoretical options, may have deep methodological consequences for the experimental work.

As far as deductive reasoning is concerned, several periods and lines of research can be identified, which reflect the status given to classical logic by psychologists. In the early days research on reasoning was focused on individuals' formal reasoning capabilities, and the unique reference was Aristotle's syllogistic. This is the case of Binet (1902) and James (1908). Piaget (Inhelder & Piaget, 1958) extended the model of reference to propositional logic but again regarded the adult human being as an implicit logician. In sum, theorists chose that portion of classical logic known to them both as a descriptive and a normative system of reference. However, the work of Wason and Johnson-Laird (1972) revealed that although individuals exhibit some logical competence overall, their answers to a variety of deductive tasks are too often at variance with logic for the Piagetian position to be defensible. At about the same time the results of the "heuristics and biases school" (Kahneman, Slovic, & Tversky, 1982) highlighted an apparent gap between individuals' probabilistic judgment and the "Bayesian" norm from which a "pessimistic" view on human rationality ensued.

Even though the introduction of an approach to human reasoning based on linguistic pragmatics (Hilton, 1995; Politzer, 1986, 2004, 2007; Politzer & Macchi, 2000) helps explain a wide variety of reasoning errors, it is clear that classical logic does not adequately describe people's deductive performance. In the 1980s and 1990s the bulk of the theoretical debate about deduction revolved around the nature of the representation and processing of the input information, assumed to be syntactic and rule-governed by some researchers (Braine & O'Brien, 1998; Rips, 1994) or semantic and based on mental models by others (Johnson-Laird & Byrne, 1991). The research agenda was to explain the individual's competence and its limitations while keeping a classical view of the normative system of reference, that is, a logic whose inferences are truth preserving, which is monotonic and whose premises are certain.

Meanwhile, two domains of research started to have a decisive impact on the investigation of deduction. One, researchers in Artificial Intelligence were studying topics or phenomena hitherto disregarded by psychologists, such as abduction or defeasibility (that is, the retraction of a conclusion on learning a new piece of information). Two, the development of research on decision making using probability theory as its main tool contributed to bring uncertainty to the forefront. Both domains aim to study thinking and reasoning in real life, where people exploit their huge knowledge bases, premises are uncertain, conclusions are defeasible, and reasoning as a whole is goal oriented. In the 1990s it became apparent that the scope of research on deductive reasoning was too narrow if not misdirected. A call for a change in the normative system of reference was made by Oaksford and Chater (1995, 2001), who argued that logic is inadequate to account for performance in reasoning tasks because reasoners use their everyday uncertain reasoning strategies, whose nature is probabilistic, so that individuals' inferential activity should no longer be assessed only with regard to logical axioms; rather, it should be assessed in reference to probability theory. This view has the advantage that it integrates the studies of reasoning and of judgment and decision making into a unified domain of research. Even though other formalisms are possible candidates for psychological modeling (see Politzer & Bonnefon, 2010) probability theory has been adopted by most researchers who endorse this new approach. A majority, following Oaksford and Chater (2007), adopt Bayesianism as a normative model[1] (although the need to choose a normative system has been questioned; see Evans, 2012).

This new approach to the study of reasoning has been regarded as a paradigm shift (Evans, 2012) and called the *new paradigm psychology of reasoning* (Over, 2009). It incorporates two major assumptions.

1. The indicative conditional[2] of natural language *if A then C* does not necessarily coincide with the material conditional of formal logic, which specifies a strong relation between A and C (A entails C). Rather, it expresses a link (possibly weak) between A and C. Consequently, the valid inferences involving the material conditional that are paradoxical for common sense are not always correct.[3]

2. The conclusion of a deductive argument may carry some uncertainty inherited from the premises' uncertainty (George, 1995, 1997; Liu, Lo, & Wu, 1996; Stevenson & Over, 2001). The uncertainty is represented by an additional truth value and the logic of reference is a tri-valued logic (Baratgin, Over, & Politzer, 2014). The uncertainty is revealed in an explicit way by probabilistic inferences that can be analyzed in the light of Bayesian coherence (Pfeifer & Kleiter, 2006).

In the new paradigm the normative model of binary logic is replaced with a Bayesian model (considered as providing a "probabilistic logic") to study the uncertain deductive inferences that people carry out in daily life.

Of course, there are various conceptions of Bayesianism. The philosophical, logical, linguistic, economic literatures and also the artificial intelligence literature offer a wide variety and the choice of one or the other form for normative or descriptive purposes has important consequences (Baratgin, 2002; Baratgin & Politzer, 2006, 2007; Elqayam & Evans, 2013).

In what follows, a relatively radical choice is made, namely the adoption of subjective Bayesian probability (de Finetti, 1964; Ramsey, 1926/1990; Savage, 1954). There are several reasons for this choice. One, this theory provides a criterion of rationality that can be used experimentally in an effective manner (see below). Two, its conception of probability as dependent on the individual's state of knowledge establishes the investigation of probability judgment as an interdisciplinary object open to psychological investigation, as de Finetti (1957, 1974) himself suggested. Three, de Finetti (1930) explicitly stated the hypothesis that reasoning is naturally probabilistic[4]:

The calculation of the probabilities is the logic of the probable. Formal logic teaches us to deduce the truth or falsity of certain consequences of the truth or falsity of certain premises; likewise the calculation of probabilities teaches us to deduce the greater or lesser likelihood or probability of certain consequences of the greater or lesser likelihood or probability of certain premises. (pp. 261–262, our translation)

Even though the subjective interpretation of probability was defended by several scientists right from the beginning of probability theory, two authors can be regarded as the "founding fathers" of subjective Bayesianism: at the end of the 1920s, Bruno de Finetti and Frank P. Ramsey independently and virtually simultaneously developed a theory founded on a subjective, or personal, interpretation of probability. Sometime later Leonard L. Savage proposed a related approach that generalizes to decision theory. However, de Finetti's theory differs from it by at least three specific characteristics:

1. "Radical" subjectivism.[5] De Finetti emphatically rejects any conception of probability that is not subjective, meaning that probability is measured by the degree of belief of a given individual (at a given instant). From his point of view probability is in all cases the expression of a subjective opinion, be it defined as the limit of a frequency or a

proportion dictated by the logical analysis of the possible cases. Many other "subjective" Bayesian theorists accept, under some conditions, the notion of physical probability viewed as irreducible (applicable when probability expresses the intrinsic property of a phenomenon all the factors of which are known or regular), and also the notion of ontological probability or objective chance (see Good, 1976; Jeffrey, 1976; Lewis, 1980). Ramsey himself defends a dualistic position for probability (see the note "chance" in Ramsey (1926/1990) and Galavotti's analysis (1995, 1997, 1999).

2.De Finetti's theory is also specific by its direct method of evaluation of probability. For Ramsey (1926/1990) probabilities and utilities are simultaneously revealed to the individual by a system of axioms that define coherent preferences. For Savage (1954) it is the agents' choices that allow them to attribute numerical values to their beliefs. In contrast, for de Finetti a person can directly assess the probability of occurrence of a particular event by considering how much she is willing to bet in favor of the realization of this event. The Dutch book procedure (that is, an assessment resulting in a sure loss in case of a series of bets) is the only criterion of rationality together with an operational method of betting that makes it possible to justify the notion of coherence. It amounts to saying that rational individuals cannot contradict themselves. This process of direct evaluation is essential from an experimental point of view because participants in experiments are explicitly required to directly express a probability.

3.Last, based on the notion of exchangeable events (meaning that their probability is invariant under permutation; see de Finetti 1931, 1937, 1979) and the subjective conception of probability, it is possible to reproduce all the classical properties and theorems that are known in the frequentist interpretation of probability. The evaluation of a probability, the unique result of the quantification of the judgment of likelihood attributed to an isolated event given by a particular individual on the basis of his or her own knowledge, coincides in every respect with what experimenters studying probability judgment ask of participants. It seems that the de Finettian model can naturally be associated with the psychological investigation because they have the same objective (Baratgin, in press).

Two Levels of Analysis: A Dual System

In the old paradigm the validity of the arguments presented to the participants was dictated by formal logic independently of the meaning of the propositions or events that constitute the arguments. The experimental study relied on the acceptance of the truth of the premises by the participant. The participants' inference was revealed by their characterization of the conclusion presented to them in terms of "true," "false," (or "one cannot know,") or by the production of their own conclusion. The main objective was to determine whether the formulas (in particular the connectives for propositional logic) were correctly interpreted (that is, in agreement with the truth tables of classical

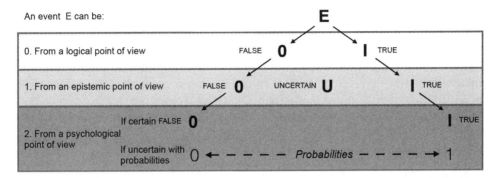

An event E can be:

0. From a logical point of view   FALSE **0** **I** TRUE

1. From an epistemic point of view   FALSE **0** UNCERTAIN **U** **I** TRUE

2. From a psychological point of view   If certain FALSE **0** **I** TRUE

If uncertain with probabilities   0 ← – – – – – *Probabilities* – – – – → 1

Figure 6.1
The three levels of knowledge (de Finetti, 1980).

logic) and whether participants complied with valid arguments of formal logic. In the new paradigm this methodology is not always applicable because in the first place one must consider which of the individual's "levels of knowledge" is being investigated. The old paradigm considers only the logical level, which is what de Finetti (1980) calls the objective level of knowledge, or level 0, where an event is either true or false in an absolute way. Even if this level is ideal from an axiomatic point of view, it is practically sterile because it has no other use than the enumeration, and the exposition in a different shape, of already known events (de Finetti, 2006). In the new paradigm it is useful to consider the two supplementary levels of knowledge of an event defined by de Finetti (1980), both of which are subjective (figure 6.1).

Level 1, the elementary level of knowledge, is the level of the individual's beliefs about the events. An event has an epistemic and subjective import. It concerns a specific object or phenomenon defined by its own characteristics known to the individual. An event is always conditioned on the individual's personal state of knowledge. For de Finetti the value *true* attributed to an event corresponds to its realization and the value *false* to the absence of its realization. From now on, we qualify an event by *true* or *false* in this sense. Now, very frequently we are uncertain about the realization of an event. This is why de Finetti defends the notion of a three-valued logic to define (conditional) events. A conditional event may be true, false, or *doubtful*. This "doubtful" value reflects a transitory epistemic state of uncertainty that depends on the individual who does not know the event's truth value at a given instant. This third value should not be compared ordinally with the absolute values *true* or *false* of binary logic; rather, is it to be conceived of as a "null" or "void" value. This leads to a three-valued logic "provisional and superimposed" on ordinary bivalued logic (de Finetti, 1995, p. 39). The individual is also assumed to be capable of executing logical operations on the propositions associated to the events. As already mentioned, an event is always conditional; when necessary, following de Finetti, we will call it a "tri-event" or a "conditional event" (noted C|A). A (conditional) event expresses a link between the occurrence of two other events: the event A and the event C. This defines a weakened form of the material conditional (for which A entails C). The conditional event is true when A and C are true, false when A is true and C is false, and null when A is false independently of the value of C. This constitutes the framework of a trivalued logic of which the system outlined by Baratgin, Over, and Politzer (2014) constitutes a possible construal. The principal property of this elementary system is the fundamental relation of the conditional event defined by de Finetti (1995) written here using his own notations:

If A then C = C|A = C&A|A   (6.1)

Level 2, the *meta-epistemic* level of knowledge, concerns the *degrees* of belief about the event, which are subjective additive probabilities. It is assumed that during the transitory state of uncertainty the individual is capable of producing a degree of belief (always dependent on his or her state of knowledge) about the doubtful event (which could lead to a multiple-valued logic—in fact, an infinity of values). The individual's judgments of probability apply in particular to the conclusion of arguments knowing the subjective probability of the premises. As far as the conditional event is concerned, the principal property of this level is that the probability of an indicative conditional of natural language, *if A then C,* is equal to the conditional probabilty of C given A, $P(C|A)$, and not to the probability of the material conditional as was assumed in the old paradigm, which is equivalent to $1 - P(A \& \text{not-}C)$. This property reads:

$P(\text{if A then C}) = P(C|A)$ (6.2)

This equation is supported by various arguments, philosophical and logical (see Adams, 1998; Edgington, 1995, for reviews) and psychological (Cruz & Oberauer, 2014; Evans & Over, 2004; Oaksford & Chater, 2007, 2009; Pfeifer & Kleiter, 2010).

The Relationship between the Two Levels of Knowledge

The passage from the elementary level, where the individual has an opinion about an event characterizable by three truth values, to a meta-knowledge level, where the individual evaluates the event's probability quantitatively, is an open field of study for psychologists—including the problem of the transformation of degrees of belief into quantitative values within the second level. Psychologists generally implicitly assume that these values are additive because, following the adoption of the Bayesian framework, they are committed to the endorsement of this property. De Finetti offers the outline of an account of the relationship between the two levels. At a first stage the intuition of the probability of occurrence of an event is qualitative and can be positioned on an ordinal scale; it is therefore susceptible of comparison with another event.

At a second stage a quantitative evaluation may follow, but it is constrained by the rules of additive probabilities, which define the static coherence (see Baratgin & Politzer, 2006). The betting schema (de Finetti, 1964) reveals the trivalent character of an event and makes it possible to obtain a "quantitative, numerical, definition of the degree of probability attributed by a given individual to a given event" (p. 101). The relations (1) and (2) can be obtained directly by the betting schema:

A tri-event corresponds, in contrast, to a bet whose validity is subordinate to some conditions which must be verified. One can bet, for example, on the victory of one of the competitors in a race which ought to take place tomorrow; if one understands that the bet is totally lost if this event does not take place, one is in the first case; one is in the second case, if one establishes that the bet is null and without effect if the race can not take place, if the competitor in question is not able to participate, or in any other eventually whatever ... The probability (subjective) attributed by the individual O to an event A (in general tri-event) is the price p at which he considers it equitable to exchange a sum pS for a sum S, the possession of which is conditioned by the verification of A (in the case of tri-events it is necessary to specify again: the payment of the stake pS is conditional on the arrival of the "hypothesis" A). (de Finetti, 1995, pp. 185–186)

The relation (2) can also be directly established based on the consideration of the three possible values of an event. The probability of the event is the probability that the event is true given that it is defined (de Finetti, 1995).

Besides these formal foundations, the relation (2) has a psychological justification. It can be interpreted as the application of the procedure devised by Ramsey, called the "Ramsey test" in the philosophical literature:

If two people are arguing "If p will q?" and are both in doubt as to p, they are adding p hypothetically to their stock of knowledge and arguing on that basis about q; so that in a sense "If p, q" and "If p, not-q" are contradictory. We say they are fixing their degrees of belief in q given p. If p turns out false, these degrees of belief are rendered *void*. If either party believes *not-p* for certain, the question ceases to mean anything to him except as a question about what follows from certain laws or hypotheses. (Ramsey, 1929/1990, p. 143)

Assuming that at the second level the degrees of belief are additive subjective probabilities (which was proposed by Ramsey, 1929/1990), the test can be interpreted as the description of the psychological process of moving from level 1 to level 2 (which is at the heart of the suppositional theory of probability; see Evans, Handley, & Over, 2003; Evans & Over, 2004; Handley, Evans, & Thompson, 2006). The individual imagines the antecedent or makes the supposition[6] that p, then evaluates the probability of the consequent q.[7] Ramsey (1929/1990, p. 143) does not go through the definition of the conditional event as in (1); rather, he describes a dynamic process of belief revision. A conditional is accepted with respect to an initial state of knowledge if this knowledge revised by the antecedent p (which is here considered as the revision message) validates the consequent.

The new paradigm has focused primarily on the indicative conditional in natural language. In doing so it has addressed the two problems of the comprehension of the conditional and the evaluation of its probability as a conditional probability. These are two different objects of study that have not been clearly distinguished so far. It seems important to distinguish de Finetti's conditional event from the probabilty of this event. It is the latter that has been studied the most through Adams's "conditional probability." However, we believe that it is important for the experimental studies to define a specific methodology for each of these two levels of representation of belief.

Comparing the Methodology for Studying the Two Upper Levels of Knowledge

Studying individuals' inferences at the level of the events and at the level of their probability may require different methodologies. For the first level the methodology is close to the traditional bivalued case provided a norm of inferential validity has been chosen and a trivalent system has been precisely defined. For the second level there are two possibilities. One is to choose a general notion of *probabilistic* validity, and, as at the first level, the experimenter observes whether participants' inferences comply with validity. The other is to exploit the informational richness of the domain and to adopt identically the methodology used to study probability judgment, which is based on the examination of the coherence of participants' evaluations. We detail this below.

The Conditional Event
Trivalued Logics and Validity   At the elementary level, by giving up classical logic, the new formalism widens the bivalued semantic framework (T = true, F = false) to adopt a trivalued one (T, F, U), where U admits of different interpretations (for a review, see Ciucci & Dubois, 2012). As is well known, in trivalent logic the concept of tautology (like its dual concept, contradiction) is equivocal, contrary to binary logic (in which a sentence is a tautology if its truth value is *true* whatever the truth value of its constituents). Similarly, the validity of an argument in binary logic is defined by the preservation of truth from premises to conclusion (which is equivalent to the preservation of nonfalsity). In trivalent logic the definition of tautology depends on the choice of the *designated* value(s), and the definition of contradiction depends on the *antidesignated* value(s) (see Rescher, 1969, pp. 66–71). In brief, before any experimental investigation, it is necessary to ponder the definition of a tautology, a contradiction, and a valid consequence. One guideline may help make a choice: it seems desirable that the new formalism makes it possible to explain the robust results of the old paradigm as a limiting case. This leads to three desiderata.

1. The inferences that are valid in the traditional sense and widely accepted by reasoners should be retained as valid by the new definition. For example, it is well

established that virtually everyone endorses the conclusion of *modus ponens,* and a clear majority endorses the conclusion of *modus tollens* (for a recent review, see Manktelow, 2012).

2. Many studies have shown that people reason nonmonotonically (Bonnefon & Hilton, 2002; Byrne, 1989; Pelletier & Elio, 1997; Politzer, 2005; Stenning & Van Lambalgen, 2008; Stevenson & Over, 1995; for a review see Politzer & Bourmaud, 2002). In particular the eight axioms of system P (Kraus, Lehmann, & Magidor, 1990; Lehmann & Magidor 1992), namely, *reflexivity, left equivalence, right weakening, cut, cautious monotonicity, equivalence, and,* and *or* seem to be endorsed by a majority of reasoners (see Benferhat, Bonnefon, & Da Silva Neves, 2004, 2005; Da Silva Neves, Bonnefon, & Raufaste, 2002). If these results are confirmed, it is desirable that these axioms constitute valid inferences according to the next definition.

3. The definition should not include the arguments that are not endorsed by the majority of the reasoners. This is the case, for instance, of the two "paradoxes" of the conditional: the negated antecedent and the affirmed consequent (for the latter, see Bonnefon & Politzer, 2011).

In the literature there are at least four possible definitions of validity that have been applied to trivalued logics.

V_T. T is the designated value. Validity is defined by the preservation of truth from the premises to the conclusion and the preservation of falsity from the conclusion to the premises. It is also the definition of validity generally adopted by trivalued logics.[8] Formally, it coincides with Kleene's trivalent system.[9] Egré and Cozic (in press) describe the main inference schemas of de Finetti's trivalent logic valid under this definition. The definition of tautology (and of contradiction) coincides with the classical definition of bivalued logic.

V_{nF}. The designated values are T and U. An argument is valid if it preserves nonfalsity from the premises to the conclusion. It is necessary that no nonfalse premise leads to a false conclusion.[10]

V_{TnF}. This definition is the conjunction of the two previous definitions. There is preservation of truth from premises to conclusion along with preservation of nonfalsity from premises to conclusion. It is impossible for the conclusion to be false without at least one premise being false. Taking the following order for the truth values: $F < U < T$, V_{TnF} prescribes that the conclusion cannot have a truth value weaker than the minimum of the truth values of the premises. This definition of validity is adopted by Blamey (2001) and McDermott (1996).[11]

V_U. A validity has been proposed under the name of "correct inference" by Hailperin (1996) for his "suppositional logic," which is identical to de Finetti's with "indeterminate" as a third value and $F < U < T$. Hailperin defines validity as the preservation of U-validity from premises to conclusion.[12]

Validities V_T and V_{TnF} are not convenient from a psychological point of view because they do not lead to the recognition of *modus ponens* as a valid argument. V_T, V_{NF}, and V_{TnF} do not respect all the axioms of system P.[13] So, validity V_U seems the most appropriate to specify de Finetti's system. However, even if it is necessary to define a validity to start experimental work on inference schemas, it is far from sufficient. Although de Finetti was a pioneer, he was not the only theorist to have defined the natural language conditional as a conditional event. The formal literature in philosophy, linguistics, and AI offers other interpretations of the conditional that can follow different semantics while remaining trivalent (see Baratgin, Politzer, & Over, 2013).[14] This is why our first experimental objective was to identify, among different trivalent logical systems that incorporate a conditional event, a system that is the closest possible description of the interpretation of natural language connectives (Baratgin, Over, & Politzer, 2014). We have considered three possible conditional events and nine possible trivalent systems.

Preliminary Results

We have carried out a series of experiments in which participants had to construct three-valued truth tables for the main connectives (negation, conjunction, disjunction, conditional, and the material conditional worded as *not-(A and not-C)*. The material consisted of chips defined by their color and shape. The color could be clearly black or clearly white or uncertain, and the shape clearly round or clearly square or uncertain. Two alternative scenarios allowed the assertion of a sentence or of a conditional bet on the same sentence. The main results are summarized below (see Baratgin, Over, & Politzer, 2013, for the indicative conditional).

1. For all the connectives considered, participants' answers in the two scenarios are very close. They treated the questions relative to the truth, the falsity, or the uncertainty of an assertion in the same manner that they treated the questions relative to winning or losing or calling off a bet on that sentence; and this applies in particular to the conditional. These results confirm and generalize the observation made by Politzer, Over, & Baratgin (2010) that a majority of participants judged that, when the antecedent A of a conditional *if A then C* is false, the sentence is neither true nor false, and when A does not occur, a bet on this conditional is neither won nor lost,[15] suggesting that for the speech acts of conditional assertion and conditional bet there is a similar, trivalent, interpretation of the conditional sentence.

2. When we consider first only the four cells of the classical bivalued truth tables, that is, restrict ourselves to the two values T and F (or win and lose), participants' answers coincided with the traditional tables for negation, conjunction, and, to a lesser extent, disjunction. In contrast, answers for the conditional were distributed essentially over two tables coinciding with the conditional and the conjunction. A conjunctive interpretation for the conditional with abstract material has already been observed (Girotto & Johnson-Laird, 2004), and our results confirm this with an even higher rate.[16]

3. Next, for the various connectives under study, we have considered all nine cells of the three-valued table defined by T, F, U (or win, lose, void), and we have analyzed each participant's truth table and compared it with the tables available in the literature (see, for example, Rescher, 1969). This analysis has produced three main results.

(a) Most tables produced by participants can be classified among a limited number of truth tables belonging to the nine systems mentioned earlier. This is remarkable, as there are theoretically three (values) to the power 5 (cells with one or two U marginal values) = 243 possible truth tables for each connective. In other words, participants treat uncertainty in a systematic manner.

(b) For negation, almost all participants have produced the table of *involutive* negation (in which the negations of U, T, and F are U, F, and T, respectively). For the other connectives the majority of the answers coincide with tables that belong to de Finetti's logic. This applies both to the assertion and to the betting conditions.

(c) Most participants have given different tables for the conditional event and the material conditional. This result, obtained with a disambiguated form of the material conditional and using three truth values, extends the observations made in the old paradigm showing that lay reasoners do not interpret the natural language conditional as a material conditional; rather the most frequent interpretation is de Finetti's conditional event.

In conclusion, these first experimental data point to de Finetti's trivalent logic as an adequate descriptive system at the first level of knowledge. This system seems to offer a good description of human evaluation of logical relations under uncertainty. This applies in particular to the conditional event and explains an emblematic result of the old paradigm, namely the "defective" truth table. However, these results have been obtained using a third truth value explicitly defined as a physical uncertainty bearing on the characteristics of an object—here a visual uncertainty. The limitation is that we still need to understand how individuals construe this third value. Further work is necessary to address this essential question.

Conditional Probability

At the second level of knowledge, the shift from formal logic to Bayesian probability can be done in two ways.

Probabilistic Validity   One way consists of replacing the traditional notion of logical validity of an argument built on truth preservation, such as, *it is impossible for the premises of a valid argument to be true when the conclusion is false,* by a notion of probability preservation. There are various choices (see Adams, 1996) such as, for instance, *it is impossible for the probability of the premises of a valid argument to be non-null when the probability of the conclusion is null.* Other definitions of probabilistic validity are focused on the probability of the conclusion, given the probability of the premises. They are concerned with the possibility of warranting a sufficiently high value (typically a threshold > ½) for the probability of the conclusion (see Adams, 1975; Coletti & Scozzafava, 2002; Gilio, 2002).[17]

The most widely accepted criterion of probabilistic validity in the new paradigm, and also the most debated in philosophy, is Adams's (1998) *p*-validity, which is defined by comparing the uncertainty of the premises and the uncertainty of the conclusion. Uncertainty is defined as the probability of falsity[18]: the uncertainty of an event A, which has a probability $P(A)$ is $P(not\text{-}A) = 1 - P(A)$. *An argument is p-valid if and only if the uncertainty of its conclusion cannot exceed the sum of the uncertainties of its premises* (Adams, 1998, p. 131). This leads to partitioning the arguments into those that are *p*-valid and those that are not (the former obey the sum condition, and the latter do not). Then, consider a *p*-valid argument: given the probabilities of its premises, the sum property allows the calculation of a limit in probability such that an individual who estimates the probability of the conclusion to be no lower than this limit can be regarded as possessing an implicit knowledge of the probability preservation of the argument. And reciprocally, a lower estimate indicates absence of such knowledge.[19]

Coherence   The second way of adopting Bayesian theory is to opt for de Finetti's (1937/1964) notion of coherence. De Finetti (1931, 1964) provides an effective method to appraise the coherence of a probability evaluation by using coherence intervals determined by the probability of the premises and the static (respecting the axioms of additive probabilities) and dynamic (revising by Bayes's rule) criteria of coherence (Baratgin & Politzer, 2006).[20] This approach is expounded by Coletti and Scozzafava (2002) and Hailperin (1996). It has been used by proponents of *probabilistic mental logic* within the new paradigm in several investigations of deduction under uncertainty (Gilio & Over, 2012; Pfeifer, 2014; Pfeifer & Kleiter, 2009, 2010, 2011; see also Cruz, Baratgin, Over, & Oaksford, 2015; Evans, Thompson, & Over, 2015; Politzer & Baratgin, in press; Singmann, Klauer, & Over, 2014) for studies using both coherence and *p*-validity.

When the concept of coherence is applied to inference schemas, it can be shown (Pfeifer & Kleiter, 2006) that again the arguments can be partitioned into two classes. For some arguments, called *probabilistically noninformative*, any assessment on the interval [0, 1] for the probability of the conclusion is compatible with the axioms of probability, whatever the probabilities of the premises. For the other arguments, called *probabilistically informative*, the interval is constrained by the probabilities of the premises. It has a lower and an upper limit [l, u], so that an individual presented with such arguments is deemed coherent if and only if his or her assessment falls inside the interval.

From a methodological point of view, it is clear that at the second level of knowledge as well, the new paradigm is far more demanding than the old one. For both approaches the probability estimates have to be fixed for the premises and measured for the conclusion, keeping in mind that for the latter the aim is to find the position of the participant's estimate with respect to a probability interval calculated by the experimenter. For coherence, the limits of the intervals can be calculated by applying the theorem of total probability in simple cases or by solving a system of linear equations (Coletti & Scozzafava, 2002). A very simple method applicable to the inference schemas commonly discussed in philosophical logic and in psychology is described by Politzer (in press).

Comparing p-Validity and Coherence   The main interest of the coherence approach stems from the fact that conceptually it justifies the idea that the subjective degrees of belief attributed to the sentences can be regarded as additive probabilities (de Finetti, 1964). That is, that an individual's degree of belief in the conclusion lies inside the coherence interval means that he or she respects the laws of probability, which is the major question of interest on the investigators' agenda. In addition this methodological approach offers a unified framework to study both deduction under uncertainty and probability judgment.

The method of coherence aims to study the participants' probability estimate of the conclusion in relation to their estimate of the probability of the premises. This method allows one to study, for each argument presented, whether participants produce a coherent evaluation. Then one can characterize the arguments according to the percentage of coherent evaluations observed across all the participants and differentiate the arguments for which the majority of the probability evaluations fall within the coherence interval (these arguments could be called "natural") from those for which this is not the case.

The aim of p-validity differs. The question it aims to answer is whether an individual's degree of belief in the conclusion is high enough to respect a criterion of probability preservation that is used to characterize the argument as deductive. This comes in the wake of the old paradigm, whose main interest was the investigation of individuals' deductive abilities. In the new paradigm it seems important to study whether reasoners judge that a high degree of belief in the premises warrants a high degree of belief in the conclusion, which is a property entailed by p-validity. Also, the p-validity measure may be used to differentiate, among probabilistically informative arguments, those that are deductive (p-valid) from those that are inductive (p-invalid; see Evans & Over, 2013).

However, the adoption of p-validity raises several difficulties. One follows from the fact that the intervals determined by the two criteria generally do not coincide. Take for instance *modus ponens*: *A; if A then C; therefore C;* and let $P(A) = r$ and $P(C/A) = q$. The intervals are [max $\{r + q - 1, 0\}$, 1] for p-validity and [rq; $rq + 1 - r$] for coherence; taking for example $q = 0.5$ and $r = 0.8$, the intervals are [0.3, 1] and [0.4, 0.6], respectively. Suppose two individuals give evaluations such as 0.35 and 0.7. They are both p-valid but incoherent, the former being too low and the latter too high. We may be satisfied that we have found two evaluations reflecting the deductive import of *modus ponens*, but both violate the laws of probability.

Another limitation of the *p*-validity measure is that as soon as the sum of the uncertainties of the premises is above some value (1 for two premises, 2 for three premises, etc.), no interval can be defined because by definition the uncertainty of the conclusion would be greater than 1, and so the probability would be negative.

Conclusion

A number of theorists have proposed Bayesianism as a norm of reference to investigate the psychology of reasoning. The ensuing new paradigm poses many new conceptual and methodological questions. The existence of a plurality of competing Bayesian formal models is one of the thorniest questions. The proposed solution is to distinguish the elementary level of knowledge (the level of the conditional event) from the meta-epistemic level (the level of the conditional probability). For the first level a comparative method that bears on the main trivalent systems has been presented. Our preliminary experimental results support the de Finettian model. They indicate participants' strong preference for the truth tables defining the connectives of de Finetti's trivalent logic. In addition this model explains straightforwardly the classical "defective" truth table of the conditional.

For the second level the proposed solution is to replace the traditional method based on validity by a method based on de Finettian coherence. At this level our results confirm the hypothesis of the equivalence between assessing the probability of the conditional event and the conditional probability of the consequent conditioned on the antecedent, and we have shown that this obtains both in the situation of a conditional assertion and in the situation of a conditional bet. Because the second level concerns probabilistic evaluation, it is relevant for deductive inferences under uncertainty. Our ongoing research focuses on lay people's confidence in the conclusion of simple inference schemas using verbal probabilities, which has not been studied so far. Our observations (Politzer & Baratgin, in press) confirm, and extend to other schemas, the results reported by probabilistic mental logic investigators.[21]

In this chapter we have considered the major change brought about by the new paradigm, which primarily consists of taking uncertainty into account. There are, however, other key features of the new paradigm that may affect the methodology. One, reasoning cannot be reduced to a neutral mechanism independent of the individual's motivation, preferences and objectives (Bonnefon, 2009; Bonnefon & Sloman, 2013). Two, reasoning depends on the context: one reasons to and from an interpretation (Stenning & Van Lambalgen, 2008). Three, it depends on the individual's knowledge base and cognitive abilities (Stanovich, 2011).

Consequently, it is important to pay heed to individual differences in the experimental studies. Individual subjective responses (judgments and decisions) may be different but nevertheless coherent because they are based on different states of knowledge. Different responses may also reflect different mental processes (Bonnefon, 2013). The various factors at the origin of the subjectivity of beliefs such as cultural background should also be taken into account. The objectives of the new paradigm meet the foundation of the Bayesian subjective theory, and more precisely, in an explicit way, the de Finettian theory. This convergence suggests that the new paradigm could be the solution to de Finetti's "true subjective probability problem," which is to understand "the ways in which probability is assessed and used by educated people, and the way in which such abilities can be improved" (de Finetti, 1974, p. 15).

Acknowledgment

Financial support for this work was provided by a grant from the ANR Chorus 2011 (project BTAFDOC).

Notes

1. The term "normative models" designates formal models that define a norm of behavioral rationality for reasoning, judgment, and decision making. These models most often pertain to the fields of probability theory, economics, philosophy, or artificial intelligence.

2. The philosophical, linguistic, and also psychological literatures offer a variety of classifications of sentences of the type *"If A, then C."* One major distinction consists of separating subjunctive conditionals, that is, contrary-to-fact conditionals, from the indicative conditionals. We are concerned only with the latter.

3. This is the case, for instance, of the inference from the consequent to the conditional: *C, therefore if A then C* (Bonnefon & Politzer, 2011; Hailperin, 1996; Pfeifer, 2014; Suppes, 1966).

4. De Finetti keeps on warning against general formulations that often are at the source of misunderstanding. This is why in what follows we refer to "de Finetti's model" to underscore the originality of his theory and specify it with regard to "Bayesian model" and even to "subjective Bayesian model."

5. The expression "radical" to qualify de Finetti's point of view was used by Jeffrey (1983) and is now used by experts in de Finetti's work (see in particular Galavotti, 2008). Note that de Finetti himself was the first to regard and call his views deeply "radical" (de Finetti, 1969, pp. 6 and 8).

6. The use of the expression "supposition" is motivated and has a long history. Bayes used it to define conditional probability, and Laplace when justifying Bayes's rule (see Edgington, 1995, p. 262 for the citations). It is for similar reasons that Hailperin (1996) called his theory "suppositional logic."

7. Notice that at the end of the quote Ramsey (1929/1990) explicitly referred to a third truth value for the conditional, qualifying the degree of belief as "void" when the antecedent is false.

8. This is the case, for instance, of Łukasiewicz's, Bochvar's, and Kleene's logics.

9. Kleene's trivalued logic coincides with de Finetti's trivalued logic except for the absence of the conditional event and for the fact that U explicitly stands for "indeterminate."

10. This validity is often defined when U is interpreted as a value that stands for inconsistency (see, for example, Avron, 1991). This is the case of paraconsistent and relevant logics.

11. Milne (2012) uses the same validity without specifying the trivalent system chosen.

12. Using the author's terminology, an event is U-valid if, in its truth table restricted to T and F, no value is equal to F and at least one value is equal to T. For instance, the conditional event A|A is U-valid because if A is T, A|A is T; and if A is F, A|A is U.

13. With a system similar to de Finetti's trivalued logic, Rescher (1969) showed that the definitions V_T and V_{nF} may lead to inconsistency when a value U or F is antidesignated.

14. This is the case, for instance, of systems that take a different definition for conjunction such as quasiconjunction where the composition of T and U yields T (Dubois & Prade, 1994, 1995; Milne, 1997). The desiderata mentioned above can be satisfied with a trivalent system other than de Finetti's and a definition of validity other than V_U. This is the case, for example, of the trivalent system SAC (see Calabrese, 2002) using V_{TnF}.

15. For a similar result, see DeRose & Grandy (1999, p. 166, note 9).

16. Fugard, Pfeifer, Mayerhofer, and Kleiter (2011) have observed a shift in interpretation toward the conditional as the sequence of the items proceeds. The fact that in the classic task the rate of conjunctive interpretation diminishes across age (Barrouillet, Gauffroy, & Lecas, 2008) suggests that it is related to cognitive load or to limited cognitive ability; indeed, it has been found that the conjunctive interpretation is more frequent among those who score lower on general ability tests (Evans, Handley, Neilens, & Over, 2007; Sevenants, Dieussaert, & Schaeken, 2011).

17. Coletti and Scozzafava (2002) and Gilio (2002) have the same interpretation of de Finetti's conditional event. However, they do not use the same definition of conjunction or the same value for the threshold.

18. The expression "uncertainty" for the probability of falsehood is somewhat ambiguous. Among various formalizations of nonadditive measures of degrees of belief (which can be regarded as generalizations of additive probability), there exist several expressions for the degree of disbelief in a hypothesis A. Bel(not-A) is called degree of "incredulity," "doubt," or "potential surprise" (for a review, see Haenni, 2009). The expression "uncertainty" is sometimes used (see Jøsang, 2001) for $1 - \text{Bel}(A) + \text{Bel}(not\text{-}A)$, which is null in the Bayesian framework. Note also that the definition of a valid inference in a nonadditive framework amounts to Adams's p-validity: A entails C if Bel(A) is smaller than, or equal to, Bel(C).

19. There is no strict coincidence between the inferential validity of the first level and the probabilistic validity of the second level. For instance, for SAC with V_{TnF} and V_U, transitivity is valid at the first level but not at the second level. (For other differences, see Milne, 2012.)

20. This method of calculation of an interval has a very old origin (for a review, see Hailperin, 1996).

21. Although the *mental probability logic* approach is methodologically close to what we are advocating here, we have some reservation with respect to the status of the coherence intervals. It seems doubtful to us that human beings possess the computational ability to assess or determine the intervals, which we view as a metacognitive ability. Rather than being part of a competence model, we view the coherence intervals as an effective tool designed by the experimenter or the theorist to examine individuals' coherence, that is, whether their evaluation belongs to the interval. This is the way they were in fact used by de Finetti. Moreover, the mental representation of probability as an interval is a case of imprecise probability incompatible with the de Finettian view.

References

Adams, E. (1975). *The logic of conditionals*. Dordrecht: D. Reidel.

Adams, E. (1996). Four probability-preserving properties of inferences. *Journal of Philosophical Logic, 25*, 1–24.

Adams, E. (1998). *A primer of probability logic*. Stanford: CLSI Publications.

Avron, A. (1991). Natural 3-valued logics. Characterization and proof theory. *Journal of Symbolic Logic, 56*, 276–294.

Baratgin, J. (2002). Is the human mind definitely not Bayesian? A review of the various arguments. *Cahiers de Psychologie Cognitive/Current Psychology of Cognition, 21*, 653–680.

Baratgin, J. (in press). Le raisonnement humain: Une approche finettienne [Human reasoning: A Finettian approach]. Paris: Hermann.

Baratgin, J., Over, D. E., & Politzer, G. (2013). Uncertainty and the de Finetti tables. *Thinking & Reasoning, 19*, 308–328.

Baratgin, J., Over, D. E., & Politzer, G. (2014). New psychological paradigm for conditionals and general de Finetti tables. *Mind & Language, 29*, 73–84.

Baratgin, J., & Politzer, G. (2006). Is the mind Bayesian? The case for agnosticism. *Mind & Society, 5*, 1–38.

Baratgin, J., & Politzer, G. (2007). The psychology of dynamic probability judgment: Order effect, normative theory and experimental methodology. *Mind & Society, 5*, 53–66.

Baratgin, J., Politzer, G., & Over, D. E. (2013). The psychology of indicative conditionals and conditional bets. Working paper. Institut Jean Nicod website: http://jeannicod.ccsd.cnrs.fr/ijn_00834585/.

Barrouillet, P., Gauffroy, C., & Lecas, J.-F. (2008). Mental models and the suppositional account of conditionals. *Psychological Review, 115*(3), 760–771.

Benferhat, S., Bonnefon, J.-F., & Da Silva Neves, R. M. (2004). An experimental analysis of possibilistic default reasoning. In D. Dubois, C. A. Welty, & M.-A. Williams (Eds.), *Principles of knowledge representation and reasoning. Proceedings of the Ninth International Conference (KR2004)* (pp. 130–140). Menlo Park, CA: AAAI Press.

Benferhat, S., Bonnefon, J.-F., & Da Silva Neves, R. M. (2005). An overview of possibilistic handling of default reasoning with experimental studies. *Synthese, 146*, 53–70.

Binet, A. (1902). *La psychologie du raisonnement: Recherches expérimentales par l'hypnotisme* [The psychology of reasoning: Experimental research by hypnosis]. Paris: Alcan.

Blamey, S. (2001). Partial logic. In D. Gabbay & F. Guenthner (Eds.), *Handbook of philosophical logic, V* (pp. 261–353). Amsterdam: Elsevier.

Bonnefon, J.-F. (2009). A theory of utility conditionals: Paralogical reasoning from decision-theoretic leakage. *Psychological Review, 118*, 888–907.

Bonnefon, J.-F. (2013). New ambitions for a new paradigm: Putting the psychology of reasoning at the service of humanity. *Thinking & Reasoning, 18*(3), 381–398.

Bonnefon, J.-F., & Hilton, D. J. (2002). The suppression of *modus ponens* as a case of pragmatic preconditional reasoning. *Thinking & Reasoning, 8*, 21–40.

Bonnefon, J.-F., & Politzer, G. (2011). Pragmatics, mental models and one paradox of the material conditional. *Cognitive Science, 37*, 141–155.

Bonnefon, J.-F., & Sloman, S. A. (2013). The causal structure of utility conditionals. *Psychological Review, 26*, 193–209.

Braine, M., & O'Brien, D. P. (1998). *Mental logic*. Mahwah, NJ: Lawrence Erlbaum Associates.

Byrne, R. M. J. (1989). Suppressing valid inferences with conditionals. *Cognition, 31*, 61–83.

Calabrese, P. (2002). Deduction with uncertain conditionals. *Information Sciences, 147*, 143–191.

Ciucci, D., & Dubois, D. (2012). Three-valued logics for incomplete information and epistemic logic. In L. Fariñas del Cerro, A. Herzig, & J. Mengin (Eds.), *Logics in artificial intelligence* (pp. 147–159). Oxford: Oxford University Press.

Coletti, G., & Scozzafava, R. (2002). *Probabilistic logic in a coherent setting*. Dordrecht: Kluwer.

Cruz, N., Baratgin, J., Oaksford, M., & Over, D. P. (2015). Bayesian reasoning with ifs and ands and ors. *Frontiers in Psychology, 6*, 192.

Cruz, N., & Oberauer, K. (2014). Comparing the meanings of "if" and "all." *Memory & Cognition, 42*, 1345–1356.

Da Silva Neves, R., Bonnefon, J.-F., & Raufaste, E. (2002). An empirical test for patterns of nonmonotonic inference. *Annals of Mathematics and Artificial Intelligence, 34*, 107–130.

de Finetti, B. (1930). Fondamenti logici del ragionamento probabilistico [Logical foundations of probabilistic reasoning]. *Bollettino dell'Unione Matematica Italiana, 9,* 258–261.

de Finetti, B. (1931). Sul significato soggettivo della probabilità [On the subjective signification of probability]. *Fundamenta Mathematicae, 17,* 298–329.

de Finetti, B. (1957). L'informazione, il ragionamento, l'inconscio nei rapporti con la previsione [Information, reasoning, and the unconscious in relation to prevision]. *L'industria, 2,* 3–27.

de Finetti, B. (1964). Foresight: Its logical laws, its subjective sources. In H. Kyburg & H. E. Smokier (Eds.), *Studies in subjective probability* (pp. 55–118). New York: Wiley. (Original work published 1937.)

de Finetti, B. (1969). Initial probabilities: A prerequisite for any valid induction. *Synthese, 20,* 2–16.

de Finetti, B. (1974).The true subjective probability problems. In C. von Holstein (Ed.), *The concept of probability in psychological experiments* (pp. 15–23). Theory and Decision Library. Amsterdam: Reidel.

de Finetti, B. (1979). Probability and exchangeability from a subjective point of view. *International Statistical Review, 47,* 129–135.

de Finetti, B. (1980). Probabilità [Probability]. In *Enciclopedia,* Vol. X (pp. 1146–1187). Torino: Einaudi.

de Finetti, B. (1995). The logic of probability. *Philosophical Studies, 77,* 181–190. (Original work published 1936.)

de Finetti, B. (2006). *L'invenzione della verità* [The invention of truth]. Milan: Cortina.

DeRose, K., & Grandy, R. E. (1999). Conditional assertions and "biscuit" conditionals. *Noûs, 33,* 405–420.

Dubois, D., & Prade, H. (1994). Conditional objects as nonmonotonic consequence relationships. *IEEE Transactions on Systems, Man, and Cybernetics, 24,* 1724–1740.

Dubois, D., & Prade, H. (1995). Logique possibiliste, modèles préférentiels et objets conditionnels [Possibilistic logic, preferential models and conditional objects]. In J. Dubucs & F. Lepage (Eds.), *Méthodes logiques pour les sciences cognitives* [Logical methods for cognitive science] (pp. 99–120). Paris: Hermès.

Edgington, D. (1995). On conditionals. *Mind, 104,* 235–329.

Egré, P., & Cozic, M. (in press). Conditionals. In M. Aloni & P. Dekker (Eds.), *Handbook of formal semantics.* Cambridge: Cambridge University Press.

Elqayam, S., & Evans, J. S. B. T. (2013). Rationality in the new paradigm: Strict versus soft Bayesian approaches. *Thinking & Reasoning, 19*(3), 453–470.

Evans, J. S. B. T. (2012). Questions and challenges for the new psychology of reasoning. *Thinking & Reasoning, 18,* 5–31.

Evans, J. S. B. T., Handley, S., Neilens, H., & Over, D. E. (2007). Thinking about conditionals: A study of individual differences. *Memory & Cognition, 35,* 1359–1371.

Evans, J. S. B. T., Handley, S. J., & Over, D. E. (2003). Conditionals and conditional probability. *Journal of Experimental Psychology: Learning, Memory, and Cognition, 29,* 321–335.

Evans, J. S. B. T., & Over, D. E. (2004). *If.* Oxford: Oxford University Press.

Evans, J. S. B. T., & Over, D. E. (2013). Reasoning to and from belief: Deduction and induction are still distinct. *Thinking & Reasoning, 19,* 267–283.

Evans, J. S. B. T., Thompson, V., & Over, D. E. (2015). Uncertain deduction and conditional reasoning. *Frontiers in Psychology, 6,* 398.

Fugard, J. B., Pfeifer, N., Mayerhofer, B., & Kleiter, G. (2011). How people interpret conditionals: Shifts towards the conditional event. *Journal of Experimental Psychology: Learning, Memory, and Cognition, 37,* 635–648.

Galavotti, M. C. (1995). F.P. Ramsey and the Notion of Chance. In J. Hintikka & K. Puhl (Eds.), *The British Tradition in the 20th Century Philosophy. Paper of the Proceedings of the 17th International Wittgenstein Symposium* (pp. 330–340). Vienna: Holder-Pichler-Tempsky.

Galavotti, M. C. (1997). Probabilism and beyond. *Erkenntnis, 45,* 243–265.

Galavotti, M. C. (1999). Some remarks on objective chance (F. P. Ramsey, K. R. Popper, and N. R. Campbell). In R. Giuntini, M. Chiara, & F. Laudisa (Eds.), *Language, quantum, music. Papers of the 10th international congress of logic, methodology and philosophy of science, Florence, Italy, August 1995* (p. 73–82). Dordrecht: Kluwer.

Galavotti, M. C. (2008). *Bruno de Finetti radical probabilist. Texts in philosophy.* London: College Publications.

George, C. (1995). The endorsement of the premises: Assumption based or belief-based reasoning. *British Journal of Psychology, 86,* 93–111.

George, C. (1997). Reasoning from uncertain premises. *Thinking & Reasoning, 3,* 161–189.

Gilio, A. (2002). Probabilistic reasoning under coherence in System P. *Annals of Mathematics and Artificial Intelligence, 34,* 5–34.

Gilio, A., & Over, D. E. (2012). The psychology of inferring conditionals from disjunctions and its probabilistic analysis. *Journal of Mathematical Psychology, 56,* 118–131.

Girotto, V., & Johnson-Laird, P. N. (2004). The probability of conditionals. *Psychologia, 47,* 207–225.

Good, I. J. (1976). The Bayesian influence, or how to sweep subjectivism under the carpet. In C. A. Hooker & W. L. Harper (Eds.), *Foundations of probability theory* (Vol. II). *Statistical inference, and statistical theories of science* (pp. 125–174). Dordrecht: Springer.

Haenni, R. (2009). Non-additive degrees of belief. In F. Huber & C. Schmidt-Petri (Eds.), *Degrees of belief* (pp. 121–159). Dordrecht: Springer.

Hailperin, T. (1996). *Sentential probability logic: Origins, development, current status, and technical applications*. Bethlehem, PA: Lehigh University Press.

Handley, S. J., Evans, J. S. B. T., & Thompson, V. A. (2006). The negated conditional: A litmus test for the suppositional conditional? *Journal of Experimental Psychology: Learning, Memory, and Cognition, 32*, 559–569.

Hilton, D. (1995). The social context of reasoning: Conversational inference and rational judgment. *Psychological Bulletin, 118*, 248–271.

Inhelder, B., & Piaget, J. (1958). *The growth of logical thinking from childhood to adolescence*. New York: Basic Books.

James, W. (1908). *Text-book of psychology, briefer course*. London: Macmillan.

Jeffrey, R. C. (1976). Judgmental probability and objective chance. *Erkenntnis, 24*, 5–16.

Jeffrey, R. C. (1983). *The logic of decision*. Chicago: University of Chicago Press.

Johnson-Laird, P. N., & Byrne, R. M. J. (1991). *Deduction*. Hove, London: Lawrence Erlbaum.

Jøsang, A. (2001). A logic for uncertain probabilities. *International Journal of Uncertainty, Fuzziness and Knowledge-based Systems, 9*, 279–311.

Kahneman, D., Slovic, P., & Tversky, A. (1982). *Judgment under uncertainty: Heuristics and biases*. Cambridge: Cambridge University Press.

Kraus, S., Lehmann, D., & Magidor, M. (1990). Nonmonotonic reasoning, preferential models and cumulative logics. *Artificial Intelligence Journal, 44*, 167–207.

Lehmann, D., & Magidor, M. (1992). What does a conditional knowledge base entail? *Artificial Intelligence, 55*, 1–60.

Lewis, D. (1980). A subjectivist's guide to objective chance. In R. Jeffrey (Ed.), *Studies in inductive logic and probability* (Vol. 2, pp. 262–292). Berkeley: University of California Press.

Liu, I. M., Lo, K. C., & Wu, J. T. (1996). A probabilistic interpretation of "if-then." *Quarterly Journal of Experimental Psychology, 49A*, 828–844.

Manktelow, K. I. (2012). *Thinking and reasoning: Psychological perspectives on reason, judgment and decision making*. Hove: Psychology Press.

McDermott, M. (1996). On the truth conditions of certain "If" sentences. *Philosophical Review, 105*(1), 1–37.

Milne, P. (1997). Bruno de Finetti and the logic of conditional events. *British Journal for the Philosophy of Science, 48*, 195–232.

Milne, P. (2012). Indicative conditionals: A request for more experiments. *Thinking & Reasoning, 18*, 196–224.

Oaksford, M., & Chater, N. (1995). Theories of reasoning and the computational explanation of everyday inference. *Thinking & Reasoning, 1*, 121–152.

Oaksford, M., & Chater, N. (2001). The probabilistic approach to human resoning. *Trends in Cognitive Sciences, 5*, 349–357.

Oaksford, M., & Chater, N. (2007). *Bayesian rationality: The probabilistic approach to human reasoning.* Oxford: Oxford University Press.

Oaksford, M., & Chater, N. (2009). Précis of *Bayesian rationality:* The probabilistic approach to human reasoning. *Behavioral and Brain Sciences, 32*, 69–84.

Over, D. E. (2009). New paradigm psychology of reasoning. *Thinking & Reasoning, 15*, 431–438.

Pelletier, F. J., & Elio, R. (1997). What should default reasoning be, by default? *Computational Intelligence, 13*, 165–187.

Pfeifer, N. (2014). Reasoning about uncertain conditionals. *Studia Logica, 8*, 1–18.

Pfeifer, N., & Kleiter, G. D. (2006). Inference in conditional probability logic. *Kybernetica, 42*(4), 391–404.

Pfeifer, N., & Kleiter, G. D. (2009). Framing human inference by coherence based probability logic. *Journal of Applied Logic, 7*, 206–217.

Pfeifer, N., & Kleiter, G. D. (2010). The conditional in mental probability logic. In M. Oaksford & N. Chater (Eds.), *Cognition and conditionals: Probability and logic in human thinking* (pp. 153–173). Oxford: Oxford University Press.

Pfeifer, N., & Kleiter, G. D. (2011). Uncertain deductive reasoning. In K. Manktelow, D. E. Over, & S. Elqayam (Eds.), *The science of reason: A Festschrift for Jonathan St B.T. Evans* (pp. 145–166). Hove, UK: Psychology Press.

Politzer, G. (1986). Laws of language use and formal logic. *Journal of Psycholinguistic Research, 15*, 47–92.

Politzer, G. (2004). Reasoning, judgement and pragmatics. In I. N. Noveck & D. Sperber (Eds.), *Experimental pragmatics* (pp. 94–115). Houndmills: Palgrave Macmillian.

Politzer, G. (2005). Uncertainty and the suppression of inferences. *Thinking & Reasoning, 11*, 5–33.

Politzer, G. (2007). Reasoning with conditionals. *Topoi, 26*, 79–95.

Politzer, G. (in press). Deductive reasoning under uncertainty: A water tank analogy. *Erkenntnis.* doi: 10.1007/s10670-015-9751-0.

Politzer, G., & Baratgin, J. (in press). Deductive schemas with uncertain premises using qualitative probability expressions. *Thinking & Reasoning.* doi: 10.1080/13546783.2015.1052561.

Politzer, G., & Bonnefon, J.-F. (2010). Two aspects of reasoning competence: A challenge for current accounts and a call for new conceptual tools. In M. Oaksford & N. Chater (Eds.), *Cognition and conditionals* (pp. 371–386). Oxford: Oxford University Press.

Politzer, G., & Bourmaud, G. (2002). Deductive reasoning from uncertain conditionals. *British Journal of Psychology, 93*, 345–381.

Politzer, G., & Macchi, L. (2000). Reasoning and pragmatics. *Mind & Society, 1*, 73–93.

Politzer, G., Over, D. E., & Baratgin, J. (2010). Betting on conditionals. *Thinking & Reasoning, 16*, 172–197.

Ramsey, F. P. (1990). Truth and probability. In D. H. Mellor (Ed.), *Philosophical papers* (pp. 52–94). Cambridge: Cambridge University Press. (Original work published 1926).

Ramsey, F. P. (1990). General propositions and causality. In D. H. Mellor (Ed.), *Philosophical papers* (pp. 145–163). Cambridge: Cambridge University Press. (Original work published 1929.)

Rescher, N. (1969). *Many-valued logic*. New York: McGraw-Hill.

Rips, L. J. (1994). *The psychology of proof*. Cambridge, MA: MIT Press.

Savage, L. J. (1954). *The foundations of statistics*. New York: Dover Publications.

Sevenants, A., Dieussaert, K., & Schaeken, W. (2011). Truth table tasks: Irrelevance and cognitive ability. *Thinking & Reasoning, 17*, 213–246.

Singmann, H., Klauer, K. C., & Over, D. (2014). New normative standards of conditional reasoning and the dual-source model. *Frontiers in Psychology, 5*, 316.

Stanovich, K. E. (2011). *Rationality and the reflective mind*. Oxford: Oxford University Press.

Stenning, K., & Van Lambalgen, M. (2008). *Human reasoning and cognitive science*. Cambridge: MIT Press.

Stevenson, R. J., & Over, D. E. (1995). Deduction from uncertain premises. *Quarterly Journal of Experimental Psychology, Section A, 48*, 613–643.

Stevenson, R. J., & Over, D. E. (2001). Reasoning from uncertain premises: The effects of expertise and conversational context. *Thinking & Reasoning, 7*, 367–390.

Suppes, P. (1966). Probabilistic inference and the concept of total evidence. In J. Hintikka & P. Suppes (Eds.), *Aspects of inductive logic* (pp. 49–65). Amsterdam: North-Holland.

Wason, P. C., & Johnson-Laird, P. N. (1972). *Psychology of reasoning: Structure and content*. Cambridge, MA: Harvard University Press.

7   Dual Reasoning Processes and the Resolution of Uncertainty: The Case of Belief Bias

Linden J. Ball and Edward J. N. Stupple

Contemporary researchers studying human thinking, reasoning, and judgment often appeal to a distinction between two forms of processing, one fast and intuitive, the other slow and reflective (e.g., Evans, 2008, 2010; Kahneman, 2011). These so-called *dual-process* accounts of high-level cognition had rather modest beginnings in reasoning research being conducted in the 1970s and 1980s by Peter Wason, Jonathan Evans, and their collaborators (e.g., Evans, 1977, 1989; Evans, Barston, & Pollard, 1983; Evans & Wason, 1976; Wason & Evans, 1975). Nowadays, the study of intuitive versus reflective reasoning processes is a major focus of many international laboratories, with dual-process concepts currently dominating a great deal of published work on thinking, reasoning, judgment, and decision making and also making inroads into related areas such as problem solving, insight and creativity (e.g., Gilhooly, Ball, & Macchi, 2015), and mental-state understanding (e.g., Wilkinson & Ball, 2013). There is, however, an emerging debate over the credibility of the dual-process distinction, particularly in terms of whether it can engender falsifiable predictions. Evans and Stanovich (2013a, 2013b) have recently defended the dual-process conceptualization of reasoning, arguing that it is supported by a considerable body of evidence obtained over many decades from a wide range of sources. One important component of Evans and Stanovich's defense is their claim that critics who latch onto the falsifiability issue are confused about different "levels" of theorizing. That is, dual-process opponents typically lodge their attacks at the level of meta-theory, yet falsifiable predictions arise at a level below this in the form of task-specific models.

In this chapter we aim to examine the credibility of task-specific, dual-process models of people's reasoning when tackling belief-oriented logical arguments. Such arguments give rise to a well-established phenomenon termed *belief bias*, which is the tendency to endorse an argument based on the believability of its conclusion rather than the argument's logical validity (see Evans et al., 1983, for pioneering research on this bias). For example, the conclusion to the following syllogistic argument is invalid, yet people tend to endorse it because of its believability: "All flowers need water. All roses need water. Therefore, all roses are flowers" (this example is adapted from one used by De Neys & Franssens, 2009). Dual-process theories have dominated attempts to provide a theoretical explanation of belief bias since Evans et al.'s (1983) seminal research, which involved stringent experimental controls that enabled the existence of belief bias to be firmly established.

In examining belief-bias theories in this chapter we focus on two main classes of dual-process explanation. First, we review accounts that emphasize a staged, *sequential* progression from fast, intuitive to slow, reflective thinking. We show how such sequential models can account for basic belief-bias findings, although we also discuss how these models struggle to capture more subtle aspects of recent data, such as the way in which people appear to have a rapid sense of *uncertainty* regarding the accuracy of their inferential decisions in cases where the logic of an argument (in terms of its validity or invalidity) conflicts with the believability of a presented conclusion (in terms of its believability or unbelievability). Second, we review accounts of belief bias that emphasize the concurrent and *parallel* operation of intuitive and reflective processes (e.g., Sloman, 1996; Stupple & Ball, 2008). Although such parallel-process models seem to fare better than sequential-process models in their capacity to account both for basic belief-bias findings and for data relating to the uncertainty that people have in their answers, we discuss below how these theories still run into major conceptual difficulties pertaining to the cognitive efficiency of running two different types of processes in parallel.

The challenges that confront both sequential and parallel dual-process conceptualization of belief bias have recently led to more nuanced, hybrid models, which encompass the operation of both sequential and parallel dual-reasoning mechanisms. A fascinating example of this approach is the model that has recently been advanced by De Neys (2012; see also De Neys, 2014; De Neys & Bonnefon, 2013), which involves a *first* stage of reasoning in which both *intuitive heuristics* (i.e., implicit processes driven by conclusion believability) and *logical intuitions* (i.e., implicit processes driven by logical considerations) function concurrently in order to deliver outputs that may concur or conflict. A conflict situation would engender uncertainty, thereby triggering a *second* stage of reflective reasoning aimed at conflict resolution and uncertainty reduction.

We suggest later in this chapter that hybrid dual-process models such as the one espoused by De Neys (e.g., 2012), although admittedly controversial, nevertheless appear to capture a diverse range of belief-bias data that derive from varied experimental manipulations and research methodologies, including neuroscientific investigations and the analysis of individual differences. We also note that De Neys's hybrid account represents a good example of the way in which generally effective thinking can occur by capitalizing on intuitive processing and avoiding the need for costly, reflective processing. In other words, rather than the cognitive system getting embroiled in resource intensive, attention-demanding, and time-consuming reflective processing for *all* reasoning and judgment situations, the cognitive system is instead well adapted to engage in reflective processing only in those situations where there is a conflict between the outputs of intuitive heuristics and logical intuitions. In this way highly constrained cognitive resources associated with reflective processing can be deployed efficiently to those reasoning situations that are characterized by uncertainty, where it is beneficial for the reasoner to attempt to resolve such uncertainty so as to determine whether or not a conclusion is normatively sanctioned (cf. Stupple & Ball, 2014). Of course, although people may be triggered to reason more reflectively because of perceived uncertainty, there is no guarantee that all individuals will have either the working-memory capacity (see Evans & Stanovich, 2013a) or the analytic tools (i.e., what Stanovich, 2009, refers to as the *mindware*) to reason out a normative solution. Indeed, various individual differences in reflective reasoning appear to be key moderator variables in accounting for the relative presence or absence of belief-bias effects, which is another theme that we pick up in later sections of this chapter. We conclude our contribution to this volume by considering some of the remaining challenges for hybrid dual-process models of belief bias and by pointing toward important new directions for belief-bias research within the dual-process framework.

Belief Bias: Key Paradigm and Core Conclusion-Endorsement Findings

Belief bias has traditionally been studied using *categorical syllogisms*. These are deductive reasoning problems that involve two premises and a conclusion, each of which features a standard logical quantifier (i.e., *all*, *no*, *some*, or *some … are not*). Above we presented an example syllogism adapted from one used by De Neys and Franssens (2009) in which the premises and conclusion involve the logical qualifier *all*: "All flowers need water. All roses need water. Therefore, all roses are flowers." A logically valid conclusion describes the relationship between the premise terms in a way that is necessarily true, which means that conclusions that are merely consistent with the premises but are not necessitated by them are invalid. In studies using categorical syllogisms, participants can be asked either to generate a logical conclusion in response to given premises or to evaluate the validity of a conclusion that is presented with the argument. Studies of belief-bias effects have almost entirely used a conclusion-evaluation paradigm. The following discussion therefore centers on findings deriving from this paradigm, although it has to be acknowledged that any truly comprehensive theory of belief bias should also capture belief-bias data that derive from the use of conclusion-production tasks.

In table 7.1 we show three other syllogisms based on those created by De Neys and Franssens (2009) in order to demonstrate the way in which logical validity and conclusion believability are systematically manipulated in standard belief-bias experiments. As can be seen from table 7.1, the crossing of conclusion validity and conclusion believability produces problem types in which validity and believability are in opposition (i.e., *conflict* items) as well as problem types in which validity and believability are

Table 7.1

Examples of the four permutations of belief-oriented syllogisms that arise from the systematic crossing of the logical status and the belief status of presented conclusions

	Logical Status	
Belief Status	Valid	Invalid
Believable	**Valid-Believable**	**Invalid-Believable**
	All birds have wings	All flowers need water
	All crows are birds	All roses need water
	Therefore, all crows have wings	Therefore, all roses are flowers
Unbelievable	**Valid-Unbelievable**	**Invalid-Unbelievable**
	All mammals can walk	All meat products can be eaten
	All whales are mammals	All apples can be eaten
	Therefore, all whales can walk	Therefore, all apples are meat products

Source: Adapted from De Neys and Franssens (2009).

in alignment (i.e., *no-conflict* items). Although contemporary belief-bias studies often aggregate data for conflict versus no-conflict problems, such aggregation obscures problem-specific effects that appear to be theoretically important. In this chapter we touch on such item-specific findings when they are particularly germane to the conceptual discussion.

In terms of findings deriving from belief-bias studies, there are three fundamental effects that are almost always observed and that require theoretical explanation. First, people reliably endorse presented conclusions that are compatible with beliefs more frequently than conclusions that contradict beliefs. This is the nonlogical belief-bias effect, which indicates that people give undue weight to conclusions that concur with their real-world knowledge and prior assumptions. Second, people reliably endorse more logically valid conclusions than invalid ones, such that at one and the same time people demonstrate evidence of logicality as well as bias. Third, there is consistent evidence that belief bias is more pronounced on invalid problems than on valid problems, giving rise to a logic-by-belief interaction in conclusion-endorsement rates. Formulating an effective explanation of the main effects of logic and belief together with the logic-by-belief interaction has been a nontrivial endeavor for researchers, with numerous competing theories having been generated over the last 30 years or so. It is looking increasingly likely that a resolution to this theoretical debate will require a radical new explanatory framework, which we are perhaps seeing signs of currently with the emergence of novel theoretical concepts such as the notion of "logical intuitions" in reasoning (De Neys, 2012).

Within a single chapter it not possible to review all of the dual-process theories of belief bias that have been formulated since the early 1980s; it is arguably also not profitable to engage in such a review given that some of the theories have been relatively short lived and have succumbed to fairly rapid falsification. Instead, we aim to survey only some of the dominant theories that have largely stood the test of time and that still feature in the contemporary literature. As noted above, we structure this review by first examining sequential dual-process models before then assessing parallel dual-process models and then completing the picture with an exploration of more recently proposed hybrid models that incorporate both sequential and parallel dual-process elements.

Belief Bias and *Sequential* Dual-Process Models

An early explanation of belief-bias effects based on the sequential operation of intuitive processes that are then followed by reflective processes is the *selective-scrutiny* model, originally espoused by Evans et al. (1983). This model assumes that believable conclusions are responded to via rapid, intuitive *heuristics* and are simply accepted, whereas unbelievable conclusions motivate more rigorous, "analytic" processing directed at testing conclusion validity. This model thereby explains the logic by belief interaction as arising from the more assiduous scrutiny that is directed at syllogisms with unbelievable conclusions because this additional analysis will allow the validity of valid-unbelievable conclusions to be determined as well as the invalidity of invalid-unbelievable conclusions.

A similar account to the selective-scrutiny model—which is also based on the idea that unbelievable conclusions stimulate an analytic reasoning process—is captured by the "mental models" theory of belief bias advanced by Oakhill and Johnson-Laird (1985; see also Garnham & Oakhill, 2005; Oakhill, Johnson-Laird, & Garnham, 1989). According to this account, conclusions that are supported by people's initial representations of given premises (i.e., their initial *mental models*) will simply be accepted if they are believable, but if a conclusion is unbelievable, then it will be tested more rigorously against alternative and potentially falsifying mental models of the premises (i.e., unbelievable conclusions serve to trigger a "selective falsification" strategy; see Roberts & Sykes, 2003). In this way the occurrence of a logic-by-belief interaction can be explained because believable conclusions will typically be endorsed, whereas unbelievable conclusions may either be endorsed or rejected, dependent on the outcome of the more diligent process of counterexample search, which will establish their validity or invalidity.

The selective scrutiny model of belief bias and the subsequent mental models account are both specified in sufficient detail to enable the derivation of predictions relating to the time that people should spend reasoning about syllogistic arguments to determine the validity or invalidity of presented conclusions. More specifically, reasoners should take reliably longer to engage with problems that have unbelievable conclusions compared to those that have believable conclusions. This is because only an unbelievable conclusion will trigger more diligent reflective processing aimed at testing conclusion validity, which would require premise rereading to refresh and revise the mental representation of the information they contain. Ball, Phillips, Wade, and Quayle (2006) tested this prediction in an eye-tracking study, which permitted detailed monitoring of the time that people spent attending to the premises and conclusions of presented problems before registering a "yes" (endorse) or "no" (reject) decision. Ball et al.'s chronometric data revealed the exact opposite of what was predicted by the selective scrutiny and mental models accounts in that it was actually the *believable* conclusions rather than unbelievable conclusions that were associated with longer problem inspection times (see also Thompson, Striemer, Reikoff, Gunter, & Campbell, 2003, for similar evidence from measures of decision latencies).

Ball et al.'s (2006) eye-tracking study also revealed the existence of a reliable interaction between logic and belief in both a global measure of overall syllogism-processing times and in a local measure of premise inspection times. The interaction indicated that people spent more time reasoning about the "conflict" syllogisms (those with invalid-believable and valid-unbelievable conclusions) relative to the "no-conflict" syllogisms (those with valid-believable and invalid-unbelievable conclusions). The interaction between logic and belief observed in eye-tracking data has also been replicated in subsequent studies using computer-based tracking techniques to monitor the processing times for syllogism components (Stupple & Ball, 2008; Stupple, Ball, Evans, & Kamal-Smith, 2011; see also Ball, 2013a, for a review of the use of eye-tracking data in research on belief bias and other reasoning biases). A further observation in Ball et al.'s (2006) eye-tracking study was that the logic-by-belief interaction that arose in premise inspection times was localized specifically at a *post-conclusion-viewing* stage. What this means is that the effect occurred only *after* participants had read a conclusion and ascertained whether it was believable or unbelievable. This finding suggests that people's second-stage reflective reasoning seems to be guided by the belief status of given conclusion in line with selective scrutiny and mental models assumptions, but, crucially, that this second-stage processing happens in a manner completely *unlike* that predicted by these theories.

The consistent evidence that conflict problems take longer to process than no-conflict problems indicates that conflict problems are in some way special, and their unique properties need to be taken seriously when formulating theoretical explanations of belief bias. Furthermore, the sequential-process accounts that we have considered so far seem to be unable to explain why conflict items are processed in such a distinct manner. One critical problem for existing sequential-process models has been identified by De Neys (2012), which is that the way in which intuitive processing precedes reflective processing means that there is no mechanism whereby the reasoner can detect that there is a conflict between the output of the intuitive process and the reflective processes until both processes have had time to run their course. Such *late* identification of conflict is, of course, possible but seems to be outside the scope of the selective scrutiny and mental model accounts. Take, for example, the case of a believable-invalid conflict problem. Under either a selective scrutiny or a mental models view, the reasoner is supposed only to engage additional reflective processing in the case of *unbelievable* conclusions, so there is simply no way for these theories to account for the triggering of further processing of *believable* conclusions that are, in fact, invalid but that people cannot ascertain are invalid until they have engaged in further processing. One way out of the conundrum would be for sequential-process theorists to accept that conflicts are detected late, that is, after the operation of both the intuitive and reflective processing stages. The long response times for conflict items could then be explained because the conflict that is identified would prompt further time-consuming processing aimed at conflict resolution. Such a model does not appear to have been mooted in the literature, although Evans (2009) has sketched what is essentially a sequential-process account that also picks up on some of these ideas. We note, however, that even a late-occurring conflict-detection mechanism seems to be insufficient to explain various processing-time effects and confidence effects discussed later on.

A more recent instantiation of the selective-scrutiny view is termed the *selective-processing* model (e.g., Evans, 2000, 2007; see also Evans, Handley, & Harper, 2001; Klauer et al., 2000; Morley, Evans, & Handley, 2004). However, this model also runs into conceptual difficulties in accounting for the increased processing times associated with conflict problems relative to no-conflict problems. According to the selective-processing model, belief bias is determined by the operation of both intuitive/heuristic and reflective/analytic components that operate within a sequential-process framework. The *default*, heuristic response is to accept believable and to reject unbelievable conclusions. This explains why belief bias arises with both valid and invalid inferences, thereby accounting for the main effect of belief that is evident in conclusion-endorsement data. The analytic component of the model is, however, needed to explain the logic-by-belief interaction. If the analytic system *intervenes* during reasoning, then an attempt is made to construct just a single mental model of the premises. This analytic process is, however, itself biased by the believability of conclusions such that for a believable conclusion a search is initiated for a single mental model that *supports* the conclusion, whereas for an unbelievable conclusion a search is initiated for a single mental model that *refutes* the conclusion. These latter assumptions provide a clear rationale for the emergence of a logic-by-belief interaction in conclusion-endorsement data. When conclusions are valid, despite unbelievable content motivating a search for a counterexample model, such a model cannot be found, which serves to limit the influence of belief bias. When conclusions are invalid, however, models exist that both support *and* refute such conclusions, thus leading to high levels of erroneous acceptance of invalid-believable items and high levels of correct rejection of invalid-unbelievable items.

Despite the considerable conceptual ingenuity of Evans's (e.g., 2007) selective-processing model, it again fails to address the issue of the increased processing times that are associated with conflict problems relative to no-conflict problems. The model provides no obvious reason for why conflict problems—especially invalid-believable ones—should be subjected to increased processing effort because an element of time-consuming analytic processing should be equally likely for *all* problem types given that a search for a confirming or disconfirming model is always pursued. Stupple et al. (2011) have noted, however, that the selective-processing model is rather more subtle than it appears at first sight because it also embodies assumptions regarding the specific nature and quality of the analytic intervention that arises for different problems types (Evans, 2007; see also Ball, 2013b). As a result, it is possible for processing latencies to vary dependent on factors such as item type, item difficulty, and reasoning ability. More specifically, Stupple et al. (2011) suggested that the particularly long latencies observed in aggregate data for invalid-believable items may well reflect the performance of a *subset* of individuals who are better able to understand the logic of these problems but whose reasoning necessitates time-consuming processing.

Some support for Stupple et al.'s (2011) proposals derives from an earlier study of individual differences in reasoning reported by Sá, West, and Stanovich (1999), who showed that high cognitive ability, dispositions toward active and open-minded think-ing, and skills in cognitive decontextualization are all associated with the avoidance of responding to syllogistic conclusions on the basis of prior beliefs. Sá et al.'s research therefore attests to the existence of a subgroup of participants who are able to avoid the biasing effects of beliefs so as to reason more effectively in accordance with logical norms. As Stupple et al. (2011) note, such a subgroup of "logical" participants would fall outside the explanatory reach of the currently formulated selective-processing model, which has essentially been devised to capture the behavior of the "average" reasoner, who demonstrates a mix of logicality and bias when responding to belief-oriented problems. Stupple et al. (2011) provided new evidence in support of their proposals in a processing-time study that clearly identified three subsets of reasoners: (1) a *low-logic* group showing high levels of belief bias and very fast response times, indicative of the operation of a heuristically driven response bias; (2) a *medium-logic* group showing moderate belief bias and slower response times, consistent with enhanced analytic pro-cessing, albeit selectively biased by conclusion believability (as per the assumptions of the original selective-processing model); and (3) a *high-logic* group showing low belief bias, where such unbiased responses were also associated with increased processing times, especially for invalid-believable conclusions.

Stupple et al.'s (2011) individual differences analyses of response times within the belief-bias paradigm are important because they clearly indicate that people of varying logical ability have different response-time profiles. Moreover, the response-time profile for the middle-ability reasoning group appears to align well with key assumptions of the selective-processing model. Arguably, too, this model embodies sufficient flexibility to be able to accommodate the response profiles of reasoners who polarize at the lower-ability and higher-ability ends of the spectrum (see Stupple et al., 2011, for an elaborated selective-processing model that can encompass individual differences in reasoning ability). A further, intriguing aspect of Stupple et al.'s study, however, was evidence showing that although the response-time difference for conflict versus no-conflict problems was particularly marked for high-ability reasoners, it was nevertheless *still* reliably present (albeit diminished) for low-ability reasoners. This finding indicates that even in low-logic individuals who are responding primarily on the basis of rapid, intuitive, heuristic processes, there is still significant evidence for them having equally rapid sensitivity to logic-belief conflicts (i.e., that the logic of the conclusion and its belief status are in opposition). Such evidence, however, is clearly very challenging for a sequential dual-process model, which would assume that logic-belief conflicts should only be identified over an extended period during which intuitive and reflective processes unfold in a staged, sequential manner. We explore this issue further in the next section.

Challenges for Sequential Dual-Process Models of Belief Bias

As outlined, sequential dual-process models of belief bias that have been developed in the literature since the 1980s have had a fair degree of success in explaining conclusion-endorsement findings, including the main effects of logic and belief that are observed in the data as well as the ubiquitous logic by belief interaction. As discussed, an important, recent instantiation of a sequential dual-process view—the selective-processing model of Evans and colleagues (e.g., Evans, 2007)—is associated with a number of assumptions concerning the nature and quality of reasoning arising with different problem types. These assumptions enable the model to go a long way toward accounting for not only conclusion-endorsement data but also processing-time data for individual problem items. Although the model has required the addition of further assumptions (e.g., Stupple et al., 2011) to enable it to explain individual differences in belief-bias effects, such auxiliary assumptions mean that the model can generally account effectively for a majority of the extant conclusion-endorsement and response-time findings. Notwithstanding the numerous strengths of the selective-processing model, however, we also suggested above that it runs into conceptual difficulties in relation to the response-time evidence indicating that reasoners—including those who typically respond very rapidly and intuitively—still seem to be *aware* that the logical status and the belief status of logic/belief conflict problems are pulling in different response directions, which manifests as longer processing times for conflict items despite highly belief-biased decisions. As De Neys (2012) puts it, the evidence suggests that people are in some way able to "detect that they are biased" (p. 29).

There are now a number of studies that support such bias detection, not only in the belief-bias paradigm but also in other paradigms that pit intuitive, heuristic responses against logical or probabilistic norms. Such studies reveal that accuracy rates on no-conflict versions of problems are typically high, and response latencies are relatively short, whereas accuracy rates are much lower on the conflict versions, and response latencies are relatively long (e.g., Bonner & Newell, 2010; De Neys & Glumicic, 2008; Pennycook, Fugelsang, & Koehler, 2012; Villejoubert, 2009). The generality of such latency findings across a wide range of tasks provides strong support for the idea that people are sensitive to normative principles even when responding rapidly and non-normatively. Such observations challenge sequential-process accounts of reasoning across all of these paradigms (see De Neys, 2012).

De Neys and colleagues have also been influential in demonstrating how a range of cognitive and neuropsychological measures that extend beyond latencies also seem to reveal implicit bias detection with conflict items. Again, such evidence arises not only with belief-oriented syllogisms but with numerous other problems that involve violations of traditional logical or probabilistic norms. Focusing specifically on the domain of belief-oriented reasoning, research using confidence measures has indicated that reasoners show decreased response confidence after responding to belief/logic conflict items relative to no-conflict items (e.g., De Neys, Cromheeke, & Osman, 2011; also see Quayle & Ball, 2000 for pioneering work on response confidence and belief bias). In addition, research examining participants' skin conductance as a way to monitor autonomic nervous system activation (De Neys, Moyens, & Vansteenwegen, 2010) has shown that responding to conflict syllogisms results in a clear electrodermal activation spike relative to responding to no-conflict syllogisms. In other words tackling belief/logic conflict items is associated with increased arousal for participants despite their overarching tendency to respond in a biased manner such that they fail to evaluate conclusions accurately (see also Morsanyi & Handley, 2012).

Brain-imaging research has further added to the body of evidence that logic/belief conflict items are processed differently from no-conflict items. For example, Goel and Dolan (2003) demonstrated that successfully overcoming belief bias during reasoning activates a specific region of the frontal lobes—the lateral prefrontal cortex (see also De Martino, Kumaran, Seymour, & Dolan, 2006; De Neys, Vartanian, & Goel, 2008; Houdé et al., 2000; Prado & Noveck, 2007; Sanfey, Rilling, Aronson, Nystrom, & Cohen, 2003). This same brain region has also been shown to be involved in responding to tasks that require inhibition of prepotent but erroneous responses (e.g., Aron, Robbins, & Poldrack, 2004). Furthermore, De Neys and Franssens (2009) have presented compelling evidence that belief bias arises from participants' failure to *complete* the inhibition process rather than from a failure to recognize the need to inhibit inappropriate beliefs in the first place. This evidence points strongly to the conclusion that people are, in fact, highly sensitive to logic/belief conflicts and that they strive to inhibit the influence of prior beliefs, albeit often unsuccessfully (see also Luo et al., 2013; Luo, Tang, Zhang, & Stupple, 2014).

In sum, the behavioral and neuroscientific evidence deriving from the belief bias paradigm seems to be at odds with established sequential dual-process assumptions in suggesting that biased reasoners detect conflict just as unbiased reasoners do, sensing that their intuitive, heuristic response is of questionable accuracy (De Neys, 2014). One of the key upshots of this consistent body of evidence indicating that biased reasoners exhibit such belief/logic conflict sensitivities is that belief bias does not arise from conflict-detection failure per se but arises from downstream processing failures to intervene on intuitively cued responses (again see De Neys, 2014; for the generalization of these conclusions to other reasoning paradigms outside of belief bias see, for example, Bonner & Newell, 2010; Denes-Raj & Epstein, 1994; Mevel, Poirel, Rossi, Cassotti, Simon, Houdé, et al., 2015; Stupple, Ball, & Ellis, 2013; Villejoubert, 2009). In the next section we briefly address whether parallel dual-process conceptualizations of belief-bias effects fare any more successfully than sequential dual-process models in accounting for findings relating to people's inherent sensitivity to belief/logic conflicts.

Belief Bias and *Parallel* Dual-Process Models

The *parallel-process* account of belief-bias effects is founded on the assumption that intuitive, heuristic and reflective, analytic processes operate *concurrently* to generate a response, with conflicts arising whenever the two processing streams produce different responses (e.g., Stupple & Ball, 2008; see also Sloman, 1996, 2002). If conflicting responses arise, then conflict resolution will be necessary in order for the reasoner to be able to determine whether to endorse or reject a presented conclusion. This conflict-resolution process will take time, thereby increasing the processing latencies for conflict items relative to no-conflict items in line with what is observed in the response-time data with belief-oriented syllogisms (e.g., Ball et al., 2006; Stupple & Ball, 2008; Stupple et al., 2011; Thompson et al., 2003; Thompson, Morley, & Newstead, 2011).

A major strength of parallel-process models of belief bias compared to sequential-process models is that they embody an assumption that people should have some sense of awareness of the conflict between the belief-based and the logic-based responses that are being cued by intuitive and reflective processes (e.g., see Evans, 2007). This assumption clearly aligns well with the observations that people have reduced confidence when tackling belief/logic conflict problems relative to no-conflict problems (De Neys et al., 2011), with such confidence reduction also manifesting as increased autonomic arousal—as determined by galvanic skin conductance measures (De Neys et al., 2010).

However, several conceptual difficulties for the parallel-process view of belief bias have been noted in the literature, which appear to undermine its viability (e.g., see Ball, 2011; De Neys, 2012). First, the evidence consistently shows that the latency effect with belief/logic conflict items is *asymmetrical*; it is invalid-believable conflict items that are particularly prone to long response times (e.g., see Stupple & Ball, 2008; Stupple et al., 2011), yet the model provides no clear-cut basis for this difference. Ball (2013b), however, has suggested that a parallel-process model could potentially be augmented to explain such response-time asymmetries by embodying similar notions to those associated with the selective processing account in relation to individual differences in counterexample search with different problem types.

Second, the parallel-process model has to assume that there is a "decision system" that reconciles the outputs from the intuitive, heuristic and the reflective, analytic processing streams. What this means in practice is that the decision system will have to wait for the results of both processing streams before producing a final response (for relevant discussions see Evans, 2007; Stupple & Ball, 2008). Without this assumption the intuitive, heuristic system would typically win out because of its processing speed, which would produce exclusively biased performance and would also not produce the bias-detection effects that are observed in people's confidence ratings and autonomic responses. However, a dual-process architecture in which the fast-processing intuitive route always has to wait for the slow-processing route seems cumbersome and inefficient and loses any adaptive advantages that may arise from having a fast processing system in the first place (cf. Evans, 2007).

Third—and for similar reasons of questionable cognitive efficiency—it seems curious to have a processing architecture whereby people analytically compute a normative response that aligns with logic *in addition to* an intuitive response that aligns with prior beliefs only for the intuitive output not only potentially to conflict with the normative decision but also often to win out in terms of the final response that is made. As De Neys (2012) explains, the parallel model essentially throws away the benefits of the intuitive route, particularly because the intuitive and the reflective routes do not always conflict, such that when there is no conflict it would be absolutely fine to rely on the intuitive route. As De Neys (2012, p. 34) puts it: "Engaging in demanding deliberate operations is redundant in this case and would be a waste of scarce cognitive resources. … Hence, what dual-process models need is a way to detect whether deliberate thinking is required without having to engage in deliberate thinking."

In sum, although parallel-process models of belief bias do seem to have some apparent advantages over sequential-process models in terms of explaining the available data deriving from a range of behavioral and neuroimaging measures (e.g., Ball, 2013a; Stupple & Ball, 2008), such models also run into major conceptual difficulties. In particular, it simply seems to be cognitively inefficient and evolutionarily maladaptive to activate simultaneous intuitive and reflective processing streams from the outset for all reasoning tasks only for the resource-demanding and time-consuming reflective process to struggle even to have its voice heard. De Neys's (2012) aforementioned point is very compelling in this respect, which is that what seems to be needed is a means for the reasoning system to determine whether reflective thinking might be beneficial—but without having to engage such reflective thinking in the first place in order to inform this decision. Such a mechanism would be maximally adaptive in terms of the efficient deployment of cognitive resources only when they are most needed. Developing a theoretical account of what such a mechanism might look like in practice is a major conceptual challenge. However, as we show in the next section, De Neys (2012, 2014) has already gone a long way toward formulating a psychologically plausible model of how conflict between normative and nonnormative reasoning responses (e.g., as arises with belief/logic conflict items) can be detected solely through automatically activated intuitive processes.

Belief Bias and *Hybrid* Dual-Process Models

As De Neys (2012) notes, the various conflict-detection studies that have been conducted in recent years—including those focused on belief/logic conflict problems—have established that despite people's failure to provide normatively correct answers on classic reasoning problems, it is demonstrably *not* the case that people simply disregard the normative aspects of problems when making their judgments. Rather, people show a high degree of sensitivity to the fact that their intuitive, heuristic answers conflict with responses that should be made on the basis of logical or probabilistic principles. The important question, of course, is how best to conceive of the nature of the normative knowledge that people appear to possess, which allows, for example, for the detection of belief/logic conflicts. De Neys (2012) lays out a model in which this normative sensitivity is, in fact, entirely *intuitive* in nature, having its basis in implicit and automatically activated knowledge rather than knowledge that is elicited and applied through effortful, explicit reasoning processes.

De Neys's (2012) account of conflict sensitivity in reasoning clearly flies in the face of conventional dual-process theorizing as exemplified in sequential-process and parallel-process models of reasoning phenomena such as belief bias. What De Neys is proposing, albeit controversially, is that people are actually "intuitive logicians" whose gut feelings in relation to the logical components of reasoning problems are cuing some sense of awareness of normative solutions. The key point here is that this activation of logical knowledge is effortless and does not require the deployment of any demanding reflective or analytic processing. One valuable upshot of De Neys' (2012) logical intuition model is that it provides exactly the kind of mechanism that enables conflict detection in reasoning—such as the detection of conflicts between belief and logic—while sidestepping the conceptual difficulties associated with standard dual-process theorizing. Thus, if the intuitive system cues both a logical response and a belief-based heuristic response, then the potential conflict can be detected *without* the prior need to engage the reflective reasoning system.

The idea that belief/logic conflict sensitivity can derive purely from the parallel operation of two different intuitive processing streams is certainly fascinating and unusual, but it is also powerful in terms of its explanatory reach. One of the strengths of this model is the way in which it can readily account for many of the belief-bias findings that we have reviewed above that derive from analyses of response times, confidence ratings, brain-activation patterns, and autonomic arousal data. For example, if the two intuitive systems are consistent in their responses then people will simply select the cued responses; the reasoning process will be relatively fast and will terminate without further reflective thinking being triggered (De Neys, 2012). However, any conflict between the logic-based and belief-based responses that arises from intuitive processing will signal the need for the reasoner to engage in reflective thinking in order to attempt to resolve the conflict. The emergence of belief/logic conflict within the intuitive system would explain the conflict-sensitivity findings that are observed in the belief-bias paradigm, which manifest very clearly in people's confidence ratings as well as electrodermal measures that index autonomic activation. In addition, the reflective processing that is triggered as a result of the conflict sensitivity arising in the intuitive system would engender increased processing times for belief-logic conflict items, as established in numerous belief-bias experiments that focus on time-based measures of responding. De Neys (2012, 2014) is at pains to stress that the fact that reflective processes are invoked in the case of belief/logic conflict items does not imply that such reflective processes will necessarily be recruited and applied successfully to enable normatively correct responses to be generated. As he notes, however, the conflict sensitivity arising in the intuitive system does, nevertheless, provide a clear "switch" mechanism that can serve to determine whether reflective engagement is required, without the need to posit the inefficient, permanent activation of the reflective system from the outset, as espoused within the standard parallel-process framework (e.g., Sloman, 1996, 2002; Stupple & Ball, 2008).

De Neys's (2012) model is essentially a "hybrid" dual-process account in that it embodies both intuitive and reflective processes, with the latter arising subsequent to an intuitive processing stage that involves the parallel operation of logical intuitions and heuristics. De Neys (2012, 2014) presents further proposals regarding the way in which the logical intuition model aligns with evidence from recent developmental studies demonstrating that both young infants and older children have an intuitive grasp of many normative principles associated with effective probabilistic and logical thinking (e.g., Brainerd, & Reyna, 2001; Kushnir, Xu, & Wellman, 2010; Morris, 2000; Perret, Paour, & Blaye, 2003; Téglás, Girotto, Gonzalez, & Bonatti, 2007; Xu & Garcia, 2008). We do not review this developmental literature here because of space limitations, but we nevertheless concede that the evidence points strongly to the conclusion that even young children have a high degree of sensitivity to key normative principles.

Another important body of literature that also appears to support De Neys's (2012) logical intuition model is that concerned with the role of meta-cognition in reasoning (see Ackerman & Thompson, 2014, for a recent review of this *meta-reasoning* literature, see Thompson, 2009, for seminal research in this area). Studies of meta-reasoning are centrally concerned with the mechanisms that *regulate* reasoning by setting goals, determining how much effort to apply, deciding among strategies, monitoring ongoing progress (e.g., assessing whether a putative solution is correct and deciding whether to initiate further processing if it is viewed as inadequate), and terminating activity. This meta-reasoning framework assumes that people do not have direct access to their underlying reasoning processes but instead base their regulation decisions on their experience with similar problems as well as on surface-level properties of the problem that is being attempted. These relatively superficial properties relate primarily to cues to problem difficulty such as the perceived coherence among problem components (e.g., Topolinski, 2014; Topolinski & Reber, 2010), the ease with which the problem can be mentally represented (e.g., Quayle & Ball, 2000; Stupple et al., 2013), and the fluency with which a solution comes to mind (e.g., Ackerman & Zalmanov, 2012; Alter & Oppenheimer, 2009).

The latter property of problems (i.e., answer fluency) is a factor that Thompson and colleagues (e.g., Thompson, 2009; Thompson, Prowse Turner, & Pennycook, 2011; Thompson, Prowse Turner, Pennycook, Ball, Brack, Ophir, et al., 2013) have suggested is especially important in mediating a judgment that they term a *feeling of rightness*. Such feeling-of-rightness judgments in turn act as meta-cognitive triggers, *terminating* processing in situations where an intuitive process has produced a rapid, intuitive answer that is attributed to be correct, or else *switching* processing from intuitive to reflective modes in situations where the initial, intuitive answer is associated with a low feeling of rightness and is therefore considered to be potentially incorrect (see also Ackerman, 2014). In this way it is possible to see very clear theoretical links between the *intuitive-to-reflective* switching role played by feeling-of-rightness judgments in the meta-reasoning framework and De Neys's (2012, 2014) claims that there is a switch mechanism that initiates reflective processing in cases where there is a perceived conflict between logical intuitions and heuristics. Thompson and colleagues have also suggested that the switch decision might be affective in nature (e.g., Thompson, 2009; Thompson & Morsanyi, 2012), which again ties in neatly with autonomic-arousal findings in conflict-detection studies (e.g., De Neys et al., 2010).

To summarize, recent evidence gives clear grounds for viewing meta-reasoning judgments as playing an important role in the monitoring and regulation of reasoning, such that initial confidence or "rightness" assessments may well act as a switch to trigger the application of downstream reflective thinking in relation to a reasoning task that is perceived to be challenging. These meta-reasoning concepts provide a useful bridge to De Neys's (2012) logical intuition model that readily applies to phenomena such as belief bias, in that conflicting intuitive responses will likewise give rise to low confidence in the rightness of the more highly activated belief-based solution (see De Neys, 2014), thereby switching the system to engage in reflective processing aimed at potential conflict resolution. This reflective processing effort takes time, even if the reasoner still ends up defaulting to a dominant belief-driven, conclusion-endorsement decision, as is typical within the belief-bias paradigm.

Conclusions

In this chapter we have systematically examined the viability of dual-process models of people's reasoning when tackling belief-oriented syllogistic arguments. Such reasoning problems are known to produce a number of established conclusion-endorsement findings, including main effects of belief and logic as well as a reliable logic-by-belief interaction, whereby belief bias is more pronounced on invalid problems than on valid problems. Beyond explaining conclusion-endorsement rates, however, theoretical developments in this area have both benefited from and been challenged by a range of recent findings that derive from novel measures, including response times, confidence ratings, brain-activation signatures, and electrodermal arousal patterns. Taken as a whole, the extant data have, in general, meant that dual-process theories of belief-bias effects that were established up until a decade ago no longer appear to be sustainable. Such theories include those that are based around a sequential, staged view of intuitive-then-reflective processing as well as those that are based around a parallel view, whereby intuitive and reflective processes are deployed concurrently. The sequential-process view struggles, in particular, to explain the increased processing times and decreased confidence ratings associated with belief/logic conflict problems versus no-conflict problems. The parallel-process view, on the other hand, seems to run into conceptual difficulties associated with considerations relating to cognitive efficiency and psychological plausibility.

The explanatory and theoretical problems that confront established models of belief bias have paved the way for the emergence of fresh and original conceptual models that seem to be better able to account for a broad range of empirical findings that have been acquired through the deployment of a variety of convergent measures. We have argued that the *logical intuition* model of De Neys (2012, 2014) represents a particularly exciting way to conceptualize the manner in which belief-bias effects emerge in reasoning. According to this hybrid, dual-process model, outputs that emerge from parallel streams of automatic, intuitive processing, where one stream is sensitive to logic and the other is based on heuristics, can give rise to either congruent intuitions or conflicting intuitions. In the case of conflicting intuitions a feeling of uncertainty will arise that will also be correlated with a state of autonomic arousal. Such events will, it is suggested, act as a switch that serves to trigger a further, downstream stage of explicit, reflective reasoning aimed at resolving the current logic/belief conflict. This reflective processing takes time and effort, thereby giving rise to the longer processing times that are observed with conflict problems relative to no-conflict problems. Whether reasoners end up resolving the conflict in favor of logic may well be dependent on a variety of factors, including the presence of appropriate mindware and dispositions to engage in reflective thinking (Stanovich, 2009), the existence of sufficient available working memory capacity (De Neys, 2006; Evans & Stanovich, 2013; Quayle & Ball, 2000), and the successful execution of belief inhibition (De Neys & Franssens, 2009; Franssens & De Neys, 2009; Houdé et al., 2000; Luo et al., 2014). As De Neys (2014) acknowledges, we are some way off from understanding the precise nature of the reflective processing failures that may mean that a biased response is *still* produced despite a logic/belief conflict having been detected at the intuitive processing stage.

De Neys's (2012, 2014) logical intuition model seems to take us a long way toward an integrative explanation of many phenomena associated with the emergence of belief-bias effects in human reasoning. As De Neys notes, however, there are important boundary conditions that need to be borne in mind when considering the scope of this novel—and admittedly rather radical—theoretical account. First and foremost, De Neys (2012, 2014) is careful to point out that his claims specifically apply to the classic tasks that have been the basis of many decades of theorizing in the reasoning, judgment, and decision-making literatures. In this chapter we have primarily focused on one such classic task, that is, belief-oriented syllogisms in which logic and belief are either set in opposition to one another (i.e., in the case of conflict items) or are congruent with one another (i.e., in the case of no-conflict problems).

Much of the research of De Neys and colleagues in relation to belief-bias effects has been based around belief-oriented syllogisms whose underpinning logic is actually fairly *elementary* in nature (see, for example, the syllogisms depicted in table 7.1). Such problems are ones that people show a strong likelihood to respond to normatively when they are presented in an *abstract* form (e.g., see Johnson-Laird & Byrne, 1991). As De Neys (2012) concedes, the elementary nature of the logical principles involved in these problems represents an intrinsic boundary condition for his logical intuition claims because it is, of course, very much to be expected that the logical intuitions will arise in situations where the logical solution is simple and likely to be automatically activated. Whether such logical intuitions arise in the case of more complex problems remains an empirical question; indeed, it is perfectly reasonable to imagine (as De Neys concedes) that the logical intuition model may not generalize to more complex reasoning tasks. That said, some of the evidence that can be taken as supportive of De Neys's model actually derives from belief-bias studies that involve complex syllogisms.

For example, the research by Stupple and colleagues (e.g., Stupple & Ball, 2008; Stupple et al., 2011) has, to date, always been based on the use of "three-model" syllogisms (Johnson-Laird & Byrne, 1991), which participants find it difficult to provide normative responses to. Even for these "difficult" syllogisms, however, there are arguments in the literature that they involve intrinsic cues to their logical status that may be identified by the intuitive processing system (in particular see Hardman & Payne, 1995 and Quayle & Ball, 2000).

To conclude, the logical-intuitions model of belief bias appears to stand as an important recent proposal that is supported by a good deal of recent data. Apart from its impressive capacity to account for many findings across a variety of different measures, a further strength of the model is that it enables the generation of testable predictions; that is, the theory appears to be falsifiable, which, as we noted in our introduction, counters at least one of the standard criticisms that tend to be lodged against dual-process accounts of reasoning. Of course, the scope of the account still needs to be determined empirically. In our view, however, the proponents of this account (e.g., De Neys, 2012, 2014; De Neys & Bonnefon, 2013) should be given due credit for remaining appropriately cautious regarding its generalizability and for being quick to acknowledge likely boundary conditions that might limit its breadth of application. We are also mindful of the calls from Singmann, Klauer, and Kellen (2014) for the application of the most rigorous scientific approach possible when examining such "extraordinary" claims as the existence of an intuitive logic (see also Klauer & Singmann, 2013). We anticipate that such rigorous testing of the logical intuition model will be forthcoming over the next few years, and we very much look forward to seeing where the latest evidence will take us in terms of advancing our theoretical understanding of belief-bias phenomena.

References

Ackerman, R. (2014). The Diminishing Criterion Model for metacognitive regulation of time investment. *Journal of Experimental Psychology. General, 143,* 1349–1368.

Ackerman, R., & Thompson, V. A. (2014). Meta-reasoning: What can we learn from meta-memory. In A. Feeney & V. A. Thompson (Eds.), *Reasoning as memory* (pp. 164–182). Hove, UK: Psychology Press.

Ackerman, R., & Zalmanov, H. (2012). The persistence of the fluency–confidence association in problem solving. *Psychonomic Bulletin & Review, 19,* 1189–1192.

Alter, A. L., & Oppenheimer, D. M. (2009). Uniting the tribes of fluency to form a metacognitive nation. *Personality and Social Psychology Review, 13,* 219–235.

Aron, A. R., Robbins, T. W., & Poldrack, R. A. (2004). Inhibition and the right inferior frontal cortex. *Trends in Cognitive Sciences, 8,* 170–177.

Ball, L. J. (2011). The dynamics of reasoning: Chronometric analysis and dual-process theories. In K. I. Manktelow, D. E. Over, & S. Elqayam (Eds.), *The science of reason: A Festschrift for Jonathan St. B. T. Evans* (pp. 283–307). Hove, UK: Psychology Press.

Ball, L. J. (2013a). Eye-tracking and reasoning: What your eyes tell about your inferences. In W. De Neys & M. Osman (Eds.), *New approaches in reasoning research* (pp. 51–69). Hove, UK: Psychology Press.

Ball, L. J. (2013b). Microgenetic evidence for the beneficial effects of feedback and practice on belief bias. *Journal of Cognitive Psychology, 25,* 183–191.

Ball, L. J., Phillips, P., Wade, C. N., & Quayle, J. D. (2006). Effects of belief and logic on syllogistic reasoning: Eye-movement evidence for selective processing models. *Experimental Psychology, 53,* 77–86.

Bonner, C., & Newell, B. R. (2010). In conflict with ourselves? An investigation of heuristic and analytic processes in decision making. *Memory & Cognition, 38,* 186–196.

Brainerd, C. J., & Reyna, V. F. (2001). Fuzzy-trace theory: Dual processes in memory, reasoning, and cognitive neuroscience. *Advances in Child Development and Behavior, 28,* 49–100.

De Martino, B., Kumaran, D., Seymour, B., & Dolan, R. J. (2006). Frames, biases, and rational decision-making in the human brain. *Science, 313,* 684–687.

Denes-Raj, V., & Epstein, S. (1994). Conflict between intuitive and rational processing: When people behave against their better judgment. *Journal of Personality and Social Psychology, 66,* 819–829.

De Neys, W. (2006). Dual processing in reasoning: Two systems but one reasoner. *Psychological Science, 17,* 428–433.

De Neys, W. (2012). Bias and conflict: A case for logical intuitions. *Perspectives on Psychological Science, 7,* 28–38.

De Neys, W. (2014). Conflict detection, dual processes, and logical intuitions: Some clarifications. *Thinking & Reasoning, 20,* 169–187.

De Neys, W., & Bonnefon, J. F. (2013). The "whys" and "whens" of individual differences in thinking biases. *Trends in Cognitive Sciences, 17,* 172–178.

De Neys, W., Cromheeke, S., & Osman, M. (2011). Biased but in doubt: Conflict and decision confidence. *PLoS One, 6,* e15954.

De Neys, W., & Franssens, S. (2009). Belief inhibition during thinking: Not always winning but at least taking part. *Cognition, 113,* 45–61.

De Neys, W., & Glumicic, T. (2008). Conflict monitoring in dual process theories of reasoning. *Cognition, 106,* 1248–1299.

De Neys, W., Moyens, E., & Vansteenwegen, D. (2010). Feeling we're biased: Autonomic arousal and reasoning conflict. *Cognitive, Affective & Behavioral Neuroscience, 10,* 208–216.

De Neys, W., Vartanian, O., & Goel, V. (2008). Smarter than we think: When our brains detect that we are biased. *Psychological Science, 19*, 483–489.

Evans, J. St. B. T. (1977). Toward a statistical theory of reasoning. *Quarterly Journal of Experimental Psychology, 29*, 297–306.

Evans, J. St. B. T. (1989). *Bias in human reasoning: Causes and consequences*. Hove: Lawrence Erlbaum Associates.

Evans, J. S. B. T. (2000). Thinking and believing. In J. Garcìa-Madruga, N. Carriedo, & M. J. González-Labra (Eds.), *Mental models in reasoning* (pp. 41–56). Madrid: UNED.

Evans, J. S. B. T. (2007). On the resolution of conflict in dual process theories of reasoning. *Thinking & Reasoning, 13*, 378–395.

Evans, J. S. B. T. (2008). Dual-processing accounts of reasoning, judgment and social cognition. *Annual Review of Psychology, 59*, 255–278.

Evans, J. S. B. T. (2009). How many dual-process theories do we need: One, two or many? In J. S. B. T. Evans & K. Frankish (Eds.), *In two minds: Dual processes and beyond* (pp. 31–54). Oxford: Oxford University Press.

Evans, J. S. B. T. (2010). *Thinking twice: Two minds in one brain*. Oxford, UK: Oxford University Press.

Evans, J. S. B. T., Barston, J. L., & Pollard, P. (1983). On the conflict between logic and belief in syllogistic reasoning. *Memory & Cognition, 11*, 295–306.

Evans, J. S. B. T., Handley, S. J., & Harper, C. (2001). Necessity, possibility and belief: A study of syllogistic reasoning. *Quarterly Journal of Experimental Psychology, 54*, 935–958.

Evans, J. S. B. T., & Stanovich, K. E. (2013a). Dual-process theories of higher cognition: Advancing the debate. *Perspectives on Psychological Science, 8*, 223–241.

Evans, J. S. B. T., & Stanovich, K. E. (2013b). Theory and metatheory in the study of dual processing: Reply to comments. *Perspectives on Psychological Science, 8*, 263–271.

Evans, J. S. B. T., & Wason, P. C. (1976). Rationalization in a reasoning task. *British Journal of Psychology, 63*, 205–212.

Franssens, S., & De Neys, W. (2009). The effortless nature of conflict detection during thinking. *Thinking & Reasoning, 15*, 105–128.

Garnham, A., & Oakhill, J. V. (2005). Accounting for belief bias in a mental model framework: Comment on Klauer, Musch, and Naumer (2000). *Psychological Review, 112*, 509–517.

Gilhooly, K. J., Ball, L. J., & Macchi, L. (2015). Insight and creative thinking processes: Routine *and* special. *Thinking & Reasoning, 21*, 1–4.

Goel, V., & Dolan, R. J. (2003). Explaining modulation of reasoning by belief. *Cognition, 87*, B11–B22.

Hardman, D. K., & Payne, S. J. (1995). Problem difficulty and response format in syllogistic reasoning. *Quarterly Journal of Experimental Psychology, 48A,* 945–975.

Houdé, O., Zago, L., Mellet, E., Moutier, S., Pineau, A., Mazoyer, B., et al. (2000). Shifting from the perceptual brain to the logical brain: The neural impact of cognitive inhibition training. *Journal of Cognitive Neuroscience, 12,* 721–728.

Johnson-Laird, P. N., & Byrne, R. M. J. (1991). *Deduction.* Hove: Lawrence Erlbaum Associates.

Kahneman, D. (2011). *Thinking, fast and slow.* New York: Farrar, Straus & Giroux.

Klauer, K. C., Musch, J., & Naumer, B. (2000). On belief bias in syllogistic reasoning. *Psychological Review, 107,* 852–884.

Klauer, K. C., & Singmann, H. (2013). Does logic feel good? Testing for intuitive detection of logicality in syllogistic reasoning. *Journal of Experimental Psychology: Learning, Memory, and Cognition, 39,* 1265–1273.

Kushnir, T., Xu, F., & Wellman, H. M. (2010). Young children use statistical sampling to infer the preferences of other people. *Psychological Science, 21,* 1134–1140.

Luo, J., Liu, X., Stupple, E. J., Zhang, E., Xiao, X., Jia, L., et al. (2013). Cognitive control in belief-laden reasoning during conclusion processing: An ERP study. *International Journal of Psychology, 48,* 224–231.

Luo, J., Tang, X., Zhang, E., & Stupple, E. J. (2014). The neural correlates of belief-bias inhibition: The impact of logic training. *Biological Psychology, 103,* 276–282.

Mevel, K., Poirel, N., Rossi, S., Cassotti, M., Simon, G., Houdé, O., et al. (2015). Bias detection: Response confidence evidence for conflict sensitivity in the ratio bias task. *Journal of Cognitive Psychology, 27,* 227–237.

Morley, N. J., Evans, J. S. B. T., & Handley, S. J. (2004). Belief bias and figural bias in syllogistic reasoning. *Quarterly Journal of Experimental Psychology, 57A,* 666–692.

Morris, A. K. (2000). Development of logical reasoning: Children's ability to verbally explain the nature of the distinction between logical and nonlogical forms of argument. *Developmental Psychology, 36,* 741–758.

Morsanyi, K., & Handley, S. J. (2012). Logic feels so good—I like it! Evidence for intuitive detection of logicality in syllogistic reasoning. *Journal of Experimental Psychology: Learning, Memory, and Cognition, 38,* 596–616.

Oakhill, J., & Johnson-Laird, P. N. (1985). The effect of belief on the spontaneous production of syllogistic conclusions. *Quarterly Journal of Experimental Psychology, 37A,* 553–570.

Oakhill, J., Johnson-Laird, P. N., & Garnham, A. (1989). Believability and syllogistic reasoning. *Cognition, 31,* 117–140.

Pennycook, G., Fugelsang, J. A., & Koehler, D. J. (2012). Are we good at detecting conflict during reasoning? *Cognition, 124,* 101–106.

Perret, P., Paour, J.-L., & Blaye, A. (2003). Respective contributions of inhibition and knowledge levels in class inclusion development: A negative priming study. *Developmental Science, 6*, 283–288.

Prado, J., & Noveck, I. A. (2007). Overcoming perceptual features in logical reasoning: A parametric fMRI study. *Journal of Cognitive Neuroscience, 19*, 642–657.

Quayle, J. D., & Ball, L. J. (2000). Working memory, metacognitive uncertainty and belief bias in syllogistic reasoning. *Quarterly Journal of Experimental Psychology, 53A*, 1202–1223.

Roberts, M. J., & Sykes, E. D. A. (2003). Belief bias and relational reasoning. *Quarterly Journal of Experimental Psychology, 56A*, 131–154.

Sá, W. C., West, R. F., & Stanovich, K. E. (1999). The domain specificity and generality of belief bias: Searching for a generalizable critical thinking skill. *Journal of Educational Psychology, 91*, 497–510.

Sanfey, A. G., Rilling, J. K., Aronson, J. A., Nystrom, L. E., & Cohen, J. D. (2003). The neural basis of economic decision making in the ultimatum game. *Science, 300*, 1755–1758.

Singmann, H., Klauer, K. C., & Kellen, D. (2014). Intuitive logic revisited: New data and a Bayesian mixed model meta-analysis. *PLoS One, 9*, e94223.

Sloman, S. A. (1996). The empirical case for two systems of reasoning. *Psychological Bulletin, 119*, 3–22.

Sloman, S. A. (2002). Two systems of reasoning. In T. Gilovich, D. Griffin, & D. Kahneman (Eds.), *Heuristics and biases: The psychology of intuitive judgment* (pp. 379–398). Cambridge, UK: Cambridge University Press.

Stanovich, K. E. (2009). Distinguishing the reflective, algorithmic and autonomous minds: Is it time for a tri-process theory? In J. S. B. T. Evans & K. Frankish (Eds.), *In two minds: Dual processes and beyond* (pp. 55–88). Oxford: Oxford University Press.

Stupple, E. J. N., & Ball, L. J. (2008). Belief–logic conflict resolution in syllogistic reasoning: Inspection-time evidence for a parallel process model. *Thinking & Reasoning, 14*, 168–189.

Stupple, E. J. N., & Ball, L. J. (2014). The intersection between Descriptivism and Meliorism in reasoning research: Further proposals in support of "Soft Normativism." *Frontiers in Psychology, 5* (Article 1269), 1–13.

Stupple, E. J., Ball, L. J., & Ellis, D. (2013). Matching bias in syllogistic reasoning: Evidence for a dual-process account from response times and confidence ratings. *Thinking & Reasoning, 19*, 54–77.

Stupple, E. J. N., Ball, L. J., Evans, J. S. B. T., & Kamal-Smith, E. (2011). When logic and belief collide: Individual differences in reasoning times support a selective processing model. *Journal of Cognitive Psychology, 23*, 931–941.

Téglás, E., Girotto, V., Gonzalez, M., & Bonatti, L. L. (2007). Intuitions of probabilities shape expectations about the future at 12 months and beyond. *Proceedings of the National Academy of Sciences of the United States of America, 104*, 19156–19159.

Thompson, V. A. (2009). Dual-process theories: A metacognitive perspective. In J. Evans & K. Frankish (Eds.), *In two minds: Dual processes and beyond* (pp. 171–195). Oxford: Oxford University Press.

Thompson, V. A., Morley, N. J., & Newstead, S. E. (2011). Methodological and theoretical issues in belief-bias: Implications for dual process theories. In K. I. Manktelow, D. E. Over, & S. Elqayam (Eds.), *The science of reason: A Festschrift for Jonathan St. B. T Evans* (pp. 309–338). Hove, UK: Psychology Press.

Thompson, V. A., & Morsanyi, K. (2012). Analytic thinking: Do you feel like it? *Mind & Society*, *11*, 93–105.

Thompson, V. A., Prowse Turner, J., & Pennycook, G. (2011). Intuition, reason, and metacognition. *Cognitive Psychology*, *63*, 107–140.

Thompson, V. A., Prowse Turner, J., Pennycook, G., Ball, L. J., Brack, H., Ophir, Y., et al. (2013). The role of answer fluency and perceptual fluency as metacognitive cues for initiating analytic thinking. *Cognition*, *128*, 237–251.

Thompson, V. A., Striemer, C. L., Reikoff, R., Gunter, R. W., & Campbell, J. D. (2003). Syllogistic reasoning time: Disconfirmation disconfirmed. *Psychonomic Bulletin & Review*, *10*, 184–189.

Topolinski, S. (2014). Intuition: Introducing affect into cognition. In A. Feeney & V. A. Thompson (Eds.), *Reasoning as memory* (pp. 146–163). Hove, UK: Psychology Press.

Topolinski, S., & Reber, R. (2010). Gaining insight into the "aha" experience. *Current Directions in Psychological Science*, *19*, 402–405.

Villejoubert, G. (2009). Are representativeness judgments automatic and rapid? The effect of time pressure on the conjunction fallacy. In N. Taatgen, & H. van Rijn (Eds.), *Proceedings of the Thirty-First Annual Conference of the Cognitive Science Society* (pp. 2980–2985). Austin, TX: Cognitive Science Society.

Wason, P. C., & Evans, J. S. B. T. (1975). Dual processes in reasoning? *Cognition*, *3*, 141–154.

Wilkinson, M. R., & Ball, L. J. (2013). Dual processes in mental state understanding: Is theorizing synonymous with intuitive thinking, and is simulation synonymous with reflective thinking? In M. Knauff, M. Pauen, N. Sebanz, & I. Wachsmuth (Eds.), *Proceedings of the Thirty-Fifth Annual Conference of the Cognitive Science Society* (pp. 3771–3776). Austin, TX: Cognitive Science Society.

Xu, F., & Garcia, V. (2008). Intuitive statistics by 8-month-old infants. *Proceedings of the National Academy of Sciences of the United States of America*, *105*, 5012–5015.

8   The Basic Assumptions of Intuitive Belief: Laws, Determinism, and Free Will

Steven A. Sloman and Aron K. Barbey

Our goal in this chapter is to address the problems of determinism and free will. But do not give up on us quite yet. We do not purport to have anything to say about whether the world is deterministic or whether there are agents in it that have free will. Our goals are much more modest: to make some suggestions about what the human cognitive system assumes about determinism and free will. We want to know not whether determinism is true and free will exists but whether people believe, at some level, in determinism and free will. Our task is much easier than that of a metaphysician because logical consistency is not axiomatic in psychology. Even if free will and determinism are in fact incompatible, people could still believe in both.

We also address a closely related question: Do people believe that the world is law governed? This is not the same as the question of determinism because laws are not necessarily deterministic: laws can be probabilistic (consider the laws of thermodynamics). It is also not the same as the question of free will because one might believe that free will is an exception to the laws that govern the world, or one might believe that free will is itself governed by those laws. Alternatively, one might not believe in free will regardless of one's position on laws.

We think of these three questions as determining a set of basic assumptions made by the cognitive system. To make inferences the cognitive system needs some direction, some driving force or principles of operation. We call this a *logic*, and all logics make assumptions. What assumptions are made by whatever logic drives human inference? Our three questions are foundational. Any logic must take a position on them. Of course, the cognitive system could take different positions at different times. Indeed, it does. We have conversations in which we imagine the world is deterministic or that it has free will, and we have conversations in which we imagine neither. But it seems plausible to us that there is a kind of basic, intuitive system for making inferences that is more consistent in the foundational assumptions it makes. It is that intuitive system—one whose operation we probably do not have conscious access to—that we focus on here.

The Space of Possibilities

We have three binary questions, so there are $2 \times 2 \times 2 = 8$ possible answers. People might believe the world is not law governed, not deterministic, and that there is no free will. Or they might believe that the world is law governed, not deterministic, and that there is no free will, or any other possible combination up to the possibility that they do believe in all three properties: that the world is law governed, that it is deterministic, and that there is free will.

These three questions are conceptually related in the sense that certain combinations of answers seem to make more sense than others. For instance, one could argue that a determinist should be more skeptical about free will than a nondeterminist. But the questions are independent in the sense that all possibilities are up for grabs; one could take any position on any question regardless of the positions one takes on the others. Indeed, although we do not review it, we suspect that every combination of positions on the corresponding metaphysical questions is represented in philosophy (e.g., Hoefer, 2010; Salmon, 1998).

Our Hypothesis

Before reviewing the evidence, we reveal our own position. On the first question, we believe that the cognitive system assumes that the world is law governed; in particular, the cognitive system assumes that the world is governed by causal laws, by mechanisms that take causes as inputs and that deliver effects as outputs.

On the second question, we believe that the cognitive system assumes that the world is deterministic. This is likely our most tendentious claim. Our claim, to be specific, is that the cognitive system believes that, whatever forces govern the world, they do so in such a way that a complete description of the world at time 1 entails a complete description of the world at time 2. There is no spontaneous noise that makes the world unpredictable if you have complete knowledge of it. I may accept quantum uncertainty, I may even accept as a matter of principle that there is randomness at the subatomic level. Our claim is that the intuitive cognitive system does not abide by this principle; it does not allow for quantum randomness (primarily because it does not understand it).

There is, however, a huge caveat to our claim: belief in determinism is conditioned on having a complete description of the world. Yet we never have such a description, and we know it. So we allow for probability despite our belief that there are deterministic laws, and we justify our use of probability in terms of uncertainty. We know we are ignorant about some things and that ignorance means that we cannot be perfect predictors. For example, we may believe that smoking causes cancer deterministically without believing that every person who smokes will get cancer. After all, a smoker could get hit by a car at a young age and never have the opportunity to develop cancer. Such an event would be unknowable, not because it is inherently unpredictable but because the world is too complicated. Given our limited cognitive resources and limited exposure to events, we cannot have enough knowledge to allow us to predict all the complex interactions that occur in the real world. So even though we believe that all events are determined by their causes, we also know that all events are impossible to predict because we are necessarily ignorant. Probability arises from ignorance of initial conditions or relevant variables or governing laws.

On the third question, we believe that people believe in free will. The cognitive system assumes that people make choices that they are responsible for, and we take that assumption to heart, even when making inferences about ourselves.

The Evidence

People Believe the World Is Law Governed

In principle, the mind need not assume the world is law governed. It could assume, for instance, that all it need do is represent the structure in experience, whatever that structure is. Indeed, this is a common assumption in psychology and even in cognitive science. Historically, psychologists have tended to think the mind imposes as little structure as possible, and so behaviorists such as Pavlov, Watson, and Skinner believed that animals of all kinds would learn whatever associations they were taught. Watson was famously bold about it:

Give me a dozen healthy infants, well-formed, and my own specified world to bring them up in, and I'll guarantee to take any one at random and train him to become any type of specialist I might select—doctor, lawyer, artist, merchant-chief, and, yes, even beggar-man and thief, regardless of his talents, penchants, tendencies, abilities, vocations, and race of his ancestors. (Watson, 1930, p. 82)

Learning was just a matter of internalizing experience according to one or another principle of learning and reward. It was not until the seminal work of Garcia (Hankins et al., 1976), Premack (Premack & Putney, 1962), and many others that the vacuity of this idea was fully realized. All animals, including people, are predisposed to learn certain kinds of relations and not others.

Garcia's work suggests that the predisposition is to learn relations that are consistent with the mechanisms that operate in their environment. So a rat will be predisposed to learn that illness is associated with a food eaten and not with a light but that a shock is associated with a light but not with food. In other words, the rat is more likely to learn the relations that make causal sense.

The tendency to rely on experience as the basis for mental representation did not die along with behaviorism during the cognitive revolution of the 1970s. Connectionism allowed for more sophisticated learning, representational, and inferential processes than behaviorism, but it retained the view that what is learned is what is experienced. In connectionist models experience is represented in terms of correlations and higher-order correlations among elements of experience. Modern neural net models that are based largely on connectionist principles retain this property. They usually have structure, but the structure does not reflect any laws that might or might not govern the world; the structure reflects neural structure, the structure of the brain.

Another type of model that is current is the probability model. Probabilistic models come in more than one variety. Some (e.g., Anderson, 1993) are intended to be direct representations of the statistics of the world. Such models explicitly reject the notion of representing laws; they only represent outcomes and claim that the cognitive system is tuned to the relative frequency of states in the world.

Other models (e.g., Tenenbaum et al., 2006) try to have it both ways. They represent statistics, but in a Bayesian way. That is, their representations are biased in favor of prior probabilities. Whether or not such models assume that the world is law governed depends on where their priors come from. If their priors are "flat" (uniform distributions sometimes called ignorance or noninformative priors), or if they are chosen because they are mathematically tractable (e.g., conjugate priors; see Griffiths et al., 2010), then they are not assuming a law-governed world. At least, they are not assuming the world is governed by any kind of natural law but rather by a law intended to ease calculation.

But if the prior is chosen to reflect causal structure (Pearl, 2000; Spirtes, Glymour, & Scheines, 1993/2000), then it is making an assumption about natural law, namely, that the world is governed by mechanisms that obey causal logic. Causal Bayes nets are Bayesian probabilistic models that represent the world using graphs intended to reflect causal structure. The structure is specifically causal in that it supports intervention: intervening to impose a value on a variable in the model changes the structure of the model so that new inferences are only made about effects of the intervened-on variable, not its normal causes (Sloman, 2005, offers a painless introduction). Causal Bayes nets actually entail a causal logic. For instance, they imply that if A causes B and B causes C, then A causes C (we do have to make additional assumptions such as the absence of a second causal path in which A inhibits C; Halpern & Hitchcock, 2013).

It is not our intention to argue that human reasoning relies on causal Bayes nets; in fact we do not believe that it does (see Sloman & Lagnado, 2015). But we do believe that some form of causal logic governs how people reason, that people are natural causal reasoners (Waldmann & Holyoak, 1992).

We reason very comfortably and naturally about causal structure and not about other kinds of logical structure. When engaging in everyday reasoning, we are bad at propositional reasoning, syllogistic reasoning, and probabilistic reasoning but good at qualitative causal reasoning (for supporting arguments, see Chater & Oaksford, 2013; Holyoak & Cheng, 2011).

Some of the evidence in favor of our position comes from demonstrations that causal knowledge trumps probabilistic knowledge in reasoning. Bes et al. (2012) report an example of such trumping. They tested the hypothesis that conditional probability judgments can be influenced by causal structure even when the statistical relations among variables are held constant. They informed participants in a series of experiments that a specific set of correlations held among three variables A, B, and C (for instance, a person's quality of sleep, level of magnesium, and muscle tone). They also offered an explanation for these correlations, a different explanation offered to each of three groups. Each explanation took the form of a causal model. One group was told the variables were correlated by virtue of a common cause ("An increase in the level of magnesium leads to an increase in the quality of sleep. An increase in the level of magnesium also leads to an increase in muscle tone"). Another group was told the correlations were due to a causal chain ("An increase in the level of magnesium leads to an increase in the quality of sleep, which in turn leads to an increase in muscle tone."), and a third to a diagnostic chain ("An increase in muscle tone leads to an increase in the quality of sleep, which in turn leads to an increase in the level of magnesium."). They were all asked the same question: they were told that Mary, 35 years old, has good quality of sleep and were asked "According to you, what is the probability that Mary has good muscle tone?" In other words, they were asked to judge a conditional probability, $P(C|A)$, whose value had not been given but could easily be calculated from the data that they were given.

The results showed that conditional probability judgments were highest when the explanation was a causal chain, next when the explanation was a diagnostic chain, and lowest after the common-cause explanation. Bes et al. also showed that the data could not be accounted for by a large class of Bayesian learning models, the most natural class to use. Specifically, they considered Bayesian models that assume that conditional probability judgments are informed not just by data but also by prior beliefs about causal structure. They also constrained the parameters of the models to a single value in the causal direction and a single value in the diagnostic direction (after all, any model with enough free parameters could fit the data perfectly after the fact). Such models predict that conditional probability predictions on the common-cause model should be between those of the chain and diagnostic models. The reason they make this prediction is that the causal chain involves two causal inferences, the diagnostic chain entails two diagnostic inferences, whereas the common cause involves one causal and one diagnostic inference.

The fact that the common-cause model engendered the lowest judgments suggests that people were not combining priors with new data in the way that Bayesianism prescribes. Instead, people were likely trying to understand the relation between the variables they were judging in qualitative terms. That relation was the most elaborate in the common-cause case because reasoning involved both a diagnostic link from C to B and a causal link from B to A. The other causal models required tracing either two causal links or two diagnostic links. In other words the conditional probability judgments were influenced by the difficulty of constructing an explanation from causal structure, not by Bayesian reasoning.

Another demonstration suggesting that causal knowledge is basic to human judgment was reported by Park and Sloman (2013). They investigated what kind of information people use to make predictions. Causal Bayes nets theory implies that people should follow a particular structural constraint called the Markov property or "the screening-off rule" when reasoning. We describe the property informally. It states that if the causes of an effect are held constant, then the effect should be treated as independent of other variables related to the effect only via the cause. For example, consider a common-cause structure in which B causes A and B also causes C. If the value of B is known, then A tells us nothing we do not already know about C. So A and C should be independent conditional on B. But a variety of previous work shows that people frequently violate this principle (Chaigneau et al., 2004; Lagnado & Sloman, 2004; Rehder & Burnett, 2005; Waldmann & Hagmayer, 2005; Walsh & Sloman, 2008).

Park and Sloman (2014) showed that this violation occurs only when B causes both A and C in the same way, when the same causal mechanism is responsible. For instance, consider the following pair of causal relations:

Smoking more than three packs of cigarettes a day often causes impairment of lung function.

Smoking more than three packs of cigarettes a day often causes damage to blood vessels.

In both cases, the causal relation is supported by similar mechanisms, namely the tar and nicotine in cigarettes does damage to the body. But now consider a different pair of causal relations:

Smoking more than three packs of cigarettes a day often causes impairment of lung function.

Smoking more than three packs of cigarettes a day often causes a financial burden on the family budget.

In this case the causal relations are supported by quite different mechanisms. The financial impact has to do with spending money, not damage to the body. Park and Sloman's main finding is that violations of screening off only occurred for the first type of example, in which mechanisms were the same, not when they were different. Specifically, when people were asked for the probability that an individual would have damage to blood vessels given that they smoked three packs a day and their lung function was normal, they gave a higher estimate than when asked for just the probability that an individual would have damage to blood vessels given that he smoked three packs a day. In contrast, when asked for the probability that an individual would have a financial burden given that he smoked three packs a day and his lung function was normal, they gave the same estimate that they gave when asked for the probability of a financial burden given that the individual had smoked three packs a day.

Our explanation for this finding is that people use knowledge about underlying mechanisms to infer latent structure to make conditional probability judgments, especially when their expectations are violated. When told that someone who smokes a lot has normal lung function, there is something to be explained. It can be explained by introducing a disabler (e.g., the person must have smoked highly filtered cigarettes) or by introducing a mediating mechanism (e.g., the person did not inhale). When the two mechanisms are the same, the disabler or mediating mechanism is likely to apply to the other effect too (if the person did not inhale, then her blood vessels are also likely not damaged). But if the mechanisms are different, then the disabler or mediating mechanism has no implications for the other effect (not inhaling does not relieve the financial burden).

What these experiments show is that people construct a causal understanding based on the evidence they are given when they are given enough information to do so. Then they make probability judgments in a way that is compatible with their causal understanding. Both the Bes et al. (2012) and Park and Sloman (2014) studies imply that the causal understanding comes first and that the cognitive system is designed to generate a causal explanation and derive probability judgments from that causal explanations. Because causal explanations require the assumption that the world is governed by causal mechanisms and that such mechanisms are manifestations of natural laws, we conclude that the cognitive system assumes the world is law governed.

Further evidence that causal understanding comes before probabilistic knowledge comes from the literature on causal learning. Only rarely and with some difficulty are people able to induce causal structure from probabilistic data. Most of the time people rely on single-event cues to causal relations, cues that are grounded in perception (White, 2014), temporal relations, instruction, or intervention (for a review, see Lagnado, Waldmann, Hagmayer, & Sloman, 2007).

Causal knowledge represents our understanding of how the world works, capturing systematic (i.e., law-governed) relationships. Rather than being reducible to associative links, causal knowledge is based on mental representations of cause-effect relations that reflect the workings of the external world. So if we are right that the logic of intuition is causal, then we can conclude that, at the most fundamental level, the human reasoning system assumes that the world is governed by natural laws, that is, causal laws.

People Are Deterministic: Probability Arises from Ignorance of Initial Conditions or Governing Laws

We propose that people are determinists for whom probability originates from ignorance about causal laws rather than representing a central feature of causal knowledge. This is an old and established position articulated by Laplace (1814/1902) among others, including Pearl (2000). We address the evidence in favor of the determinist assumption and the origins of probability separately.

Evidence in Favor of Determinism   Accumulating evidence indicates that people have a deterministic representation of causal laws. Causal determinism is the idea that all events have causes (Friedman, 1982; Hirschfeld & Gelman, 1994). It predicts that people will infer unobserved causes whenever events appear to occur spontaneously. There is considerable evidence that both adults and children do this (Chandler & Lalonde, 1994; Friedman, 1982). For example, Schulz and Sommerville (2006) report a series of experiments demonstrating that children infer unobservable causes whenever events appear to occur spontaneously. In these experiments a toy is placed on a device that illuminates only when the switch on the device is turned ON (deterministic condition) or when the switch is turned ON or OFF (stochastic condition) and children are asked to perform an action to enable or disable the device. Rather than inferring probabilistic causation, their actions indicate that children assume that causes produce their effects deterministically. They believe, in the stochastic condition, that an inhibitory cause is at work (e.g., there is a device in the experimenter's hand that blocks the effect of the switch). Thus, the reported findings indicate that children resist believing direct causes can act stochastically, instead preferring a deterministic representation of causal laws.

Other evidence in favor of determinism is less direct. English speakers recognize there is a difference in meaning between causes and enabling conditions. Yet the two are identical in their effect on probability; both increase the probability of their associated effect. The difference between them cannot be described in terms of probability (Cheng & Novick, 1991; Wolff et al., 2010) and so must have some other basis. A likely possibility is that they differ in the role they play in causal mechanisms (Sloman et al., 2009). Such differences in role are consistent with a deterministic model.

Another argument comes from learning. Reasoners often infer a causal relation from a single observation (e.g., Ahn et al., 1995; Schlottmann & Shanks, 1992; White, 2014). But, if causal assertions are probabilistic, single observations should rarely suffice to establish cause and effect because probabilistic representations tolerate exceptions. Single observations may be sufficient in the context of extremely strong prior beliefs that require very little additional information to compel a conclusion. But learning from a single instance occurs more often than such a view would lead one to expect.

Furthermore, people are happy to conceive of interventions that initiate a causal chain in a deterministic way (Gopnik et al., 2004; Hagmayer & Sloman, 2009; Pearl,

2000; Tenenbaum & Griffiths, 2001). Interventions are actions that, on most views, cannot even be assigned a probability (Spohn, 1977) and are generally modeled using deterministic operators.

Evidence That Probability Is Associated with Ignorance   There is no direct evidence that people believe that probability originates in ignorance; perhaps such evidence is impossible. Nevertheless, the claim is at least consistent with data showing that changing people's feelings of ignorance influences their judged probability. The Ellsberg paradox (Ellsberg, 1961) is that people prefer to bet on an urn of known distribution than one of unknown distribution. In the simplest case, you are offered two urns, one with 50 balls of one color (say red) and 50 balls of a different color (say blue). The other urn also has 100 balls and all of them are either red or blue, but you are not told how many there are of each color. You are told you can pick an urn, one ball will be selected from that urn, and if it is red, you will win $50. Which urn do you choose? Given the reasonable assumption of symmetry—that anything relevant to the likelihood of picking a red ball applies equally to picking a blue ball—the effective probability of winning from either urn is 0.5. The probability of winning from either urn is identical, and yet most people choose the first urn, the urn of known composition. The greater knowledge (of the composition of balls) associated with the first urn makes people feel more confident or certain about the first urn and therefore more willing to bet on it.

Heath and Tversky (1991) showed that willingness to bet on an event is proportional to one's sense of competence regarding the event, where competence refers to what one knows relative to what could be known. So less ignorance is associated with a greater willingness to bet, and willingness to bet is an operational definition of subjective probability (Savage, 1954).

More evidence that feelings of ignorance are associated with judged probability was reported by Fox and Tversky (1995). They showed that the Ellsberg paradox occurred only in within-participants designs, not when people were stating their willingness to bet on each urn separately. They call this "comparative ignorance"; it is the comparison to another event that induces the sense of ignorance. Similar demonstrations of reduced-probability judgment come from experiments varying whether more knowledgeable individuals happen to be in the room. Simply stating that an expert is present makes people feel less confident, presumably by making them feel less knowledgeable.

There Is Free Will

Evidence from psychology indicates that the belief in free will originates from the subjective experience of choosing and acting. Monroe and Malle (2010) investigated the psychological foundations of free will and provided evidence that people do believe in free will and conceptualize it as a choice that satisfies their desires, free from constraints.

Even when faced with evidence to suggest that free will is an illusion (i.e., that behavior is caused by neural impulses that precede our impression of agency), people nonetheless defend the notion that free will is central to human thought and behavior. Indeed, Nichols (2004) provides evidence to suggest that free will originates in childhood from the attribution of choice and agency to human actors. For example, children attribute agency to a human actor but not to an object, believing that the human actor could have acted differently (based on free will), whereas the object must behave deterministically (Nichols, 2004).

Recent psychological evidence further indicates that the belief in free will is central to moral judgment, providing a basis for holding people responsible for their actions (Mele, 2009). In a series of experiments Clark et al. (2014) provide evidence that the belief in free will is modulated by the desire to hold others morally responsible for their wrongful behaviors. For example, participants report greater belief in free will after considering an immoral action than a morally neutral one, consistent with the desire to punish morally wrongful behavior. The belief in free will may therefore be motivated by the human desire to blame and punish others for wrongful behavior.

More indirect evidence concerning people's assumptions about free will comes from work on intervention that was alluded to above. Interventions afford different inferences than observations (Meek & Glymour, 1994; Pearl, 2000; Spirtes et al., 1993/2000). In particular, they afford different diagnostic inferences about the causes of an event. For instance, observing someone drink 10 shots of whiskey in a row does make it more likely that the person is an alcoholic; intervening to make the person drink 10 shots of whiskey provides no such evidence. People are exquisitely sensitive to this difference in the diagnostic inferences that are afforded by observation versus intervention; they draw different conclusions in the two cases when other things are equal (Hagmayer & Sloman, 2009; Sloman & Lagnado, 2005; Waldmann & Hagmayer, 2005).

Intervention is not easy to define (Woodward, 2003), and what constitutes an intervention may well be different in different situations. We would all agree that an experiment is an example of an intervention, but we might disagree about whether there are "natural" experiments (does a hurricane provide an unconfounded test of a tree's resilience?). We can state one sufficient condition for an intervention: an action taken by virtue of an agent's own free will. The locus of the action is then an intervention. The data showing that people make appropriate inferences from intervention suggest that people can and do represent interventions in a way distinct from observations. To the extent that such interventions arise from free will, representing them presupposes the existence of free will. In that sense, people do act as if they and others have free will.

Conclusion

This chapter is speculative in that all of our conclusions are defeasible. None of the evidence is overly compelling. But it does paint a sensible portrait of human cognition.

We propose that human intuition is premised on three assumptions: that people believe the world is law governed, deterministic, and that free will exists. These are all commonsensical assumptions that we suspect would fly unnoticed at most dinner table conversations. We are well aware that uncertainty abounds and that people realize that. But the cognitive system could easily attribute that uncertainty to ignorance. It is hard to make predictions, especially about the future, and it is hard to know things, especially about the world. There is just too much to know.

What is more surprising to us is how much cognitive theory does not respect these simple, basic assumptions. Probabilistic models are fine either as approximations or as models of knowledge and inference in the face of uncertainty. But they should be grounded in a representation that, with sufficient knowledge, would be law governed and deterministic. We would go so far as to suggest that the basic problem for cognitive psychology is to understand how causal laws are represented and processed.

The implications of belief in free will are less obvious and less manageable using the mechanistic tools of cognitive science. One view is that free will arises from the quantum nature of the brain, but we fail to see how this solves any problems. At a minimum, belief in free will suggests that we need to take seriously representations that distinguish intervention from observation (Pearl, 2000).

References

Ahn, W. K., Kalish, C. W., Medin, D. L., & Gelman, S. A. (1995). The role of covariation versus mechanism information in causal attribution. *Cognition, 54,* 299–352.

Anderson, J. R. (1993). *Rules of the mind.* Hillsdale, NJ: Lawrence Erlbaum Associates.

Bes, B., Sloman, S., Lucas, C. G., & Raufaste, E. (2012). Non-Bayesian inference: Causal structure trumps correlation. *Cognitive Science, 36,* 1178–1203.

Chaigneau, S. E., Barsalou, L. W., & Sloman, S. A. (2004). Assessing the causal structure of function. *Journal of Experimental Psychology. General, 133,* 601–625.

Chandler, M. J., & Lalonde, C. E. (1994). Surprising, miraculous, and magical turns of events. *British Journal of Developmental Psychology, 12,* 83–95.

Chater, N., & Oaksford, M. (2013). Programs as causal models: Speculations on mental programs and mental representation. *Cognitive Science, 37,* 1171–1191.

Cheng, P. W., & Novick, L. R. (1991). Causes versus enabling conditions. *Cognition, 40,* 83–120.

Clark, C. J., Luguri, J. B., Ditto, P. H., Knobe, J., Shariff, A. F., & Baumeister, R. F. (2014). Free to punish: A motivated account of free will belief. *Journal of Personality and Social Psychology, 106,* 501–513.

Ellsberg, D. (1961). Risk, ambiguity, and the savage axioms. *Quarterly Journal of Economics, 75,* 643–669.

Fox, C. R., & Tversky, A. (1995). Ambiguity aversion and comparative ignorance. *Quarterly Journal of Economics, 110*(3), 585–603.

Friedman, W. J. (1982). *The developmental psychology of time*. New York: Academic Press.

Glymour, C., Scheines, R., & Spirtes, P. (2001). Causation, prediction, and search (2nd ed.). Cambridge, MA: MIT Press. (Original work published 1993.)

Gopnik, A., Glymour, C., Sobel, D. M., Schulz, L. E., Kushnir, T., & Danks, D. (2004). A theory of causal learning in children: Causal maps and Bayes nets. *Psychological Review, 111*, 3–32.

Griffiths, T. L., Chater, N., Kemp, C., Perfors, A., & Tenenbaum, J. B. (2010). Probabilistic models of cognition: Exploring representations and inductive biases. *Trends in Cognitive Sciences, 14*, 357–364.

Hagmayer, Y., & Sloman, S. A. (2009). Decision makers conceive of their choices as interventions. *Journal of Experimental Psychology. General, 138*, 22–38.

Halpern, J. Y., & Hitchcock, C. (2013). Compact representations of extended causal models. *Cognitive Science, 37*, 986–1010.

Hankins, W. G., Rusiniak, K. W., & Garcia, J. (1976). Dissociation of odor and taste in shock-avoidance learning. *Behavioral Biology, 18*, 345–358.

Heath, C., & Tversky, A. (1991). Preference and belief: Ambiguity and competence in choice under uncertainty. *Journal of Risk and Uncertainty, 4*, 5–28.

Hirschfeld, L. A., & Gelman, S. A. (1994). *Mapping the mind: Domain specificity in cognition and culture*. Cambridge, New York: Cambridge University Press.

Hoefer, C. (2010). Causal determinism. *The Stanford Encyclopedia of Philosophy* (Spring 2010 Edition), Edward N. Zalta (ed.), http://plato.stanford.edu/archives/spr2010/entries/determinism -causal/.

Holyoak, K. J., & Cheng, P. W. (2011). Causal learning and inference as a rational process: The new synthesis. *Annual Review of Psychology, 62*, 135–163.

Lagnado, D. A., & Sloman, S. (2004). The advantage of timely intervention. *Journal of Experimental Psychology: Learning, Memory, and Cognition, 30*, 856–876.

Lagnado, D. A., Waldmann, M. R., Hagmayer, Y., & Sloman, S. A. (2007). Beyond covariation: Cues to causal structure. In A. Gopnik & L. Schulz (Eds.), *Causal learning: Psychology, philosophy, and computation* (pp. 154–172). Oxford: Oxford University Press.

Laplace, P. S. (1902). *A philosophical essay on probabilities*. New York: John Wiley & Sons. (Original work published 1814.)

Meek, C., & Glymour, C. (1994). Conditioning and intervening. *British Journal for the Philosophy of Science, 45*, 1001–1021.

Mele, A. R. (2009). *Effective intentions: The power of conscious will*. New York: Oxford University Press.

Monroe, A. E., & Malle, B. (2010). From uncaused will to conscious choice: The need to study, not speculate about people's folk concept of free will. *Review of Philosophy and Psychology, 1*, 211–224.

Nichols, S. (2004). Folk concepts and intuitions: From philosophy to cognitive science. *Trends in Cognitive Sciences, 8*, 514–518.

Park, J., & Sloman, S. A. (2013). Mechanistic beliefs determine adherence to the Markov property in causal reasoning. *Cognitive Psychology, 67*, 186–216.

Park, J., & Sloman, S. A. (2014). Causal explanation in the face of contradiction. *Memory & Cognition, 1*, 1–15.

Pearl, J. (2000). *Causality: Models, reasoning, and inference*. Cambridge: Cambridge University Press.

Premack, D., & Putney, R. T. (1962). Relation between intersession interval frequency of competing responses and rate of learning. *Journal of Experimental Psychology, 63*, 269–274.

Rehder, B., & Burnett, R. C. (2005). Feature inference and the causal structure of categories. *Cognitive Psychology, 50*, 264–314.

Salmon, W. C. (1998). *Causality and explanation*. New York: Oxford University Press.

Savage, L. J. (1954). *The foundations of statistics*. New York: Wiley.

Schlottmann, A., & Shanks, D. R. (1992). Evidence for a distinction between judged and perceived causality. *Quarterly Journal of Experimental Psychology. A, Human Experimental Psychology, 44*, 321–342.

Schulz, L. E., & Sommerville, J. (2006). God does not play dice: Causal determinism and preschoolers' causal inferences. *Child Development, 77*, 427–442.

Sloman, S. A. (2005). *Causal models how people think about the world and its alternatives*. Oxford: Oxford University Press.

Sloman, S., Barbey, A. K., & Hotaling, J. M. (2009). A causal model theory of the meaning of cause, enable, and prevent. *Cognitive Science, 33*, 21–50.

Sloman, S. A., & Lagnado, D. A. (2005). Do we "do"? *Cognitive Science, 29*, 5–39.

Sloman, S. A., & Lagnado, D. (2015). Causality in thought. *Annual Review of Psychology. 66*, 223–247.

Spirtes, P., Glymour, C. N., & Scheines, R. (1993/2000). *Causation, prediction, and search*. New York: Springer-Verlag.

Spohn, W. (1977). Where Luce and Krantz do really generalize Savage's decision model. *Erkenntnis, 11*, 113–134.

Tenenbaum, J. B., & Griffiths, T. L. (2001). Generalization, similarity, and Bayesian inference. *Behavioral and Brain Sciences, 24*, 629–640, discussion 652–791.

Tenenbaum, J. B., Griffiths, T. L., & Kemp, C. (2006). Theory-based Bayesian models of inductive learning and reasoning. *Trends in Cognitive Sciences, 10*, 309–318.

Waldmann, M. R., & Hagmayer, Y. (2005). Seeing versus doing: Two modes of accessing causal knowledge. *Journal of Experimental Psychology: Learning, Memory, and Cognition, 31*, 216–227.

Waldmann, M. R., & Holyoak, K. J. (1992). Predictive and diagnostic learning within causal models: Asymmetries in cue competition. *Journal of Experimental Psychology. General, 121*, 222–236.

Walsh, C., & Sloman, S. A. (2008). Updating beliefs with causal models: Violations of screening off. In Gluck, M. A., Anderson, J. R., & Kosslyn, S. M. (Eds.), *Memory and Mind: A Festschrift for Gordon H. Bower* (pp. 345–357). Hillsdale, NJ: Lawrence Erlbaum Associates.

Watson, J. B. (1930). *Behaviorism.* New York: W.W. Norton & Co.

White, P. A. (2014). Singular clues to causality and their use in human causal judgment. *Cognitive Science, 38*, 38–75.

Wolff, P., Barbey, A. K., & Hausknecht, M. (2010). For want of a nail: How absences cause events. *Journal of Experimental Psychology. General, 139*, 191–221.

Woodward, J. (2003). *Making things happen: A theory of causal explanation.* Oxford: Oxford University Press.

III Epistemic Context in Reasoning

9 Pragmatic Approaches to Deontic Conditionals: From Rule Content to Rule Use

Denis Hilton

In this chapter I review a recent program of research on the pragmatics of deontic conditionals and outline a framework for describing how they are interpreted and reasoned about. I begin by noting that much work on deontics in the psychology of reasoning has focused on deontic rule contents, for example, in evaluating the different effects of deontic and indicative conditionals on reasoning (e.g., in the Wason selection task). I argue that, in addition to rule content, we need to consider rule-use in order to understand the meanings of deontic conditionals, as deontic rule contents can be used in both performative and constative (indicative) ways. I focus extensively on institutional deontics, which are frequently used in impersonal performatives, which create rights and duties in a social group. I show that rights and duties are perceived as instruments of societal exchange and differentiate them from everyday interpersonal deontics such as advice, warnings, and instructions. Rights and duties form shared background knowledge in acculturated members of a society, and this knowledge is used in the interpretation of deontic statements such as *permits, obligates,* and *forbids* in a way that is not fully captured by existing models. I conclude by reviewing various pragmatic approaches to the interpretation of deontic conditionals that fail to capture the phenomena of interest, before outlining the elements of a speech act approach that promises to do so.

Constative versus Performative Use of Conditionals

Austin (1961) made an important distinction between constative and performative uses of language. Constatives make statements about the world that are either true or false (e.g., indicative statements, counterfactuals), whereas performatives are intended make things happen (e.g., through promises, advice, warning, commands, interdictions). Below, I suggest that this distinction can help illuminate questions in the psychology of how conditional statements are interpreted and reasoned about. At first sight, the constative-performative distinction might look to be the same as that drawn by Manktelow and Over (1991) between indicative (e.g., *If water is heated to 100° then it will boil*) and deontic conditionals (e.g., *If you tidy your room then you can go out and play*). There is indeed a relation, but there is also an important difference. Manktelow and Over (1991) focused on the *contents* of indicative versus deontic rules and indeed showed that deontic rules led to much more "logical" performance on the Wason selection task. However, I focus below on the *uses* of rules—whether causal or deontic in content—in performatives that aim to improve the world in some way. The key distinction here is whether the speaker is using language to describe the world (constatives) or to change it (performatives).

Note that deontic rule contents can figure naturally in both performative and constative uses (Hilton, Kemmelmeier, & Bonnefon, 2005), but in interestingly different ways. For example, in a performative use, an airport security director at a French airport could issue the directive *If baggage is suspect then it will be searched*, thus creating a rule to be followed by those under his authority. However, in a constative use the French national passenger safety agency could state a rule such as *If baggage is suspect then it will be searched* as a hypothesis to be tested. This could be done in one of two ways. In the first the safety agency could check the rule books in various French airports to see if this rule is included in the standard security procedures. In the second the safety agency could send passengers with suspect equipment in their baggage through security controls in various French airports to see if the suspect baggage is indeed checked. Here an institutional rule *If baggage is suspect then it will be searched* is first created by the performative utterance, which then serves as a causal mechanism that influences operatives' behavior. This deontic rule can then become the object of a constative, which can be true or false, and this can happen in one of two ways. First, we can say whether it is true or false that the rule in question figures in the rule books of the security procedures at Toulouse airport (existence criterion). Second, if the rule does exist, we can check whether it is indeed followed or not by checking whether suspect baggage is taken out or not during security checks (effectiveness criterion). For example, if the rule in fact exists in some rule book, we can say that there is indeed such a rule at Toulouse, but if the operatives do not follow it, we would say that it is ineffective. Conversely, if it is not inscribed in the rule book but the operatives search suspect baggage anyway, we would be tempted to say that not only does it not exist but that it would be superfluous. It would be a waste of time and energy to implement it if the intended behavior is already occurring.

In contrast, rules that describe physical causal mechanisms do not lend themselves to being used as performatives: they exist independently in the physical world and neither need human injunction to be created nor can they be influenced by them. We would conclude that someone who stipulated to a kettle: *If you reach 100° then you will boil!* would be odd indeed. A litmus test for deciding whether a rule describes a physical-causal or social-deontic mechanism comes from whether it can be sensibly used in the second person—as a form of address to another social being. Previous research has often tried to skate over this distinction by using ambiguous formulations of deontic conditionals that are unclear as to whether they are indicative or performative in mood. This is sometimes done by stating the rule in the third person, for example, "If workers are repairing a road, then a traffic police officer must be directing the traffic." This statement is ambiguous between an indicative *must* (implying a causal relation between road repairs and the presence of a police officer) and a deontic *must* (implying that road repairs require a police officer to be put in place). However, the performative mood can be made clear by expressing statements in the second person, for example, "If you know that workers are repairing the road, then assign a traffic police officer." When it becomes clear that the point of the rule is to get someone else to do something, then the deontic interpretation of the rule is privileged.

Whereas constative statements are simply true or false, deontic statements can be true (if they are in the rule book) or be respected (if they are successful at influencing behavior). For example, if we use modal operators such as *may* or *must*, they need to be disambiguated in context in order to specify whether the usage is epistemic or deontic (e.g., Lyons, 1977; Johnson-Laird, 1978). For example, uses of the strong deontic modal do not logically imply what is, in fact, the case (cf. Lyons, 1977). To say, on the basis of DNA evidence, that a man *must* (epistemic) be the father of a child is to imply logically that he is in fact the father. In contrast, to say, on the basis of child protection laws, that a man *must* (deontic) support his child financially is not to imply logically that he will in fact do so. Obligations of this kind can be violated without falsifying the truth of the speaker's statement. Although these laws may in fact be the case (having been legislated in the Child Protection Act), they may also be ineffective—they may fail to induce the intended behavior.

Of course, indicative causal conditionals can be used *indirectly* for performative uses in the second person by drawing the addressee's attention to the likely consequences of actions, for example: *If you pull the cat's tail, it will probably bite you.* Here this might be interpreted as a useful piece of advice: *I wouldn't pull the cat's tail if I were you.* But the deontic value of the statement is not located in the utterance itself (a simple indicative causal conditional) but in the conversational implicatures drawn from the Gricean conversational contract that guarantees its usefulness as a contribution to a cooperative conversation. In particular, the deontic introduction (going from *is* to *ought*) is made through calculating a conversational implicature on the assumption that the speaker has a cooperative intention and is respecting the maxims of conversation (Grice, 1975), which are expressed in imperative form (roughly speaking: Be accurate! Be informative! Be relevant! Be clear!). This will be so only if the information can be exploited to improve the addressee's well-being in some way. In other words, the causal conditional must be able to impact the hearer's utility, perhaps through indicating relevant causal mechanisms to her that will help her avoid a nasty surprise (Elqayam et al., in press). Because the information would become relevant only if the speaker were indeed to pull the cat's tail, it is plausible to assume that the speaker is commenting on the desirability of doing so. As a proof of the importance of taking into account the conversational point made by uttering a causal conditional, note that, in contrast, stating the same rule in the third person form seems more readily able to carry a simple constative function of prediction: *If he pulls the cat's tail, it will probably bite him.* Additional conversational context may, however, license a deontic interpretation if it is "heard" as being directed at the addressee. For example, if said to a babysitter for the child in question, the above constative (indicative conditional) might also carry a conversational implicature, *You should take care not to let him pull the cat's tail,* destined to influence the babysitter's behavior.

In sum, deontic rule contents can naturally form the object of both performatives (whereby an authority makes them come about in a given jurisdiction, e.g., by creating a law, giving permission, or imposing an obligation) and constatives (whereby an observer tests whether they exist and/or generate behavior in a given jurisdiction). Causal rule contents cannot naturally be the object of performatives, although they are naturally the object of constatives. In contrast, causal statements that are uttered in the second person may make *indirect* suggestions (advice, instructions, warnings, etc.) that are destined to influence a hearer's behavior. However, physical causal conditionals cannot be used as performatives (e.g., in giving orders, advice, warnings) destined to influence the causal mechanism in question. Whereas causal conditionals can convey deontic conclusions *indirectly* through generating conversational implicatures (see the discussion of deontic introduction above), performative conditionals explicitly introduce deontic elements that justify deontic conclusions (Searle, 1964). For example, deontic conditionals explicitly specify a permission or an obligation through the use of a deontic verb in the antecedent (e.g., "If I *promised* him to go round tomorrow" or "If the service contract *requires* me to go round tomorrow"; both imply the conclusion that I *ought* to go round tomorrow).

For these reasons I contrast two kinds of rule content (causal vs. deontic) and two kinds of rule use (constative vs. performative) below. Note that in the usage adopted here, "indicative" refers to the constative form of rule use (i.e., the use of a rule to make a descriptive statement about the world) and not to a form of rule content.

Impersonal Performatives

Having distinguished deontic from causal contents in conditional rules, I now wish to make a more specific distinction between two subtypes of deontic content, namely impersonal rules that concern a *class of people* rather than a *specific individual.* Institutional deontics such as *Taxpayers who donate to charity are permitted to claim a tax rebate* (Bucciarelli & Johnson-Laird, 2005) may be considered to be forms of "social laws" (Cosmides, 1989) that differ from "social contracts" proposed between two individuals, for example, *If you wash my motorcycle then you may ride it* (e.g., Fiddick & Grafman, 2008). Such social contracts are often based on deals made by two individuals on a specific occasion, such as a wife proposing to her husband, *If I prepare the dinner, you must take the children to the zoo on Sunday* (see Legrenzi, Girotto, & Politzer, 1996, for a comprehensive theoretical analysis). In contrast to social contracts that refer to such specific deals, *institutional deontics* refers to formal, impersonal rules issued by an authority, such as *If you are licensed by the District, then you may sell hotdogs on the Mall*. Here the "you" can be read impersonally, and as applying to all people who possess this license. Institutional deontics of this kind are likely to be found in rule books produced by formally constituted social groups (nations, companies, or clubs) and are often expressed using formal rules that are expressed in the third person. Unambiguous institutional deontics are often phrased in the language of universal laws and regulations, such as *Passengers of international flights have the duty to hold an identity card*. They have the characteristics of laws (Hart, 1961) in that they are not addressed to a specific person, they are standing orders independent of any particular time with sanctions for violators, and they can only be felicitously established as a rule by an authority.

Properties of Prototypic Institutional Deontics: The Case of Rights and Duties

Hilton, Charalambides, and Hoareau-Blanchet (in press) suggest that rights and duties are prototypic kinds of institutional deontics. They are script-like, widely shared by acculturated members of society, and can be viewed as social contracts that constitute symmetric pairs of obligations and permissions. In the case of rights an individual has a *permission* to exercise her right, but the social group is under an *obligation* to give that individual her rights if she asks for them. In the case of duties the individual is under an *obligation* to fulfill her duty, but the social group is *permitted* to dispense with this requirement if it so wishes. Just as an individual can waive her right to a benefit if she does not want it, a dispensation can be conceived as the group giving itself permission not to take a benefit if it does not wish to. However, a party that does not respect its obligations under this arrangement (e.g., by taking a right without being qualified to do so or by failing to perform a duty when required to do so) may be considered a cheat (cf. Cosmides, 1989). The strong intuition that those who violate institutional deontics are cheats locates rights and duties within the general realm of social contracts, albeit with specific characteristics that we detail below.

Hilton, Charalambides, and Hoareau-Blanchet (in press) suggest that in reasoning from deontic rules about what is permissible and impermissible, linguistic knowledge of the meanings of deontic verbs (*permits, obligates,* etc.) needs to be combined with socially shared script-like knowledge (Abelson, 1981; Schank & Abelson, 1977) about *rights* and *duties* that hold in institutions or similar social groups. This world knowledge emerges from role expectations about behavior as a function of an individual's position in a social structure (e.g., a family, a restaurant, a public railway, a government). As Sarbin writes:

Two general kinds of expectations are found: rights and obligations. Rights are role expectations in which the actor of the role anticipates certain performances from the actor of the reciprocal role; e.g., the child's right to be protected by his mother. Obligations (or duties) are role expectations in which the actor of a role anticipates certain performances *directed toward* the actor of the reciprocal role: e.g. the mother's obligation to provide protection for the child. (1954, p. 226)

It is important to note that the child's right to expect protection and the mother's duty to give it emerge from *general socially shared* expectations and not from a *particular private* deal struck between two parties and thus do not fit the form of "social contract" commonly studied in the psychology of reasoning (e.g., Cosmides, 1989).

This scripted knowledge about widely accepted rights and duties functions as *content knowledge* shared by certain social groups. For example, it is widely assumed among French university students that "People who have been made redundant have the right to benefit from unemployment benefits" and that "Passengers of SNCF trains have the duty to punch their tickets" (Hilton, Charalambides, & Hoareau, in press). We may consider these rights and duties as socially shared knowledge structures that are subsumed under the pragmatic reasoning schemas of permission and obligation, which describe "abstract knowledge structures induced from ordinary life experiences" (Cheng & Holyoak, 1985). Pragmatic reasoning schemas concerning permission and obligation encapsulate general knowledge about these relations, defined in terms of rules of inference that connect precondition-action pairs about what may, need not, must, and must not happen in certain cases. Permission schemas imply that the precondition (e.g., making donations to charity) must be satisfied for the action in question to be taken (e.g., claiming tax reductions), and obligation schemas imply that when a precondition holds (e.g., taking public transport), an obligation arises (e.g., to hold a valid ticket). If rights imply permission schemas, then this implies that cases of *not-A & B* will be perceived as violations, and if duties imply obligations, then cases of *A & not-B* will be perceived as violations (Cheng & Holyoak, 1985; Cheng, Holyoak, Nisbett, & Oliver, 1986).

The approach advocated by Hilton, Charalambides, and Hoareau-Blanchet (in press) requires making a distinction between deontic verbs (or operators, such as *permits, obligates, forbids*) and the contents of deontic rules (e.g., rights vs. duties). An advantage of the approach to defining deontic understanding in terms of basic deontic operators, such as *may* and *must*, and cognate ones such as *it is allowed that* or *it is required that*, is that these deontic primitives can be used as definitions of larger knowledge structures such as social contracts and pragmatic reasoning schemas (cf. Manktelow & Over, 1991). However, this raises the critical question of how deontic expectations activated by a deontic verb (e.g., *are obligated*) are interpreted in the context of a rule content that *prima facie* appears to contradict it. For example, for many, the assertion that *Taxpayers who make donations to charity are obligated to claim a tax rebate* will seem odd if claiming tax rebates were seen as an individual's right. This is because a right becomes a duty if its holder is obliged to exercise it, bringing with it a loss of autonomy on the part of the holder of that right. In most Western societies, this would seem to be an inappropriate imposition on the rights of taxpayers. It is important to note that although *may* can indeed be strengthened into *must* without anomaly from a logical point of view, such strengthening may be considered felicitous only when accompanied by a social justification (e.g., if the taxpayers in question are members of a political party that wants to bankrupt the state). Another example would be considered socially appropriate is the paternalistic case where instituting obligations to exercise rights would be perceived to be necessary where it is felt that those holding the right cannot be left to choose for themselves (e.g., as when children are obliged to go to school).

Although we accept that the patterns of inference about permissibility under rights and duties can be well modeled by the pragmatic reasoning schemas of permission and obligation (Holyoak & Cheng, 1995), we propose that there is more to knowledge of rights and duties than just inferential schemas. We have already seen, for example, that causal and deontic rules differ in important ways. Below, we look at further ways in which our approach to the analysis of rights and duties goes beyond that of simple pragmatic reasoning schemas. First, we propose that rights and duties are closely associated to the notion of social value inherent in social contract theory (Cosmides, 1989) and more generally to the notion of utility in decision-theoretic approaches to deontics (e.g., Over, Manktelow, & Hadjichristidis, 2004). Second, we propose that attention has to be paid to the unconventional ways in which rules whose contents are normally understood as rights or duties can be *used* (e.g., using a duty-content script such as the SNCF ticket-punching rule to express a permission, such as *Passengers on SNCF trains may punch their tickets before boarding their train*). Third, we review ways in which differences between institutional and interpersonal deontics are marked in language. Finally, we suggest that understanding of the interests that are at stake is necessary for the proper interpretation of deontic verbs such as *may, should, must, is obliged to,* and so on when used in the context of institutional deontics.

Rights and Duties as Instruments of Societal Exchange

Rights are acquired and even won (as in the right to free speech or the right to vote). They are often given as a sign of social worth, as it is natural to speak of "winning" and "deserving" rights, as in having "earned" the right to go out and celebrate once a big job of work has been done or of having "lost" the right to ask someone a favor if we have let them down. We speak easily of exchanging rights, such as when we buy and sell film and property rights. In English, the notion of "duty" also seems closely tied to the notion of value; for example, customs and excise duties are taxes imposed by a government on imported goods. Similar associations exist in other languages (as well as in old forms of English). In French, the verb *devoir* can be used to express both obligation and debt, an association present in its Latin root, *debere,* which comes from the amalgamation of *de* + *habere* (to have a debt). This historical association is also found in the English modal verb *ought,* which comes from an old English form for *owed* (Mackie, 1977).

Rights and duties may both be instituted to enable smooth social ordering and thus benefit the common good. However, at the local level, rights are likely to be seen as benefiting the individual concerned (they transfer a benefit from the group to the individual), whereas duties will be seen to benefit the group (they require an individual to transfer a benefit to the group). The English language's vocabulary of revenue transfers reflects this distinction. In Britain, welfare "rights" typically describe payments made by the state to disadvantaged groups such as the unemployed, whereas import "duties" are typically collected by the state in the form of surcharges on certain kinds of goods (e.g., alcohol, cigarettes) that incoming travelers must pay on arrival at a port of entry.

The notions of value attached to rights and duties gives them properties that go beyond the inferential patterns associated with the pragmatic reasoning schemas of permission and obligation. Rights and duties are used by Cheng and Holyoak when studying the rules intended to activate pragmatic reasoning schemas (Cheng & Holyoak, 1985; Cheng, Holyoak, Nisbett, & Oliver, 1986; Holyoak & Cheng, 1995) in that in their experiments an impersonal authority (e.g., an immigration service) gives a rule to a class of individuals (e.g., visitors to a country) concerning requirements (e.g., for admission). Despite the centrality of value to deontic reasoning, Cheng and Holyoak (1985) eschew reference to notions of value and utility in their formulation of pragmatic reasoning schemas, which in their view include nondeontic domains such as causation and prediction (where the notion of value plays no intrinsic role). For example, Cheng and Holyoak (1989, p. 288) argue explicitly against the position that such social regulations involve exchanges of benefits between an institution and an individual. They contend that such cases as their immigration rule are not real social exchanges because they do not involve significant costs and benefits for the individuals or groups concerned. However, it seems plausible to us that the cost and pain of having an inoculation can be regarded as a fair and reasonable "price" to be demanded of an individual by a country she wishes to visit, especially if it protects the people of that country. We therefore agree with Cosmides (1989; see also Kilpatrick, Manktelow, & Over, 2007) that Cheng and Holyoak's (1985) "rationale" manipulation may have implicitly introduced just these kinds of costs and benefits into their experimental materials. For example the "rationale" for their immigration rule—that inoculation prevents the spread of disease in a country—clearly evokes a benefit for that country.

Hilton, Charalambides, and Hoareau-Blanchet (in press) hypothesized that rights and duties, unlike pragmatic reasoning schemas, express a logic of societal exchange and, in particular, asymmetric directions of transfer of benefits between a society and individuals. This is because everyday conceptions of rights include the assumption that they involve transfers of benefits from the social group to a subcategory of individuals who satisfy a certain requirement, whereas duties imply transfers of benefits in the opposite direction from an identified subcategory to the larger group. Hilton, Charalambides, and Hoareau-Blanchet (in press) tested the above ideas by presenting a number of rules expressed in the form of A *(a class of individuals) have the right/have the duty to B* to our target population of French students. The rule contents were likely to be widely known to this group, and the aim was to see whether the students could classify them as rights or duties (they were given the option of choosing "have the right" or "have the duty" as the most appropriate filler expression for the space left for the deontic expression). A second aim was to test the prediction that those rules of the form A *(a class of individuals) deontic verb B* that were perceived as rights would be perceived as primarily *favoring* A (the individuals), and that those that were classified as duties would be perceived as primarily *favoring* B (the social group or authority stipulating the rule).

The results showed that participants had little difficulty selecting the deontic expression that went best with the rule in question. Thus, a large majority (80%) chose the verb *have the right* for the unemployment rule (e.g., "People who have been made redundant have *the duty/ the right* to claim unemployment benefits") and the verb *have the duty* for the transport pass rule (e.g., "Users of public transport have *the duty/the right* to hold a transport pass"). In addition, those rules that were classified as rights were indeed perceived as primarily favoring the individuals targeted by the rule, whereas rules that were classified as duties were perceived as favoring the group or authority stipulating the rule. The findings indicate that rights and duties are more than just reasoning schemas of permission and obligation (Cheng & Holyoak, 1985). In particular, rights and benefits appear to be specific kinds of social contract that fall under the general category of *social laws* (Cosmides, 1989). However, these social laws constitute social contracts that go beyond private social exchanges and cannot be captured in purely individualistic approaches to utility currently proposed in decision-theoretic accounts of deontic reasoning (Over, Manktelow, & Hadjichristidis, 2004; Perham & Oaksford, 2005) or pragmatic theories of the use and interpretation of conditional statements (Hilton, Kemmelmeier, & Bonnefon, 2005). For this reason we prefer to say the rights and duties reflect a logic of societal exchange (e.g., between an individual and a group or between two groups) rather than a social exchange between two individuals.

Nonconventional Uses 1: Giving Permissions in the Context of Duties

More importantly, understanding why a statement might be made helps us to reinterpret a rule content that is typically thought of as a duty to be, in fact, the granting of a permission, or a rule content that is typically thought of as a right as, in fact, the imposing of an obligation. As a first example, let us take the script content "Passengers of SNCF trains have *the duty/the right* to punch their tickets," which was widely recognized as a duty by participants, as they almost all selected "have the duty" as the more appropriate verb for this rule when it was presented without further context (Hilton, Charalambides & Hoareau, in press). In France, it is widely understood that punching the ticket before boarding the train makes the ticket valid for the journey being taken, and moreover removes the right of the passenger to ask for a refund for the ticket (for example if he or she is unable to take the journey). This scripted knowledge seems to intrude on their interpretation and reasoning about statements that use a duty rule content but state permissions (e.g., not to do what the rule content suggests must be done) and that may be thought of as a dispensation, which is a permission not to do something that is normally required (Beller, 2008). Nevertheless, many participants given this kind of dispensation without further context still consider that a violation of the duty specified in the rule content constitutes a violation of the rule. For example, Hilton, Charalambides, and Hoareau-Blanchet (in press, study 2) found that, when given dispensations using rule contents that are commonly recognized as duties (e.g., *Passengers of SNCF trains need not punch their tickets*), a considerable proportion (38%) of participants considered that cases of *A & not-B* (Passengers who do not punch their tickets) constitute violations of the rule. This contradicts the meaning of the deontic verb (*need not*), which allows such cases.

Note that the above results were obtained without any contextual explanation of *why* the scripted duty-content rule (punching train tickets) was expressed with a verb that expresses a permission not-to (e.g., the dispensation, *need not*). However, what if the context alters the normal relation between punching the ticket and taking a trip on a SNCF train and explains why the dispensation was given? For example, suppose that the SNCF has decided to give a free day's travel on its trains in order to compensate commuters for the disruption caused by a recent rail strike. Here, it may make sense for the company to issue the directive *Passengers of SNCF trains need not punch their tickets*). Hilton and Charalambides (unpublished data) tested this hypothesis by presenting participants with a context that explained the reasons for the dispensation: "In an attempt to compensate for the unfortunate consequences of the train drivers' strike last month, the SNCF is offering free travel today on all the railway lines of the Ile-de-France region. The head of the train station makes the following vocal announcement in order to inform passengers that: *Passengers on SNCF trains need not punch their tickets*." Here, only 6% of participants considered cases of *A & not-B* (Passengers who do not punch their tickets) to constitute violations of the rule. In addition, in this context, substantially more participants considered the rule to benefit the passengers (44%) than the SNCF (6%), consistent with the view that the rule was now perceived as a right. Here, the contextual explanation of why passengers are permitted not to stamp their tickets seems to be sufficient to cue almost all the participants into recognizing that the rule is being used to state a right and not a duty (which would normally be the case).

Nonconventional Uses 2: Imposing Obligations in the Context of Rights

We use the example of a tax adviser to illustrate the importance of knowing *why* a speaker is making a deontic statement in order to interpret it properly. In the context of financial advice, *If you donate to charity you must claim a tax rebate* the *must* seems most sensibly interpreted here as the tax advisor's insistence that claiming the rebate is very much in his client's interest. Indeed, if a tax adviser gave this advice to his client because he thought that it would be in the interests of society that his client claims a tax rebate, his client could reasonably accuse him of not doing his job properly. Although a tax law may state that having the right to claim a tax rebate is optional (i.e., the beneficiary *may* do so), the tax adviser who says that *If you donate to charity you must claim a tax rebate* is likely to be insisting on his client's perceived obligation to himself to take advantage of this right. He is not insisting on his client's legal obligation to do so, as might be the case if he were to say in the same meeting *If you donate to charity then you must file your tax return by April 21st* because the government requires up-to-date records to be kept about donations to registered charities.

When expressed in the impersonal form, most people consider rights to benefit the addressees if they are given no further information. Consider the rule: *Taxpayers who donate to charity … claim a tax rebate.* Most (99%) perceived this rule to be a right when presented with no further context, 70% perceived it to be benefiting the individual, whereas only 2% considered that this rule benefited the group, and 29% that it benefited both the individual and the group (Hilton, Charalambides, & Hoareau, in press). However, a majority of participants (63%) who read the rule *Taxpayers who donate to charity must claim a tax rebate* considered that taxpayers who donated to charity but who did not claim a tax rebate were breaking the rule, suggesting that they interpreted this rule as an obligation. This may be thought of as a kind of *deontic strengthening*, where a right is strengthened into an obligation. The rule still remains a right, but the addressee is now obliged to exercise that right.

What might be the circumstances in which it would make sense to create an institutional performative that requires a class of people to claim tax rebates? Hilton and Charalambides (unpublished data) hypothesized that this could happen where it was in the larger group's interest that a subgroup exercised their rights. For example, they presented the unemployment benefit rule in the following setting:

In Montalaguay most of the people who are unemployed don't claim unemployment benefits. It is indeed perceived as a sign of lack of moral integrity and as a sign of submissiveness to colonial values, to ask for help from the State. Unfortunately, many children suffer from malnutrition because their parents, who are poverty-stricken, would rather not get involved with the claiming procedures. In addition to the catastrophic consequences for the children, this situation gives a very negative image of the Montalaguay government to the rest of the international community. The latter imagines, quite wrongly, that the government is deliberately stopping its citizens from receiving the financial support they're in need of and is threatening to boycott Montalaguay products. The Montalaguay government, which is very alarmed by the speeches and negative judgments of the international community, has thus made the following official announcement to the people of Montalaguay: People who have been made redundant ... claim unemployment benefits. (Hilton & Charalambides, unpublished data)

Here, 50% chose to insert *must* as the modal verb (compared to 13% who inserted *have the duty to* in the equivalent task without context in Hilton, Charalambides, & Hoareau, in press), and 21% considered the rule to benefit the group and 49% to benefit both the individual and the group (compared to 5%). This strongly suggests that in this context the majority (70%) of our participants perceived this rule as expressing a duty at least partially for the benefit of the group, even if at the level of rule content it is scripted as a "right," and out of context only 5% see it as at least partially benefiting the group. Once the context changes the interests that are perceived to be at stake, it seems that many are able to state an obligation using a rule content that is normally recognized as a right.

Address and Vocabulary in Distinguishing Institutional from Interpersonal Performatives

In institutional performatives, expressions such as *may, must, are permitted to, are required to, have the right to,* and *have the duty to* are normally used. Note that interpersonal performatives often sit uneasily with the impersonal form of address characteristic of institutional performatives: whereas it sounds natural to say *If you give me an ice cream you may borrow my bike*, it seems slightly odd (and somewhat officious) to say *All those who give me an ice cream are permitted to borrow my bike*. Another intriguing difference between the language of institutional performatives and that of everyday performatives can be observed in the language of interdiction. In everyday forbidding, addressees are told what they *must not* do. However, it seems less natural for institutional interdictions to inform the public about what they *have the duty not to do* or what they are *obliged not to do* but more natural to inform them about what they *do not have the right* to do or *are not allowed* to do or *may not* do.

Institutional performatives that establish rights and duties have several characteristics that differentiate them from interpersonal deontics such as advice and instructions. For example, when we really mean business we do not use deontic modals such as *should* or *ought to* in institutional deontics (Mackie, 1977). In contrast, *should* and *ought* are often used in personal recommendations such as in Bucciarelli and Johnson-Laird's (2005 p. 160) example, *If you are in Venice then you ought to see the Accademia* (see Elqayam et al., in press, for further examples of this kind). For example, a tax lawyer would typically use the deontic modal *should* to advise his client, *If you support a charity you should claim a tax rebate*, because he saw this as being in the interests of his client. The above example gives another illustration of the importance of distinguishing rule content from rule use. When stated as an institutional deontic, the taxpayers' rule is clearly perceived to be a right. Thus, Hilton, Charalambides, and Hoareau-Blanchet (in press) found that people clearly preferred to interpret the rule *Taxpayers who support a charity have the right/have the duty to claim a tax rebate* as a right which benefits the individuals in question (taxpayers who support a charity) and could legitimately be expressed as *Taxpayers who support a charity may claim a tax rebate*. However, a tax adviser might not only say to his client that he *should* claim a tax rebate, but he might be even more forceful and suggest that *If you support a charity you must claim a tax rebate*, as he considers that it is very much in his client's interest to do so. But by the use of *must* he does not mean that his client is *obligated* (in the sense of being required by law) to claim a tax rebate. In the tax lawyer context a client can freely consider the advice, evaluating whether it is really in her interests or not, and then decide what she *should* do in the light of her personal interests.

Of course, there are intermediate cases of the institutional deontic, as when an institution makes information available with the aim of encouraging behavior that is in the public interest. An interesting case is provided by official institutions that provide advice for the public. For example, the Department of Health in Britain publishes a booklet, "Keep Warm, Keep Well: A Guide for Families," with, for example, advice on how to avoid colds and flu, "The whole family should wash their [*sic*] hands regularly with soap and water" (see the UK Department of Health website). We do not count such examples as central cases of the institutional deontic, as the institution is here standing in for a doctor and giving advice that it *hopes* the public will follow to everyone's benefit. The use of *should* is felicitous in this advice context, whereas the use of *must* or *is obliged to* could be seen as infelicitous (i.e., overly strong and forcing the point) in such a case, as no one will be liable for punishment if they do not follow the recommendations. In contrast, it would be felicitous for the Department of Health to lay down a law that cooks in a restaurant "must" or "are obliged to" wash their hands, and it may reserve the right to close down an establishment and punish its owners if this law is not respected. Such health and safety regulations *are* central examples of the institutional deontic.

Pragmatic Approaches to Deontic Conditionals

The work of Cosmides (1989) and Manktelow and Over (1991) ushered in an epoch of research on deontic conditionals (for examples, see Gigerenzer & Hug, 1992; Politzer & Nguyen-Xuan, 1992; Lieberman & Klar, 1996). Much work focused on the ability of deontic conditionals to facilitate "correct" responses (even when using unfamiliar rule content) on the Wason selection task compared to indicative conditionals. A major "pragmatic" approach to explaining facilitation effects on the Wason selection task is the *pragmatic reasoning schemas* perspective. Given that rights are permissions and duties are obligations, the pragmatic reasoning schemas perspective would identify the *not-A & B* cases as violations for rights and *A & not-B* cases as violations for duties (Cheng & Holyoak, 1985; Cheng, Holyoak, Nisbett, & Oliver, 1986; Holyoak & Cheng, 1995), as observed by Hilton, Charalambides, and Hoareau-Blanchet (in press). However, the pragmatic reasoning schemas approach has problems in explaining other results observed by Hilton, Charalambides, and Hoareau-Blanchet (in press). First, the same studies demonstrated a strong and consensual tendency to view rights as benefiting the individual (A) and duties as benefiting the collective (B). A pragmatist is often thought of as someone who takes costs and benefits into account in his decisions, but pragmatic reasoning schemas appear to eschew such considerations. Second, the pragmatic reasoning schemas approach does not address the question of how deontic expressions (such as *may, must, must not, may not*) are interpreted in the context of rule contents that are normally understood as rights or duties.

The Pragmatic Modulation of Deontic Expressions

Bucciarelli and Johnson-Laird (2005) did address the question of how deontic expressions are defined and suggested that "pragmatic modulation" may be used to select the specific meaning of an ambiguous deontic expression, such as *permits, obligates,* and *forbids*. They argue that people have an understanding of deontic modal verbs such as *permit* or *may, must* or *obligate, permit not* or *need not,* and *forbid* or *must not,* and the semantic interpreter will use these understandings to yield a set of mental models that depict interpretations of what is deontically permissible and impermissible. According to this analysis, an expression such as *A permits B* may have a weak interpretation (where all combinations are permissible, *A & B, A & not-B, not-A & B,* and *not-A & not-B*) or a strong interpretation (where *not-A & B* is considered impermissible and the other three cases are considered permissible). Bucciarelli and Johnson-Laird map out the possible interpretations (whether strong or weak) of deontic words, and the suggestion that "pragmatic modulation" involves using world knowledge to select between strong and weak interpretations of deontic expressions. However, this approach raises the psycholinguistic question of how primitive deontic expressions are to be used and understood in contexts where larger molecular deontic scripts are activated. Bucciarelli and Johnson-Laird (2005, experiment 1) used rights and duties as example rules, but focused exclusively on the effect of deontic modal verbs on representation and reasoning about mental models.

Hilton, Charalambides, and Hoareau-Blanchet (in press) systematically varied the type of deontic verb (e.g., *may, must, need not,* and *must not*) and type of rule content (*right* vs. *duty*) used in deontic statements. They found that rule content (*right* vs. *duty*) affected interpretation of strong and weak meanings of *may* and *must,* and this can be taken as supporting the verb-based strategy (semantic analysis plus pragmatic modulation). However, they also found that rule content (*rights* vs. *duties*) influenced permissibility judgments in other cases, leading to apparent semantic contradictions.

For example, duty-content rules predict semantic contradictions for all four deontic verbs for judgments of the permissibility of cases of *A & not-B*. For example, as predicted by this world-knowledge approach, Hilton, Charalambides, and Hoareau-Blanchet (in press) found that a substantial number of participants (38%) who read duty content rules such as *"Users of public transport need not hold a transport pass"* considered that people who use public transport but do not hold a transport pass (*A & not-B*) are violating the rule. This interpretation contradicts Bucciarelli and Johnson-Laird's semantic analysis of "permit not" (need not), which stipulates that *A & not-B* is permissible. Similarly, it was predicted and found that using *must* in a rights-based content (e.g., *People who have been made unemployed must claim unemployment benefit*) led a substantial number of participants (63%) to consider that *A & not-B* is permissible, thus contradicting the semantic meaning of *must*. These results pose a significant challenge to Bucciarelli and Johnson-Laird's analysis as they cannot be handled by their notion of "pragmatic modulation" as currently formulated. World knowledge leads to interpretational effects that are incompatible with the proposed semantic analysis and thus cannot admit of pragmatic modulation.

Salience, Conversational Focus, and Directionality

One of the interesting aspects of Bucciarelli and Johnson-Laird's (2005) framework is that it succeeds in showing that deontic expressions that comport negations (whether implicit ones such as *forbid* or explicit ones such as *permit not*) will make the *A & not-B* cases salient, as indexed by the order in which cases are listed by participants. For example, in response to negative deontics (*forbids, permits not*), Bucciarelli and Johnson-Laird found that participants were more likely to list negative cases (e.g., *A & not-B*) first. Hilton, Charalambides, and Hoareau-Blanchet (in press) found a broadly similar pattern of results in response to a question asking participants to designate cases that were most relevant to explaining the rule. These patterns may help explain why authorities use these negative forms rather than other formally equivalent deontic expressions (Bucciarelli & Johnson-Laird, 2005). For example, people are not usually told that they have the duty not to smoke in restaurants or are obliged not to lean out of train windows but rather that they do not have the right to smoke in restaurants or that they are not allowed to lean out of train windows. Here, "pragmatics" means something like the aspect of an event that the speaker wishes to draw attention to.

The notion of conversational aspect is central to Schmeltzer and Hilton's (2014) analysis of the "pragmatic well-formedness" of conditionals. For example, when we consider the sentences *If it is possible that the operation will succeed, then have the operation* and *If it is not certain that the operation will succeed, then have the operation*, both are syntactically and semantically well formed, but the first seems to be an appropriate thing to say, whereas the second seems somehow anomalous. In other words, the choice of words that puts the accent on the chances of failure sits oddly with the encouragement to take the operation. And if one heard a doctor give this anomalous-sounding advice, one might be forgiven for wondering what exactly he was trying to tell us to do as well as questioning his rationality (or at least wondering how good his English is).

People's intuitions about pragmatic well-formedness are strong and rapidly formed and are based on the causal directionality of the relation between the antecedent and the consequent. Thus, it seems natural to focus on the consequence that naturally follows from the precondition (e.g., the possibility of sunshine seems to favor hanging washing outside to dry, whereas the possibility of rain does not). This dictates the polarity of the quantifier used to characterize that possibility (e.g., *possible* or *not certain*). Moreover, this relation is not calculated using expected probability and utility, as might be expected by a decision-theoretic approach (e.g., Over, Manktelow, & Hadjichristidis, 2004). For example, Schmeltzer and Hilton (2014) showed that incomplete conditional sentences that begin with the antecedent *If it is possible that the operation will succeed* tend to be completed with consequents that encourage performing the action in question (e.g., *then have the operation*). In contrast, those that begin with the antecedent *If is not certain that the operation will succeed* tend to be completed by consequents that discourage it (e.g., *then do not have the operation*). In terms of the probabilities referred to in the antecedent, both sentences seem to have the same "objective" meaning because *possible* and *not certain* denote similar probabilities of the operation succeeding in these examples. However, the first explicitly focuses on the chances of success, whereas the second implicitly focuses on the chances of failure (Moxey & Sanford, 1993).

In a follow-up study, participants were asked to make judgments of the coherence of these sentences. These judgments were strongly influenced by the directionality of the causal relation (e.g., whether the precondition focused on the operation's chances of success or failure) and the polarity of the quantifier (*possible that, not certain that*) used to describe these chances independently of the probability of the outcome. Moreover, electrophysiological evidence suggests that the causal direction and polarity information are computed more rapidly than probability information in sentence processing (El-Yagoubi, Hilton, Schmeltzer, & Wawrzyniak, unpublished data). In sum, it seems that although the pragmatic well-formedness of a sentence is a psychologically real phenomenon in the comprehension of such advice conditionals, it is calculated without recourse to the kind of probability information that would be used in a decision-theoretic approach to evaluate these conditionals. Thus, pragmatic well-formedness effects seem to be best subsumed under the rubric of *conventional implicatures*, which seem to be calculated more automatically than *conversational implicatures* derived from assumptions about the speaker's purpose in making an utterance that flouts conversational maxims (Grice, 1975; see also Hilton, 2009).

Toward a Speech Act Theory of Conditional Assertion and Interpretation

Below I consider how a pragmatic speech act theory (Searle, 1965) of conditional use and interpretation may be able to give a fully successful account of the way deontic conditionals are used and understood in everyday communication and reasoning. Such a theory must specify who says what to whom, how, and to what effect. For example, outside of the institutional deontic, it can be felicitous to use a deontic obligation word such as *must* in a rule that describes a right in advice situations. For instance, this may be the case where the speaker is not an institutional authority but a parent of the addressee. A mother might urge her son to do the right thing by saying *If you support charities, then you must claim a tax rebate.* This is because the mother wants the best for her son and insists that he claim a tax rebate because she thinks it is in his interest. This is quite different from the typical institutional deontic use of "must," whereas the research reported above shows that the activity in question is normally required in order to benefit the institution that imposes the rule. Alternatively, the felicity of the utterance may depend on the circumstances of the addressee. For example, a mother saying to her poor son *If you are unemployed then it is your duty to claim unemployment benefit* may be felicitous if he has a starving family to support. However, the formally similar utterance *If you support charities, then it is your duty to claim a tax rebate* seems infelicitous if one assumes that the addressee is already quite wealthy.

In addition to specifying who says what to whom to what effect, a speech act theory of conditionals may also address the question of *how* the rule is expressed. Manner matters (Grice, 1975). For example, note that substituting a formal term for an informal one seems to change the nature of the remark being made. Thus, whereas "If you go to Venice you will have to visit the Accademia" sounds like a piece of advice (because you will enjoy it), the statement "If you go to Venice you will be obliged to visit the Accademia" sounds like the statement of a duty, that you will have to go there (even against your will). Further research will be needed to map out the conditions under which all these modal utterances are felicitous.

The Role of Utilities in a Speech Act Theory of Conditionals

We expect that the core of this speech act theory of conditionals will contain a calculus of utilities concerning what is preferable for whom. Specifying utilities in this way will avoid the problem of circularity in the account given for deontic terms. For example, the pragmatic reasoning schemas of Cheng and Holyoak (1985) themselves contain *may* and *must*, not to mention *if* and *only if*, and there should be a noncircular account of these terms (Evans & Over, 2004; Over et al., 2004). The same point could be made about the use of "permissible" and "impermissible" as labels on the mental models of Bucciarelli and Johnson-Laird (2005). It seems likely that such an analysis of utilities will be needed in a noncircular theory of what is permissible and impermissible in the kind of deontic reasoning we have studied. Such an approach would be consistent with decision-theoretic approaches to deontic reasoning (Over et al., 2004; Perham & Oaksford, 2005).

Finally, it seems likely that the rational self-interest model that underlies current decision-theoretic approaches to conditionals (Evans & Over, 1996) will have to be extended to incorporate collective benefits in the way that is done by economic game theory (cf. Cosmides, 1989). This point can be illustrated by a bank robbery example. Although observers will understand and normally pardon a bank worker for handing over the money when threatened with her life (she was acting in her self-interest), the outcome (contributing to a robbery) will nevertheless meet social disapproval as being the kind of event that relevant collectives (the bank, society in general) generally wish to avoid. In contrast, her institutional obligation to hand over money to the bank client does not depend on self-interested calculations of this type—in fact, this transaction can be and often is performed by an unthinking banking machine on insertion of the appropriate banking card and code number. And in handing over the money, the bank worker (or bank machine) will be contributing to an outcome (the orderly protection and use of personal property in the form of bank savings) that will meet social approval as being the kind of event that the relevant collectives (the bank, society) generally wish to promote. An important question for future research will therefore be to establish how and why such utilities motivate an authority to give permission and impose duties and then to guide individuals in their interpretations of those rights and duties.

Finally, understanding how a deontic statement is interpreted is vital for understanding how it will be evaluated. For example, the perception of violation may depend on the perspective taken (cf. Gigerenzer & Hug, 1992; Manktelow & Over, 1991; Politzer & Nguyen-Xuan, 1992). It is possible to imagine that a rule such as *All those who are unemployed must receive benefits* could be interpreted as an instruction that a government official in the social security office might give to a junior official. In this case failing to award benefit when it is justified could be perceived as an error on the part of the junior official, making *A's that are not B's* violations of the rule, involving costs for the social security office's reputation for fairness. Alternatively, it is quite possible for recipients to inadvertently receive state benefit through administrative error, meaning that *not-A's that are B's* are violations because they are mistakes, and not because the recipients in question have cheated. This could involve a cost to the department's reputation for rigor. Hence, an important question for future research will be to test whether embedding these rules in contexts that invited such interpretations would also change the experimental participants' judgments of permissibility as well as their perception of costs and benefits for their organization.

Conclusion

We have examined various pragmatic approaches to understanding deontic conditionals. Previous work on pragmatic reasoning schemas (Cheng & Holyoak, 1985; Holyoak & Cheng, 1995) proposes that these reasoning schemas apply equally to causal and deontic rule contents and fails to address the question of how deontic verbs are interpreted in the context of rules that a priori contradict them (as when *may not* is used in the context of a duty). In normal circumstances *not-A & B* cases may be expected to be considered violations of rules perceived as rights, and *A & not-B* cases of rules perceived as duties, in line with these content-dependent pragmatic reasoning schemas perspective (Cheng & Holyoak, 1985). However, these pragmatic reasoning schemas are not content dependent, as it is possible to imagine contexts of use where a script-based content that is normally associated with a duty (e.g., punching tickets in order to board a train in France) is used to express a permission (e.g., *Passengers on SNCF trains need not punch their tickets*). Duties that normally state obligations can in certain contexts be used to state permissions, and rights that normally state permissions can be used to state obligations. Recognition of this possibility underscores the importance of distinguishing the content of institutional deontic rules from their performative use. Finally, the pragmatic reasoning schemas approach cannot account for the fact that causal statements can be used indirectly to convey deontic advice and warnings.

Bucciarelli and Johnson-Laird (2005) have proposed a model for interpreting deontic verbs, but their suggestion that ambiguity in deontic expressions can be handled by pragmatic modulation seems insufficiently specified to capture the ways in which deontic verbs can interact with deontic rule contents (Hilton, Charalambides, & Hoareau, in press). Focusing and directionality effects seem to be important to understanding important pragmatic aspects of the meaning of deontic conditionals (Bucciarelli & Johnson-Laird, 2005; Schmeltzer & Hilton, 2014) but do not seem to mobilize reasoning about the involved actors' interests. A pragmatic theory that explicates the role of utilities in a social matrix will be necessary for an understanding of how humans communicate, understand, and reason about institutional deontics in context. It will need to address effects such as salience, linguistic polarity, and relevance. Such a theory must be capable of explaining who says what to whom, how, and why.

Acknowledgments

I thank Laetitia Charalambides, Bertrand Faure, David Over, and Christophe Schmeltzer for helpful discussions that have influenced much of what is written in this chapter. The research reviewed here was supported by the Agence Nationale de la Recherche (Grant NT05–4_45862).

References

Abelson, R. P. (1981). Psychological status of the script concept. *American Psychologist, 36*(7), 715.

Austin, J. L. (1961). *How to do things with words*. Oxford: Clarendon Press.

Beller, S. (2008). Deontic norms, deontic reasoning, and deontic conditionals. *Thinking & Reasoning, 14*, 305–341.

Bucciarelli, M., & Johnson-Laird, P. N. (2005). Naïve deontics: A theory of meaning, representation and reasoning. *Cognitive Psychology*, *50*, 159–193.

Cheng, P. W., & Holyoak, K. J. (1985). Pragmatic reasoning schemas. *Cognitive Psychology*, *17*, 391–416.

Cheng, P. W., & Holyoak, K. J. (1989). On the natural selection of reasoning theories. *Cognition*, *33*(3), 285–313.

Cheng, P. W., Holyoak, K. J., Nisbett, R. E., & Oliver, L. M. (1986). Syntactic versus pragmatic approaches to training deductive reasoning. *Cognitive Psychology*, *18*, 293–328.

Cosmides, L. (1989). The logic of social exchange: Has natural selection shaped how we humans reason? Studies with the Wason selection task. *Cognition*, *31*, 187–276.

Elqayam, S., Thompson, V. A., Wilkinson, M. R., Evans, J. S. B., & Over, D. E. (in press). Deontic introduction: A theory of inference from *is* to *ought*. *Journal of Experimental Psychology: Learning, Memory, and Cognition*.

Evans, J. S. B. T., & Over, D. E. (2004). *If*. Oxford: Oxford University Press.

Evans, J. S. B. T., & Over, D. E. (1996). *Rationality and reasoning*. Hove, UK: Psychology Press.

Fiddick, L., & Grafman, J. (2008). *Separating pacts from precautions*. Unpublished manuscript, Dept. of Psychology, James Cook University, Queensland, Australia.

Gigerenzer, G., & Hug, K. (1992). Domain-specific reasoning: Social contracts, cheating, and perspective change. *Cognition*, *43*(2), 127–171.

Grice, H. P. (1975). Logic and conversation. In P. Cole & J. L. Morgan (Eds.), *Syntax and semantic 3: Speech acts* (pp. 41–58). New York: Academic Press.

Hart, H. L. A. (1961). *The concept of law*. Oxford: Oxford University Press.

Hilton, D. J. (2009). Conversational inference: Social cognition as interactional intelligence. In F. Strack & J. Förster (Eds.), *Social cognition: The basis of human interaction* (pp. 71–92). New York: Psychology Press.

Hilton, D. J., Charalambides, L., & Hoareau-Blanchet, S. (in press). Reasoning about rights and duties: Mental models, world knowledge and pragmatic interpretation. *Thinking & Reasoning*.

Hilton, D. J., Kemmelmeier, M., & Bonnefon, J.-F. (2005). Putting ifs to work: Goal-based relevance in conditional directives. *Journal of Experimental Psychology*, *134*(3), 388–405.

Holyoak, K. J., & Cheng, P. W. (1995). Pragmatic reasoning with a point of view. *Thinking & Reasoning*, *1*(4), 289–313.

Johnson-Laird, P. N. (1978). The meaning of modality. *Cognitive Science*, *2*, 17–26.

Kilpatrick, S. G., Manktelow, K. I., & Over, D. E. (2007). Power of source as a factor in deontic inference. *Thinking & Reasoning*, *13*(3), 295–317.

Legrenzi, P., Politzer, G., & Girotto, V. (1996). Contract proposals: A sketch of a grammar. *Theory & Psychology*, *6*(2), 247–265.

Lyons, J. (1977). *Semantics* (Vol. 1). Cambridge: Cambridge University Press.

Mackie, J. L. (1977). *Ethics: Inventing right and wrong*. Harmondsworth: Penguin.

Manktelow, K. I., & Over, D. E. (1991). Social roles and utilities in reasoning with deontic conditionals. *Cognition*, *39*, 85–105.

Moxey, L. M., & Sanford, A. J. (1993). *Communicating quantities: A psychological perspective*. Hove, UK: Lawrence Erlbaum Associates.

Over, D. E., Manktelow, K. I., & Hadjichristidis, C. (2004). Conditions for the acceptance of deontic conditionals. *Canadian Journal of Experimental Psychology*, *58*, 96–105.

Perham, N. R., & Oaksford, M. (2005). Deontic reasoning with emotional content: Evolutionary psychology or decision theory? *Cognitive Science*, *29*, 681–718.

Politzer, G., & Nguyen-Xuan, A. (1992). Reasoning about conditional promises and warnings: Darwinian algorithms, mental models, relevance judgments or pragmatic schemas? *Quarterly Journal of Experimental Psychology*, *44A*, 401–421.

Sarbin, T. (1954). Role theory. In G. Lindzey (Ed.), *Handbook of social psychology* (pp. 223–258). Cambridge, MA: Addison-Wesley.

Schank, R. C., & Abelson, R. P. (1977). *Scripts, plans, goals and understanding: An enquiry into human knowledge structures*. Hillsdale, NJ: Lawrence Erlbaum Associates.

Schmeltzer, C. S., & Hilton, D. J. (2014). To do or not to do? A cognitive consistency model for drawing conclusions from conditional instructions and advice. *Thinking & Reasoning*, *20*(1), 16–50.

Searle, J. (1964). How to derive "ought" from "is." *Philosophical Review*, *73*, 43–58.

Searle, J. (1965). What is a speech act? In M. Black (Ed.), *Philosophy in America* (pp. 221–239). London: Allen and Unwin.

10   Why Don't People Produce Better Arguments?

Hugo Mercier, Pierre Bonnier, and Emmanuel Trouche

That people produce arguments of low quality has been a recurring complaint from scholars of informal reasoning (Kuhn, 1991; Perkins, Farady, & Bushey, 1991) and formal reasoning (Evans, 2002), and from social psychologists (Nisbett & Ross, 1980). One of the main issues is that people tend to produce arguments that are one-sided (the one side always being their side) (Baron, 1995; Nickerson, 1998) and that they have trouble finding arguments for any other position (e.g., Kuhn, 1991). However, this is not the only problem: these biased arguments are often weak, making only "superficial sense" (Perkins, 1985, p. 568), as if people were content with the first argument that crossed their mind (Nisbett & Ross, 1980, p. 119).

These conclusions, reached in the study of arguments actually produced by participants, are bolstered by reasoning's failure to correct participants' intuitions in many tasks (e.g., Frederick, 2005; Wason, 1966). When people persist, after several minutes of reasoning, in providing the wrong answer to a simple logical or mathematical problem, it means not only that they mostly looked for arguments supporting their initial, wrong intuition but also that they were satisfied with the arguments they found—arguments that were necessarily flawed given the nature of the tasks.

Moreover, recent research bearing on confidence in reasoning has revealed that participants are often very confident in their arguments, even when they are faulty (De Neys, Cromheeke, & Osman, 2011; Shynkaruk & Thompson, 2006). In particular, a study by Trouche, Sander, and Mercier (2014) asked participants, for a standard reasoning problem, to evaluate not only their confidence in their answers but also their confidence in the reasons for their answers. The participants who gave the intuitive but wrong answer were highly confident not only in their answer but also in the—necessarily faulty—reasons for the answer.

The objective of this chapter is to explain why people seem to produce such weak arguments. In the first section we lay out how two theories of the function of reasoning—the classical theory and the argumentative theory—account for reasoning's apparent limitations. The argumentative theory, we contend, can easily explain some of these apparent limitations, such as the *myside bias*, as well as more obviously adaptive features of reasoning such as the ability to properly evaluate others' arguments. However, it is less clear how the argumentative theory can be reconciled with the low quality of arguments produced by reasoning.

Here we offer an explanation that rests on the dialogic nature of argumentation. When people produce arguments in a dialogue, they can rely on their interlocutor to explain why they find the argument defective or weak. Relying on interlocutor feedback is often more effective than trying to anticipate what a better argument would be. We then describe in more details how this account explains the various types of argument failures.

Two Theories of Reasoning and Two Features of Argument Production

It is useful to distinguish two well-established traits of argument production: the tendency to find arguments that support the reasoner's side and the tendency to be satisfied by relatively weak arguments. The first of these traits has been the focus of intense study, generally under the name of confirmation bias (Nickerson, 1998). Although this research has soundly established that reasoning is biased, the word "confirmation" is a misnomer: reasoning does not seek to confirm everything, only beliefs the reasoner endorses. By contrast, when reasoning bears on beliefs the reasoner disagrees with, it produces counterexamples, counterarguments, and other ways to falsify the beliefs (see Mercier & Sperber, 2011). As a result, it is more accurate to talk of a *myside bias* (Mercier, in press).

The myside bias flies in the face of the classical theory of (the function of) reasoning. Most scholars who have speculated about the function of reasoning postulate that it has a chiefly individual function: to correct the reasoner's mistaken intuitions, thereby guiding her toward better beliefs and decisions (Evans, 2008; Kahneman, 2003; Stanovich, 2004). To perform this function properly, reasoning should either impartially look for reasons why the reasoner might be right or wrong or, even better, preferentially look for reasons she might be wrong. Because of the myside bias, reasoning does the exact opposite, behaving in a way that is difficult to reconcile with the classical theory of reasoning.

The second apparent limitation of reasoning—that it tends to produce relatively weak arguments—has been the focus of less intense scrutiny. Yet it is no less problematic than the myside bias. If people applied very high-quality criteria to their own arguments, the effects of the myside bias would be much softened. In some cases—for instance in logical or mathematical tasks—people would have to admit that there are no good reasons for the intuitive but wrong answer, and they would be forced to change their mind. Again, the classical theory of reasoning should predict the exact opposite: in order to make sure our intuitions do not lead us astray, reasoning should check that we have good reasons for them, not just any reason.

To reconcile these two traits of reasoning—the production of biased *and* weak arguments—with the classical theory of reasoning, psychologists often invoke cognitive limitations such as low working memory (e.g., Evans, 2008). However, cognitive limitations cannot be the main explanation for these traits because reasoning exhibits them only when it produces arguments. When reasoning evaluates other people's arguments, it becomes (relatively) objective and exigent. It is objective because it accepts strong arguments and then leads the reasoner to change her mind. It is exigent because it rejects weak arguments. The good performance in reasoning tasks following group discussion demonstrates both traits of argument evaluation: if it were not (relatively) objective, people would reject the arguments for the good answer; if it were not (relatively) exigent, people would just as likely be convinced by arguments for the wrong answer (e.g., Laughlin & Ellis, 1986; Moshman & Geil, 1998). A large literature in social psychology also shows that when they care about the conclusion of an argument, people change their mind more in response to strong than to weak arguments (see Petty & Wegener, 1998). Other experiments have shown that participants are not easily swayed by straightforward fallacies but that they react appropriately to sounder versions of the same arguments (Hahn & Oaksford, 2007). Given that people are often careless in the evaluation of their own arguments—they produce relatively weak arguments—but that they judge other people's arguments more stringently, we suggest calling this property of reasoning *asymmetric argument evaluation* (Trouche, Johansson, Hall & Mercier, in press).

The argumentative theory of reasoning was developed as an alternative to the classical theory of reasoning (Mercier & Sperber, 2011). Instead of postulating an individual function for reasoning, it grounds the evolution of reasoning in the logic of the evolution of communication. Humans' reliance on communication creates selection pressures for mechanisms that protect receivers from the potentially harmful information communicated by senders (Sperber et al., 2010). To protect themselves from misleading messages, people evaluate both the content and the source of communicated information: if either is found wanting, then the information is rejected. However, these mechanisms have stringent limits: sometimes people would be better off accepting information that flies in the face of their existing beliefs and that is communicated by a source not entirely trusted—sometimes others, even others we do not fully trust, know better than we do. This "trust ceiling" affects both senders and receivers: senders fail to transmit their message, and receivers miss out on potentially valuable information. A solution is for senders to provide reasons supporting the message they want to transmit. Receivers can then evaluate these reasons and, if they are deemed sufficient, change their mind, not because they trust the sender but because it makes more sense for them to accept her message than to reject it. According to the argumentative theory, reasoning evolved chiefly to enable argumentation.

This hypothesis readily accounts for some features of reasoning described above. First, it is essential that reasoning should be able to reject weak arguments—otherwise manipulation would be altogether too easy, and people would be better off not listening to any arguments. This is what we observe: when it matters, people are not easily swayed by poor arguments.

When it comes to producing arguments, conviction is most likely to be achieved by finding arguments that support the reasoner's side or go against her interlocutor's. The myside bias is a normal feature of reasoning when it is understood as performing an argumentative function.

However, if the function of reasoning is to convince, one might expect reasoning to produce strong arguments. As we have seen, this is often not the case. We presently offer an explanation for this feature of reasoning that relies on the fit between features of reasoning and dialogical contexts—as the argumentative theory would predict. In the next section we introduce the relevant properties of dialogic contexts in the general case before exploring in more detail the case of argumentation.

Repair in Communication

Even though an interlocutor can understand what a speaker means without accepting the message, conversational contexts entail a very high overlap of interests: the speaker wants the interlocutor to understand what she means, and the interlocutor wants to understand what the speaker means. As a result, the burden of making communication as efficient as possible does not fall only on the speaker but is shared by the interlocutor, a division of labor that has been studied in linguistics (e.g., Clark & Wilkes-Gibbs, 1986; Sacks, Schegloff, & Jefferson, 1974; Schegloff, Jefferson, & Sacks, 1977; Schegloff & Sacks, 1973).

In the following example the first speaker, A, wants the interlocutor, B, to understand that he is referring to the Ford family (from Sacks & Schegloff, 1979, p. 19; cited in Levinson, 2006).

A: … well I was the only one other than the uhm tch Fords? Uh, Mrs. Holmes Ford? You know, uh, the … the cellist?

B: Oh yes. She's … she's the cellist.

A: Yes. Well she and …

A starts by simply saying "the Fords," but he does not stop there. He then points out one member of the family in particular ("Mrs. Holmes Ford") before specifying her occupation ("the cellist"). This repair was likely initiated by the lack of expected positive feedback following the first attempt to refer to the Ford family. Simply by failing to communicate that she understood who the Fords are, B made it clear that she required more information to understand the referent of A's utterance (see, e.g., Goodwin, 1981).

Whether they are "self-repairs" (as in the present example) or "other repairs," repairs are ubiquitous in verbal communication. Such repairs can follow genuine failures, for instance when the speaker chooses the wrong word. The present example might seem to reflect a failure as well: failure of the speaker to choose the optimal way to refer to the Ford family from the start. However, as argued by the linguists who have studied these repairs, such interactions should instead be understood as reflecting the efficient working of communication.

For A to find the best way to refer to the Ford family, he needs to know how B knows the Fords. This information could be easily accessible—if, for instance, A knows very well that B is a good friend of the Fords—in which case A would have no trouble referring to them. Or it might be nearly impossible to access: maybe A knows neither B nor the Fords very well and has only vague hunches about how much they know each other. In this case, A has three possible strategies. The first strategy is to think long and hard about whether B knows the Fords or not, digging into his memory and his inferential abilities to make the best possible guess. The second strategy is to provide B with an exhaustive list of the information he has about the Fords to maximize the chances that B understands whom A is talking about. The third strategy is the one A chooses: to start with the common way of referring to a family in this context and proceed to offer more clues to who they are until B indicates that she understands.

Although the first two solutions seem superficially more efficient—there might be fewer conversational turns—they are in fact more costly: A either has to take time and energy to figure out something that would be trivially revealed in the course of the conversation, or A has to make a long speech that might be irrelevant if B recognizes the Fords immediately. By contrast, a few conversational turns offer a very economical alternative: what looks like a failure might in fact be the most efficient system given the constraints.

"Repair" in Argumentation

How does this logic apply to argumentation? If the argumentative theory of reasoning is correct, argumentation solves a problem that affects both senders and receivers, so that senders have an incentive to communicate the best available reasons for their messages, and receivers have an incentive to understand what these reasons are. As in other forms of communication, the alignment of interests is not perfect, but it is strong.

As a result, the logic described above also applies to argumentation. Instead of laboring to find the strongest possible argument from the start, interlocutors can make the best of the interactive context and refine their arguments as the exchange unfolds. Indeed, argumentation should rely even more on feedback than other forms of communication. Finding good arguments is likely to be harder than finding, say, the best way to refer to someone. Fortunately, the difficulty of the task is mitigated by the richness of the feedback. Instead of a mere indication of understanding or failure of understanding, interlocutors who reject an argument often state their reasons for doing so, offering the speaker an opportunity to understand and address these reasons.

Take the following excerpt from a discussion among three students on the topic of nuclear power (from Resnick, Salmon, Zeitz, Wathen, & Holowchak, 1993, p. 350):

C4: Well, uh is, is nuclear, I'm against it. . . Is nuclear power really cleaner that fossil fuels? I don't think so.

A5: You don't think, I think …

B6: In terms of atmospheric pollution I think that … the waste from nuclear power, I think it's … much less than fossil fuels … but the waste that there is of course is quite dangerous.

C7: It's gonna be here for thousands of years, you can't do anything with it. I mean, right now we do not have the technology as …

B8: Acid rain lasts a long time too you know.

C9: That's true, but if you reduce the emissions of fossil fuels, which you can do with, uh, certain technology that we do have right now, um, such as scrubbers and such, you can reduce the acid rain, with the nuclear power you can't do any, I mean nuclear waste you cannot do anything with it except …

B10: Bury it.

A11: M-hm …

C12: Bury it and then you're not even sure if its ecologically … um … that the place you bury it is ecologically sound.

B13: I, I think if … if enough money is spent it can probably be put in a reasonably safe area.

Here we can see instances of "self-initiated repair," for instance at *C7*, when *C* specifies that what he meant was that it is impossible to get rid of nuclear waste for good. A rebuttal to a counterargument can be seen as a form of "other-initiated repair," as for example in *C7/B8/C9*, when *C* addresses *B*'s counterargument by spelling out his argument in more detail: it is not only that nuclear wastes are long lasting but that, given current technology, the damage created by other wastes is shorter lived than that of nuclear waste.

C could have tried to anticipate the counterargument offered by *B*. However, in doing so he would have been likely to think of counterarguments that *B* would never have thought of, or that she would not subscribe to, and to miss the counterargument she actually offered. In most cases such anticipation has high costs—cognitively—and little benefits because the interlocutor will give her counterarguments herself. So why bother?

People's ability to adapt and refine their arguments in the course of a discussion has been observed in various contexts such as discussions of contentious topics (Kuhn &

Crowell, 2011; Resnick et al., 1993), of logical tasks (Trognon, Batt, & Laux, 2011; Trognon, 1993), and of classroom tasks (Anderson, Chinn, Chang, Waggoner, & Yi, 1997). However, it remains an understudied topic.

The Various Ways in which Arguments Can Fail to Convince

The explanation above is not very fine-grained: it accounts for the overall limited quality of arguments, especially those most studied by psychologists, which correspond to what should only be the first turn of a discussion. To better understand what it means that people produce relatively weak arguments, it is useful to look in more detail at the various ways in which arguments can fail to convince their intended audience.[1] We suggest that there are two main stages at which this can happen.

The first stage bears on the intrinsic quality of the argument: is it a reason at all to support the conclusion? An argument that is found wanting at this point can be called *defective*. In turn, an argument can be defective in different ways, which can be categorized as *external* and *internal*. When an argument is externally defective, the audience either disagrees with or simply misses a premise (often an implicit premise). Here are two examples:

(1) *Laura:* You should go see this movie. It's by Stanley Kubrick.

George: I don't know who he is.

(2) *Laura:* You should go see this movie. It's by Stanley Kubrick.

George: I don't really like his movies.

In (1), the argument fails because Laura did not anticipate that George would not know the implicit premise (movies by Stanley Kubrick are worth watching), in (2) it fails because George disagrees with this premise.

By contrast, an internal failure happens when the argument is inherently flawed, as in (3):

(3) *Laura:* You should go see Dr. Strangelove rather than *Eyes Wide Shut*. It's by Stanley Kubrick.
George: But *Eyes Wide Shut* is also by Kubrick!

In this case nothing can be done to salvage the argument. In a simple logical or mathematical task, all the arguments for any wrong answer must fail internally.

Even an argument that is not found to be defective can fail to convince at the next stage of argument evaluation because it is *too weak*. For instance:

(4) *Laura:* You should go see *Eyes Wide Shut*. It's by Stanley Kubrick.
George: I love Kubrick, but I really hate Tom Cruise, so I think I'll pass.

Here Laura's argument is accepted by George, but it is not sufficient to convince him. The argument is simply not strong enough to change his mind. Even though we call this a failure here for simplicity, arguments that are found to be weak range from the too weak to have any effect to the nearly strong enough to tip the scales. In the latter case, adding even a relatively weak argument might suffice, so that even though the initial argument failed to completely convince the interlocutor, it will have played the major role when she eventually changes her mind.

With the exception of the internal failures, all the other types of failures reflect a lack of perspective taking: the speaker fails to properly anticipate that the interlocutor does not hold a given belief, or has a stronger belief in the conclusion than anticipated. These failures are related to a more general and well-studied failure of perspective taking known as the *curse of knowledge* (Birch & Bloom, 2007), the *false consensus effect* (Krueger & Clement, 1994), or simply *egocentrism* (Nickerson, 1999).

As argued above, these failures (again, with the exception of the internal kind) are often not very costly. They do not mean the conversation is over: more arguments can be added. In particular, external failures can be fixed by trying to change the interlocutor's mind about the problematic premise. Here, Laura could inform George that Kubrick is a widely respected director—in (1)—or try to convince George of the value of Kubrick's movies—in (2).

Failed Argument or Successful Explanation?

We now argue that in some cases these failures are only apparent, not real failures: it depends on what the objective of putting forward the argument is. One way in which reasoning can solve "trust bottlenecks" is by allowing people to provide arguments in order to convince others, as explained above. However, reasoning can also help alleviate problems of trust by enabling people to justify their decisions.

Figuring out why people do the things they do can be fiendishly difficult. When we fail to reconstruct the reasons for a given behavior, it will appear irrational. If we based our evaluation of others on the unaided understanding of their behavior, we would often be led to conclude that they are not very competent and therefore not very reliable or trustworthy. Reasoning can help solve this problem by letting people explain their apparently irrational behaviors. As in the case of argumentation, interlocutors can then evaluate these reasons to see if they are indeed good reasons. This solution is efficient because (a) it is much easier for the person who engaged in a given behavior to provide a reason for it than it is for most observers, and (b) it is easier for the observer to evaluate a reason provided to her than to figure it out on her own.

There are, however, crucial differences in the way rational explanations of behavior (or of thoughts), on the one hand, and arguments on the other ought to be evaluated (for another take on this issue, see, e.g., Bex, Budzynska, & Walton, 2012). For an explanation to be good, it has to make sense from the point of view of the speaker. By contrast, an argument has to be good from the point of view of the interlocutor. Accordingly, for explanations external failures are not failures anymore. Consider this variation on (3):

(3') *Laura:* I think I will go see this movie. It's by Stanley Kubrick.
George: I don't really like his movies.

Here George's reaction should not be understood as a refutation of Laura's explanation but simply as a statement of opinion. To the extent that George can easily fill in the implicit premise—that Laura likes Kubrick—then he should not find the explanation defective, even if he disagrees with the premise.

Similarly, explanations are less likely to be found to be too weak: they do not have to be strong enough to overcome the interlocutor's belief but simply strong enough to warrant the speaker's belief. Again, consider a variation on the preceding dialogue:

(4') **Laura:** I think I will go see *Eyes Wide Shut.* It's by Stanley Kubrick.
George: I love Kubrick, but I really hate Tom Cruise.

As long as George does not have a reason to think that Laura shares his distaste for Cruise or that she has a stronger reason to not see this movie, then he should find the explanation sound.

In many psychological experiments, reasoning might be triggered more as a way of justifying the participant's position, making sure that she stands on rational grounds, rather than as a way of trying to convince someone. For instance, in a typical reasoning task, people do not really care if others hold the same beliefs regarding the right answer. If they are motivated to reason, it is more likely to be as a way to ensure that they can provide an explanation for their answer, to show that it is rational. Even when participants are explicitly asked to defend their opinions on, say, public policy, as in Kuhn (1991), they do not actually face someone who disagrees with them and whom they would really like to convince. Although argument failures are to be expected even in a genuine argumentative discussion—for the reasons exposed above—people might be more motivated to engage in some perspective taking, and therefore avoid some argument failures, when they really aim at convincing someone of the argument's conclusion rather than of their own rationality.

Are Others Better at Detecting Internal Argument Failures?

The production of externally defective arguments and arguments too weak to convince on their own is caused by the difficulties of perspective taking: the speaker either cannot anticipate or does not make the effort to anticipate the beliefs of the interlocutor. This is exactly what one should expect to happen in interactive contexts, when it often makes more sense to let the interlocutor inform the speaker of her beliefs than to force the speaker to anticipate them. Moreover, if the goal of the speaker is to explain her position rather than to convince, most of these "failures" are not failures at all.

Internal failures are not so easily explained. They reflect ignorance about the world (rather than about the interlocutor's beliefs), failures of inference, or failures of memory. For instance, in most reasoning tasks people provide arguments that are internally faulty, and they fail to make the inferences that would enable them to realize this. Crucially, arguments that fail internally are also poor explanations: although they show that the speaker had *a* reason for her position or behavior, they reveal that this was a poor reason even from the speaker's perspective.

Even though internal failures cannot be explained as part of a well-functioning division of cognitive labor between speaker and interlocutor, they would not have to be especially mysterious if it were not for one of their features. After all, every cognitive system is bound to make some mistakes, as it does not have infinite resources. What makes internal argument failures interesting, however, is the asymmetry mentioned above: people seem to be better at spotting such failures in others than in themselves. We suggest that there are two types of explanations. The first is simply of a difference in background beliefs or in their accessibility between the speaker and the interlocutor. In example (3), George might have more knowledge about Kubrick, or he might have just been thinking about who the director of *Eyes Wide Shut* is. There is no reason, however, that interlocutors should, on average, be more likely to have access to the relevant beliefs than speakers. What explains the asymmetry is that when a speaker accesses the relevant beliefs, then he does not produce the argument at all, so there is no observable behavior. By contrast when the interlocutor does, then we can observe the defective argument being corrected.

The second explanation is more interesting. A speaker produces an argument that is, in fact, internally defective. At first, the interlocutor might simply find it too weak to change his mind but not defective. He would then likely engage in a search for counterarguments in order either to justify not changing his mind or to convince the speaker to change her mind (or both). In the process he might find arguments that support his point of view without attacking the speaker's initial argument— as in (4), for instance. But he might also find arguments that specifically target the speaker's initial argument. Such arguments are likely to reveal the defect in the initial argument—as in (3), for example. In this case the apparent difference in the way speakers and interlocutors evaluate arguments—that the speaker found the argument good enough to produce whereas the interlocutor found it defective—does not reflect a difference in evaluation *stricto sensu* but a difference in evaluation stemming from the production of a counterargument by the interlocutor. The fact that the interlocutor is more likely to find such a counterargument is a simple consequence of the myside bias.

Conclusion

Researchers who have studied argument production generally agree that quality is not very high: people routinely produce arguments that are weak or easily countered. This is a problem for the classical theory of reasoning: if reasoning's goal is to improve individual cognition, it should make sure we have good reasons for our beliefs or decisions. But this also seems to be an issue for the argumentative theory of reasoning: if the function of reasoning is to convince, then would it not be better to produce strong, convincing arguments?

We have argued that, counterintuitively, not aiming at very strong arguments is the best strategy for a device working in an interactive, cooperative context. Finding arguments that will appeal to a particular interlocutor entails having a solid grasp of the interlocutor's beliefs, making it an arduous, cognitively costly task. Instead of trying to anticipate the interlocutor's belief, it is possible to start an argumentative discussion by offering an argument that passes some minimal threshold of quality and wait for the interlocutor's feedback. If the argument does not convince the interlocutor, he will often provide the speaker with an explanation. This enables the speaker to mend her argument, to adjust it to the interlocutor's relevant beliefs.

In this perspective most argument failures are better seen as steps in a normal process of interaction. Moreover, if the goal of the reasoner is to justify her behavior rather than to convince the interlocutor of a given conclusion, then most argument failures are not even failures to begin with: they are perfectly acceptable explanations.

The only exceptions are internally defective arguments, arguments that contradict beliefs the speaker ought to have considered. These arguments not only are unconvincing but also make for poor explanations. Particularly puzzling is the asymmetry in the evaluation of internally defective arguments: why would the interlocutor be in a better position to spot such failures than the speaker, given that the argument clashes with the speaker's beliefs? We suggested that interlocutors might not, at first, be more likely to spot the defect in the argument, but that in the process of looking for a counterargument, they might find one that reveals the defect in the argument. The search for counterargument is guided by the myside bias; therefore, the argumentative theory can also account for the asymmetry in the evaluation of internally defective arguments.

Although the study of argumentative discussion, with the interactions they entail, is fraught with methodological difficulties, it is the best place to reach a genuine understanding of reasoning's strengths and (supposed) failures.

Acknowledgments

We would like to thank Steve Oswald for his very useful feedback.

Note

1. We are not taking a normative stance here and making, for instance, a distinction between whether the argument should convince based on its soundness or validity or whether it should merely "persuade." We aim at describing psychological mechanisms, and conviction is obtained when, or to the extent that, the interlocutor changes her mind. Accordingly, we urge the reader not to think of any normative framework in reading what follows: when we introduce "intrinsic quality," it will refer only to the way it is defined here, not to the more notion of logical validity.

References

Anderson, R. C., Chinn, C., Chang, J., Waggoner, M., & Yi, H. (1997). On the logical integrity of children's arguments. *Cognition and Instruction, 15*(2), 135–167.

Baron, J. (1995). Myside bias in thinking about abortion. *Thinking & Reasoning, 1*, 221–235.

Bex, F., Budzynska, K., & Walton, D. (2012). Argument and explanation in the context of dialogue. In T. Roth-Berghofer, D. B. Leake, & J. Cassens (Eds.), *Proceedings of the 7th international workshop on explanation-aware computing* (pp. 6–10).

Birch, S. A., & Bloom, P. (2007). The curse of knowledge in reasoning about false beliefs. *Psychological Science, 18*(5), 382–386.

Clark, H. H., & Wilkes-Gibbs, D. (1986). Referring as a collaborative process. *Cognition, 22*(1), 1–39.

De Neys, W., Cromheeke, S., & Osman, M. (2011). Biased but in doubt: Conflict and decision confidence. *PLoS One, 6*(1), e15954.

Evans, J. S. B. T. (2002). Logic and human reasoning: An assessment of the deduction paradigm. *Psychological Bulletin, 128*(6), 978–996.

Evans, J. S. B. T. (2008). Dual-processing accounts of reasoning, judgment and social cognition. *Annual Review of Psychology, 59*, 255–278.

Frederick, S. (2005). Cognitive reflection and decision making. *Journal of Economic Perspectives, 19*(4), 25–42.

Goodwin, C. (1981). *Conversational organization: Interaction between speakers and hearers.* New York: Academic Press.

Hahn, U., & Oaksford, M. (2007). The rationality of informal argumentation: A Bayesian approach to reasoning fallacies. *Psychological Review, 114*(3), 704–732.

Kahneman, D. (2003). A perspective on judgment and choice: Mapping bounded rationality. *American Psychologist, 58*(9), 697–720.

Krueger, J., & Clement, R. W. (1994). The truly false consensus effect: An ineradicable and ego-centric bias in social perception. *Journal of Personality and Social Psychology*, *67*(4), 596.

Kuhn, D. (1991). *The skills of arguments*. Cambridge: Cambridge University Press.

Kuhn, D., & Crowell, A. (2011). Dialogic argumentation as a vehicle for developing young adolescents' thinking. *Psychological Science*, *22*(4), 545.

Laughlin, P. R., & Ellis, A. L. (1986). Demonstrability and social combination processes on mathematical intellective tasks. *Journal of Experimental Social Psychology*, *22*, 177–189.

Levinson, S. C. (2006). On the human "interaction engine." In N. J. Enfield & S. C. Levinson (Eds.), *Roots of human sociality* (pp. 39–69). Oxford: Berg.

Mercier, H. (in press). Confirmation bias—Myside bias. In R. Pohl (Ed.), *Cognitive illusions* (2nd ed.). Hove, UK: Psychology Press.

Mercier, H., & Sperber, D. (2011). Why do humans reason? Arguments for an argumentative theory. *Behavioral and Brain Sciences*, *34*(2), 57–74.

Moshman, D., & Geil, M. (1998). Collaborative reasoning: Evidence for collective rationality. *Thinking & Reasoning*, *4*(3), 231–248.

Nickerson, R. S. (1998). Confirmation bias: A ubiquitous phenomena in many guises. *Review of General Psychology*, *2*, 175–220.

Nickerson, R. S. (1999). How we know—and sometimes misjudge—what others know: Imputing one's own knowledge to others. *Psychological Bulletin*, *125*, 737–759.

Nisbett, R. E., & Ross, L. (1980). *Human inference: Strategies and shortcomings of social judgment*. Englewood Cliffs, NJ: Prentice-Hall.

Perkins, D. N. (1985). Postprimary education has little impact on informal reasoning. *Journal of Educational Psychology*, *77*, 562–571.

Perkins, D. N., Farady, M., & Bushey, B. (1991). Everyday reasoning and the roots of intelligence. In J. Voss, D. Perkins, & J. Segal (Eds.), *Informal reasoning and education* (pp. 83–105). Hillsdale, NJ: Lawrence Erlbaum Associates.

Petty, R. E., & Wegener, D. T. (1998). Attitude change: Multiple roles for persuasion variables. In D. Gilbert, S. Fiske, & G. Lindzey (Eds.), *The handbook of social psychology* (pp. 323–390). Boston: McGraw-Hill.

Resnick, L. B., Salmon, M., Zeitz, C. M., Wathen, S. H., & Holowchak, M. (1993). Reasoning in conversation. *Cognition and Instruction*, *11*(3/4), 347–364.

Sacks, H., & Schegloff, E. A. (1979). Two preferences in the organization of reference to persons in conversation and their interaction. In G. Psathas (Ed.), *Everyday language: Studies in ethnomethodology* (pp. 15–21). New York: Irvington.

Sacks, H., Schegloff, E. A., & Jefferson, G. (1974). A simplest systematics for the organization of turn-taking for conversation. *Language*, *40*(4), 696–735.

Schegloff, E. A., Jefferson, G., & Sacks, H. (1977). The preference for self-correction in the organization of repair in conversation. *Language, 53*(2), 361–382.

Schegloff, E. A., & Sacks, H. (1973). Opening up closings. *Semiotica, 8*(4), 289–327.

Shynkaruk, J. M., & Thompson, V. A. (2006). Confidence and accuracy in deductive reasoning. *Memory & Cognition, 34*(3), 619–632.

Sperber, D., Clément, F., Heintz, C., Mascaro, O., Mercier, H., Origgi, G., et al. (2010). Epistemic vigilance. *Mind & Language, 25*(4), 359–393.

Stanovich, K. E. (2004). *The robot's rebellion.* Chicago: University of Chicago Press.

Trognon, A. (1993). How does the process of interaction work when two interlocutors try to resolve a logical problem? *Cognition and Instruction, 11*(3&4), 325–345.

Trognon, A., Batt, M., & Laux, J. (2011). Why is dialogical solving of a logical problem more effective than individual solving? A formal and experimental study of an abstract version of Wason's task. *Language & Dialogue, 1*(1), 44–78.

Trouche, E., Johansson, P., Hall, L., & Mercier, H. (in press). The selective laziness of reasoning. *Cognitive Science.*

Trouche, E., Sander, E., & Mercier, H. (2014). Arguments, more than confidence, explain the good performance of reasoning groups. *Journal of Experimental Psychology. General, 143*(5), 1958–1971.

Wason, P. C. (1966). Reasoning. In B. M. Foss (Ed.), *New horizons in psychology: I* (pp. 106–137). Harmandsworth, UK: Penguin.

11   Individual Differences in Reasoning beyond Ability and Disposition

Jean-François Bonnefon and Éric Billaut

Works of fiction often depict powerful reasoners, characters whose reasoning performance is essential to advance the plot. These reasoners tend to have rich, quirky personalities, ranging all the way from the chaotic genius to the psychorigid prodigy. On one end we find the likes of Sherlock Holmes or Gregory House, MD. Both are intense, narcissistic characters with no regard for orderliness, social conventions, or healthy habits regarding drug usage. On the other end we find the likes of Nero Wolfe or Spencer Reid from *Criminal Minds*. In opposition to chaotic characters such as Holmes, these orderly heroes value neatness and routine procedures and are often not too keen on physical contact or emotional display. Interestingly, the personalities of all these reasoners seem to affect the way they make inferences. The shabby, bohemian French detective Adamsberg solves mysteries in a dreamlike manner that infuriates some of his colleagues—but the sharp Mr. Spock solves mysteries in a crisp, logical manner that, yes, also infuriates some of his crewmates.

How does this play out in real life? Do different personality types reason differently? To answer this question, we must take baby steps because there is not a lot of research we can build on, even though there is a large literature devoted to individual differences in reasoning. From this literature we know for a fact that different people reason differently. That is, there are stable differences between the conclusions that different people reach from the same data that cannot be attributed to inadvertent mistakes and mere lapses of attention. The challenge, then, is to identify the individual characteristics that track these different conclusions. It so happens that the psychology of reasoning has focused on a rather narrow band of such characteristics (linked to the ability and disposition to engage effortful mental processing) and has not much addressed other personality traits.

In the first part of this chapter we briefly review the historical and epistemological forces that pushed for this research agenda. We then consider the huge territory that this agenda left unexplored: Can we identify personality traits that track stable differences in reasoning? We consider for inspiration the research concerned with the thinking styles associated with psychological disorders and the research on the personality correlates of language use and comprehension. We then explore further the projection effects that occur when reasoners project personality-based beliefs on the world around them. We offer two detailed examples of this phenomenon: the projection of personality signatures (statements describing individual triggers) on conditional syllogisms and the projection of the *belief in the just world* trait on problems featuring two conditional rules that must be conjoined in a single causal model.

Individual Differences in Reasoning

The closest most people come to a reasoning task is when they take an intelligence test. Lay persons can easily be forgiven, then, if they think that the tasks used in the psychology of reasoning must be very similar to the problems featured in an IQ test. There are very important differences, however, between reasoning tasks and IQ tests (Stanovich, West, & Toplak, 2011). The chief difference is that IQ tests are meant to capture subtle differences among people, whereas reasoning tasks typically focus on the modal response, less attention being given to the psychometric properties of the tasks (Bonnefon, Eid, Vautier, & Jmel, 2008; Bonnefon & Vautier, 2008; Bonnefon, Vautier, & Eid, 2007).

This focus on the modal response is probably the legacy of the *deduction paradigm* that had dominated the field for decades (Evans, 2002). One of the key assumptions of the deduction paradigm was that people broadly reasoned the same; that is, they engaged the same cognitive processes and reached the same conclusions, barring an occasional lapse of attention or an unusual construal of the problem. Accordingly, theories of reasoning typically have aimed at accounting for the modal response, treating other responses as measurement noise.

Fast forward to the present. Nowadays, students of reasoning pay close attention to the fact that different reasoners give different responses to the same problem and to the individual characteristics that predict which response will be given by a reasoner (De Neys & Bonnefon, 2013). Most articles, however, seem to focus on just two individual characteristics: the ability and the disposition to engage cognitive effort.

Look at the belief-bias task, for example. Since its introduction 30 years ago (Evans, Barston, & Pollard, 1983), the belief-bias task has become one of the great classics of the psychology of reasoning, and we assume that the reader is familiar with it. If not, here is a quick explanation: in the *belief-bias task*, reasoners are presented with syllogisms and must detect whether or not the conclusion logically follows from the premises. The catch is that the conclusion may be true or false *in the real world*, a fact that is irrelevant to the task but hard not to be influenced by. For example, consider this: "All flowers need water. Roses need water. Therefore, roses are flowers." The conclusion is true in the real world, but it does *not* follow logically from the premises. To say that it logically follows is to succumb to the belief bias.

Not all people succumb to belief bias, so it is interesting to look at what individual characteristics may predict one's susceptibility to the bias. You might think that dozens of personality traits would be candidates for such an investigation. Not so. A search for "individual differences" and "belief bias" in the Web of Science, followed by a systematic inspection of the results, returned 15 empirical articles measuring the correlation between belief bias and one or more individual differences variables. In 14 articles out of 15, these individual differences variables were linked to the ability (e.g., working memory span) or disposition (e.g., need for cognition) to engage effortful cognitive processing (De Neys, 2006; Evans, Handley, & Bacon, 2009; Evans, Handley, Neilens, & Over, 2010; Handley, Capon, Beveridge, Dennis, & Evans, 2004; Hirayama & Kusumi, 2004; Kokis, MacPherson, Toplak, West, & Stanovich, 2002; MacPherson & Stanovich, 2007; Morsanyi & Handley, 2012; Newstead, Handley, Harley, Wright, & Farelly, 2004; Sa, West, & Stanovich, 1999; Stanovich & West, 1998, 2008; Trippas, Handley, & Verde, 2013; West, Toplak, & Stanovich, 2008). Only one article focused on a variable unrelated to cognitive ability or disposition, that is, trait anxiety (Vroling & de Jong, 2010), and it actually did not find an effect of this trait, although we will have more to say about anxiety and reasoning later on.

What is the reason for such an exclusive focus on cognitive ability and disposition? Our answer requires a story, whose stage is the *great rationality debate* and whose hero is the *dual-process model*. To summarize, the great rationality debate (i.e., are humans rational, and can psychology experiments prove otherwise?) first brought to attention the cognitive ability of participants in reasoning experiments before the dual-process model emphasized that the ability to sustain effortful cognitive processing was useless without the motivation to engage this kind of processing in the first place. We now offer a more detailed narrative of these theoretical moments.

The Great Rationality Debate

Back in the 1960s and up to the 1990s, psychologists came up with deceivingly simple problems that (almost) everyone got wrong. Peter Wason invented several such problems in the field of reasoning (e.g., the selection task, Wason, 1968), and Daniel Kahneman and colleagues created hugely influential examples in the field of probabilistic judgment and decision making (Kahneman, Slovic, & Tversky, 1982). These results ignited the "great rationality debate" (Stanovich, 2009, 2011): Can studies of reasoning and decision making prove humans to be irrational?

A fairly common move in this debate was to claim that psychologists assessed reasoning performance against the wrong norm and that a norm existed for which the modal response to their tasks was actually correct. In the selection task, for example, 90% of reasoners give a response that is incorrect from the perspective of formal logic. It so happens that one can consider a Bayesian normative framework in which this response is actually correct (Oaksford & Chater, 1994). Thus, depending on the norm one decides to apply, reasoners fail at deductive logic or succeed at optimal data selection. Which is true?

This is when individual differences came under the spotlights, thanks to Keith Stanovich and Richard West. Whatever norm applied, they argued, reasoners were more likely to succeed when their cognitive ability (measured by standardized tests) was greater (Stanovich & West, 2000). Imagine a reasoning problem in which 90 reasoners give response A and 10 reasoners give response B. Now imagine that the 10 reasoners who gave response B actually form the top decile of the sample in terms of cognitive ability. Which seems the more reasonable: that these 10 reasoners gave the correct response where the other 90 failed, or that these 10 reasoners failed where the other 90 succeeded? Unless the task has some very unusual features, one would probably expect that they succeeded where others failed.

This elegant idea fueled a number of studies that attempted to correlate the cognitive ability of a reasoner to her propensity to give this or that response to various reasoning tasks (e.g., Stanovich & West, 2008). Accordingly, the first systematic wave of studies that looked at individual differences in reasoning focused on cognitive ability. This phenomenon was only amplified by the advent of the dual-process model, which we now consider.

The Dual-Process Model

Dual-process models are found across the whole spectrum of higher-order cognition (Evans, 2008; Greene, 2013; Kahneman, 2011), but they became influential in the domain of reasoning after (and perhaps because of) the great rationality debate. Dual-process models typically distinguish between *Type 1* (fast, automatic) and *Type 2* (slow, effortful) processes. Because Type 1 and Type 2 processes may deliver different outputs from the same input, the possibility arises that different individuals give different responses to the same problem because they engaged different cognitive processes. Not all dual-process models are the same, and they may emphasize different aspects of Type 1 or Type 2 processes as a function of their domain of application. For example, in the domain of moral cognition, strong emphasis is put on the emotional character of Type 1 responses (Greene, 2013), whereas this is not usually the case in the domain of reasoning.

Individual differences are a key element of all dual-process models because of the very characterization of Type 2 responses, which are assumed to require time and cognitive effort. As a consequence of this characterization, researchers have several tests at their disposal in order to prove that a response results from Type 2 processing (for example, giving reasoners a limited time to respond or giving them an interfering task). Some of these tests are correlational. For example, one can show that response A takes longer to produce than response B and argue on this basis that response A is more likely to result from Type 2 processing than response B.

In the same vein, and most importantly for our current purpose, one can show that response A is given by reasoners of greater cognitive ability than response B and argue on this basis that response A is more likely to result from Type 2 processing than response B. The rationale for this conclusion is the following: producing Type 2 responses requires sustained mental effort; people of higher cognitive ability are more apt to sustain mental effort; therefore, people of higher cognitive ability are more likely to produce Type 2 responses.

Cognitive ability is not enough, however. Although high cognitive ability may ensure that Type 2 processing is successful, it does not ensure that it is engaged. A high-ability reasoner who would never engage effortful cognitive processing could be indistinguishable from a reasoner of very low ability in terms of their ultimate likelihood to produce a Type 2 response. Accordingly, influential dual-process scholars recommended considering ability (to sustain effortful processing) and disposition (to engage effortful processing) as the two key individual parameters determining the probability of a Type 2 process. Evans (2007) noted that high-ability reasoners were more likely to give a Type 2 response but refrained from assuming that they did so because they were more likely to engage Type 2 processing. In contrast, he suggested that high-ability participants were more likely to give a Type 2 response given that they engaged in Type 2 processing in the first place but that the probability of engaging in Type 2 processing may be entirely driven by dispositional or motivational factors. In like vein Stanovich and West (2008) suggested that differences in cognitive ability correlated with the probability of sustaining effortful Type 2 processing but not with the probability of detecting the need for Type 2 processing, which was more likely to correlate with dispositional factors such as need for cognition or actively open-minded thinking.

In sum, the advent of the dual-process model gave a new twist to the question of whether different people reason differently. Whereas the great rationality debate focused on cognitive ability, the dual-process model considered both the ability and the disposition to engage in effortful processing. These two individual traits undoubtedly play very important roles in reasoning, just as the dual-process model plays an important role in theories of reasoning. We should be careful, however, to leave room for other individual differences in reasoning. As we saw earlier, of 15 articles exploring individual differences in belief bias, 14 focused on cognitive ability or disposition. This clearly results from the fact that the dual-process model dominates the field of individual differences in reasoning. Our claim is not that this domination is undeserved but to warn against some of its undesirable side effects. One of these side effects is apparently to discourage research on individual differences unrelated to the ability and disposition to engage in effortful processing. We now consider three possible directions for this type of research: psychological disorders, language use and interpretation, and personality projection.

Psychological Disorders

Earlier in this chapter, we mentioned that a search for "individual differences" and "belief bias" in the Web of Science returned only one article that did not focus on cognitive ability or disposition. This paper (Vroling & de Jong, 2010) instead studied the effect of trait anxiety on reasoning about threat and safety contents. The objective of this research was to test the hypothesis that a threat-confirming reasoning style might play a role in the development of anxious psychopathology. Although there was nothing in the data to support this hypothesis, it is worth noting here because it is representative of a wider research trend. This research trend is concerned with the reasoning style that results in, or results from, some form of psychological disorder.

Let us consider, for example, the *jump-to-conclusion* style of reasoning (or hypothesis testing). People are said to jump to conclusions when they draw conclusions based on too little evidence, especially when the little evidence they have favors their prior beliefs. This reasoning style is considered to play a central role in the formation and maintenance of delusions (for reviews see Dudley & Over, 2003; Garety & Freeman, 1999). Outside the domain of pathological delusions, jumping to conclusion is also observed in the reasoning of highly anxious or obsessive individuals drawn from nonclinical samples (Bensi & Giusberti, 2007; Bensi, Giusberti, Nori, & Gambetti, 2010). Note here that research on jumping to conclusions was conducted in both clinical and nonclinical samples. This is important because it means that a reasoning style that tracks a psychological disorder can also track a relevant personality trait in the general population, which confirms the possibility of investigating nonpathological personality correlates of reasoning.

Another instance of this research trend is concerned with the reasoning style attached to autism spectrum disorder as well as with the milder autistic traits found in the general population. Reasoners with autistic spectrum disorder seem to experience difficulties when reasoning with imperfect rules, that is, rules that allow for exceptions. Reasoners with autistic spectrum disorder are less likely to retrieve exceptions from memory when using a causal rule and less likely to take such exceptions into consideration when these are explicitly brought to their attention (McKenzie, Evans, & Handley, 2010, 2011; Pijnacker et al., 2009). This reasoning feature is also observed (among others) in typically developing populations for individuals who display high but subclinical autistic traits (Fugard, 2009).

In sum, there is a substantial body of data on the reasoning patterns found in clinical or subclinical samples. It is harder, however, to find studies investigating the reasoning patterns linked to other, nonclinical individual traits. In the next section we seek inspiration from a closely related field that has moved beyond clinical samples: the personality correlates of language use and comprehension.

Language Use and Comprehension

One possible source of inspiration for research on the personality correlates of reasoning is found in studies on individual differences in language use and comprehension. Indeed, experimental studies of language use and comprehension can come extremely close to qualifying as experiments on reasoning. This is often the case with experimental pragmatics, with its focus on the extrasemantic inferences that are required to interpret an utterance (Noveck & Reboul, 2008).

Just as with studies of individual differences in reasoning, one can find studies of language use and comprehension that focus on psychological disorders and clinical samples. For example, a substantial literature has documented the pragmatic deficits (or weaknesses) demonstrated by persons with autistic disorder or high-functioning autism (Loukusa & Moilanen, 2009). Comparable studies exist that documented the pragmatic comprehension and production deficits of persons with Parkinson's disease (e.g., Hall, Ouyang, Lonnquist, & Newcombe, 2011; Holtgraves & McNamara, 2010) as well as many other conditions.

Studies of the individual correlates of language use and comprehension have gone beyond clinical samples, though, and have explored personality traits in the typical range. Research on natural language use as a marker of personality, in particular, is concerned with finding correlations between personality traits (e.g., the *big five* factors, or the *dark triad* traits) and the preferential use of specific linguistic content (e.g., positive emotion words, swear words) or style (e.g., first-person pronouns, negations). In their review of this literature Ireland and Mehl (2014) point to several findings that can provide inspiration for studying personality correlates of reasoning.

Consider first the link between personality and tentative language, that is, the use of uncertainty terms (*maybe, doubt*) or hedges (*I think, I believe*). People who score high on Extraversion use less tentative language (especially uncertainty terms), whereas Agreeable people seem to use more tentative language (especially hedges). Can these different propensities translate into different approaches to uncertain information? We can only speculate at this stage, but it sounds possible that people with a low propensity to inject uncertainty into their sentences may give a lesser role to uncertainty in their reasoning.

Now consider a different example, the link between personality and positivity, that is, the use of positive verbs, qualifiers, and emotion words ("I enjoyed a lovely day with my friends" vs. "I hate these jerks with a vengeance"). Positivity is a robust signature of Agreeableness, whereas negativity is linked to Neuroticism. Again, we may speculate that a preference for using positive (vs. negative) words may translate into a higher propensity to endorse positive (vs. negative) premises or conclusions. Neurotic reasoners may find it easier to accept as premises or conclusions the kind of negative sentences they are likely to use themselves, whereas agreeable reasoners may find these same sentences less compellingly true.

The general idea that we are offering is that personality may drive a predisposition to endorse specific premises or conclusions. Based on data on personality and language use, we could predict neurotic reasoners to be predisposed to endorse negative emotion conclusions, whereas agreeable reasoners would be predisposed to discard these same conclusions. In contrast, agreeable reasoners would be predisposed to endorse uncertain conclusions, whereas extraverted reasoners would be predisposed to discard uncertain conclusions.

We do not know of any data that tested these predictions, which are highly speculative. What we do in the rest of this chapter, though, is to decline our general idea in two specific examples and provide relevant data. First, we show that reasoners are predisposed to endorse conditional rules that match their personality signatures (i.e., personal triggers described as *conditionals*). Second, we show that reasoners are prediposed to endorse causal models that fit their personal *belief in a just world* trait (e.g., causal models that minimize the probability of an injustice). In the conclusion of this chapter we return to our general claim and its relation to studies of language comprehension.

Projecting Personality Signatures

Our general claim is that personality traits can predispose reasoners to endorse specific conclusions when these conclusions are congruent with the personality of the reasoner. Our first example is based on the phenomenon of social projection (Robbins & Krueger, 2005). Social projection is what you do when you assume that others want and feel the same things that you want and feel. For example, if you are a liberal, you may be likely to overestimate the support for liberal parties in the population (Koudenburg, Postmes, & Gordjin, 2011).

Social projection is not restricted to beliefs and attitudes, though. It can also work on subtler individual characteristics. For example, think of your triggers, these little things that set you off in a specific way. Maybe you feel annoyed when people speak loudly; maybe you lose control in the presence of your favorite food; or maybe you feel bad when you take advantage of someone. Personality psychologists call these triggers your *personality signatures* (Mischel, 2004). Personality signatures are sets of conditional statements that describe your reactions given one of several triggers, for example:

(1a) If I were to steal from someone, I would feel very bad.

(1b) If I were to take advantage of someone, I would regret my behavior.

Knowing the triggers of others can be good for you. For instance, having an accurate knowledge of a friend's triggers makes for a deeper and more peaceful relationship (Friesen & Kammrath, 2011). But interestingly, people seem to project their triggers onto others, just as they project their preferences or attitudes (Kammrath, 2011). Let us consider more closely what this means. Imagine an individual whose personality signatures (i.e., triggers) include (1b) above. That means that (1b) is an adequate description of how this individual would react if she were to take advantage of someone. Now, if this individual projects her personality signatures on others, then she must believe that (1b) is true to some extent of other people.

This belief should have direct consequences on the way our individual reasons about other people: given the information that Alice took advantage of Bob, our individual will be more likely to think that Alice regrets her behavior than to think otherwise. More generally, any personality trait that can be described with personality signatures should affect reasoning in a predictable way. If you possess a personality trait that includes a personality signature of the form *if x, then y*, then you should be more likely to conclude *y* given *x*, than someone who does not possess the same personality trait.

This claim was tested extensively in Bonnefon (2010). The principle of the test is described in figure 11.1 (left panel). An online repository of personality scales was searched for scales that included at least two items in conditional form. For example, the *Fairness* scale included these two items:

(2a) Would feel very badly for a long time if I were to steal from someone.

(2b) Would not regret my behavior if I were to take advantage of someone impulsively.

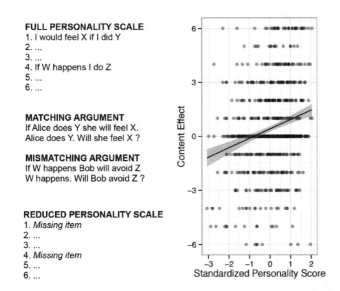

Figure 11.1

The general method used for generating the personality signature materials (left). The stronger the personality trait, the larger the content effect on inferences that match or mismatch this personality trait (right).

The idea then was to use these items as premises in a conditional inference task. One item was used to create an argument that matched the personality trait, for example:

(3) If Alice steals something, she feels very badly. Alice steals something. Will she feel badly?

The other item was used to create an argument that mismatched the personality trait; for example:

(4) If Bob takes advantage of someone, he feels proud. Bob takes advantage of someone. Will he feel proud?

Participants expressed their confidence in the conclusion of the two arguments, and the difference between their confidence in (3) and their confidence in (4) served as an index for the content effect. Most importantly, the prediction was that this content effect would be stronger for participants who scored higher on the Fairness personality scale, from which the items were extracted. To check this prediction, participants also took a subset of the Fairness personality scale, that is, the scale minus the two items used in the reasoning task.

As shown in figure 11.1 (right panel), this prediction was grounded in the data. The higher participants scored on personality scales such as Fairness, Responsive Joy, or Self-Control, the more strongly they agreed with conditional arguments that described people reacting in line with this personality trait.

A very similar method was developed by Fumero, Santamaria, and Johnson-Laird (2010), although the goal of this study was to measure reasoning performance rather than personality projection. Conditionals were extracted from items in the NEO-PI-R and used in a reasoning task, in a very similar way as in Bonnefon (2010). The results of Fumero et al. (2010) were less conclusive than those of Bonnefon (2010), but this discrepancy can probably be attributed to differences in statistical power because the former study involved a few dozens of participants, compared to many hundreds in the latter.

In sum, we have argued in this section that a great many personality traits can affect conditional reasoning as soon as these traits can be expressed by personality signatures. Generally speaking, reasoners are predisposed to accept conditional rules that match their own personality signatures, which propensity predisposes them to accept conclusions derived from these conditional rules. In the next section we continue to explore projection mechanisms, only with a more specific reasoning task and a specific personality trait.

Projecting the *Belief in a Just World*

In this section we illustrate how a basic reasoning problem, that of forming a causal model out of two rules, can be guided by a personality trait. The personality trait we are interested in is the *belief in a just world* (Lerner, 1980). People who score high on this trait are more likely to think bad people (Hafer & Bègue, 2005). In this section we show that, given an ambiguity between two causal models, people who score high on the *belief in a just world* tend to form a causal model of the world where the probability of an injustice is minimized.

Our demonstration focuses on the problem of forming a single model out of two causal rules of the form *if c_1 then e,* and if c_2 then e'. Figure 11.2 (left panel) displays the two causal models that can be formed out of these two rules. In the Collider model, c_1 and c_2 are construed as two alternative causes of the common effect e: each of them is sufficient to produce e. This model can be logically expressed as *if (c_1 or c_2), then e'.* In contrast, in the Enabler model, c_1 and c_2 are no longer construed as alternative, sufficient causes of e, although their joint occurrence still produces e. This model can be logically expressed as *if (c_1 and c_2) then e'.* In the terminology of Sloman, Barbey, and Hotaling (2009), c_1 as well as c_2 genuinely *cause* e in the Collider model, but c_1 and c_2 merely *enable* e in the Enabler model.

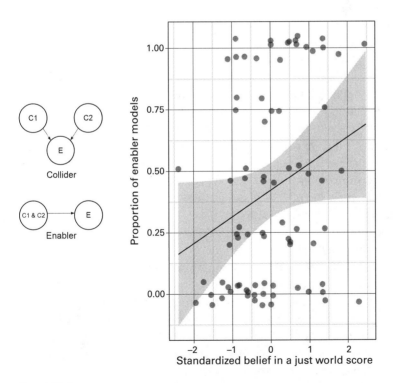

Figure 11.2
The Collider and Enabler models of two causal conditionals (left). Proportion of Enabler models when the rules threaten the *just world*, as a function of the standardized BJW score (right).

There is a rich experimental literature in the psychology of reasoning showing that depending on the linguistic contents of the rules, people may build either the Enabler or the Collider model (e.g., Politzer & Bonnefon, 2006; Stevenson & Over, 1995). Interestingly for our current purpose, several authors (Demeure, Bonnefon, & Raufaste, 2009; Stenning & van Lambalgen, 2005, 2008; Stevenson & Over, 2001) pointed out that some contents are ambiguous enough that they could equally well suggest Collider *or* Enabler models. As an illustration, consider the following rules:

(5a) If Bruno is modest, he will get in trouble.

(5b) If Bruno is sincere, he will get in trouble.

How should these two rules be combined? If combined in a Collider model, either modesty or sincerity will get Bruno in trouble. If combined in an Enabler model, only the conjunction of modesty and sincerity will get Bruno in trouble. Without further information, it is anybody's guess whether the two rules should be combined into a Collider or an Enabler model.

This ambiguity leaves room for the projection of one's *belief in a just world*. Indeed, the *just world* requires that good things happen to good people and that bad things happen to bad people. The causal rules in (5) suggest otherwise and thus threaten the *just world*. However, and this is the key idea in this section, their threat to the *just world* is greater if they are combined in a Collider model than if they are combined in an Enabler model. Accordingly, people who want the world to be just should prefer the Enabler model, which poses less of a threat to their ideal.

What we mean by "a lesser threat" is that the probability of an injustice is always lower when two unjust rules are combined in an Enabler model than when they are combined in a Collider model. We define an injustice as a good thing happening to a bad person or a bad thing happening to a good person. Bruno getting in trouble for being modest is an example of the latter kind of injustice. We define unjust rules as causal relations of the form *if c then e* where either (a) c is the mark of a bad person and e is a good outcome, or (b) c is the mark of a good person and e is a bad outcome. The rules in (5) illustrate the latter case.

It can be shown that under reasonable assumptions, the probability of an injustice is always greater in the Collider model than in the Enabler model. This can be understood intuitively. In the Collider model of (5), Bruno can get in trouble for being modest or for being sincere, whereas in the Enabler model of (5), Bruno will get in trouble if he is both sincere and modest. As long as these two characteristics are not perfectly correlated, there are more possible worlds in which Bruno gets into trouble in the Collider model than in the Enabler model. Accordingly, if reasoners project their *belief in a just world* on the problem, their propensity to form the Enabler model should increase as a function of their *belief in a just world* score.

To test this hypothesis, we asked 80 participants (62 women and 18 men, mean age 21) to paraphrase four pairs of unjust rules similar to (5). Some of these rules described bad events happening to good people, as in (5); others described good events happening to bad people, as in (6):

(6a) If Alice is selfish, she will be successful.

(6b) If Alice is dishonest, she will be successful.

All target items were constructed on this model. That is, they stated that someone would get a positive outcome (being successful) by behaving in an unsavory way (being arrogant, insincere, selfish, dishonest); or they stated that someone would get a negative outcome (being in trouble) by being modest, sincere, generous, or honest. Two versions of the questionnaire were constructed, which featured different combinations of behaviors and outcomes.

The task was to construct a paraphrase of these two rules by filling out a sentence with AND or OR, for example: "If Alice is selfish ... dishonest, she will be successful." Completing the sentence with AND was considered as reflecting the Enabler model, whereas completing the sentence with OR was considered as reflecting the Collider model.

Once the participants were done with the paraphrases, they took a six-item questionnaire aimed at measuring their *belief in a just world* (BJW). The scale was a French translation of that of Dalbert (1999). It consisted of six statements that participants had to assess on a 6-point scale anchored at *strongly disagree* and *strongly agree*: "I think basically the world is a just place"; "I believe that, by and large, people get what they deserve"; "I am confident that justice always prevails over injustice"; "I am convinced that in the long run people will be compensated for injustices"; "I firmly believe that injustices in all areas of life (e.g., professional, family, politic) are the exception rather than the rule"; "I think people try to be fair when making important decisions."

The *belief in a just world* (BJW) score of each participant was obtained by summing up the rating this participant gave to the six questions of the scale, with a greater score corresponding to a greater belief ($\alpha = 0.69$). As shown in figure 11.2 (right panel), there was a significant correlation between the BJW score and the number of *and* paraphrases (from zero to four), $r(79) = 0.32$, $p = 0.004$. The 26 participants who used three or four *and* paraphrases had an average BJW score of 19.3 (SD = 4.2), whereas the 54 participants who used 0 to 2 *and* paraphrases had an average BJW score of 17.1 (SD = 4.5), a significant difference, $t(78) = 2.1$, $p = 0.04$.

In sum, we showed in this section that a high *belief in a just world* predisposed reasoners to build causal models that minimized the probability of an injustice. That is, faced with the same information (two unjust conditional rules), different reasoners built different causal models, and the causal models they built were linked to their score on the personality trait known as the *belief in a just world*. This is a second illustration of our general claim that personality traits predispose reasoners to endorse specific premises and conclusions. We now come back to this general claim in the conclusion of this chapter.

Conclusion

Different people reason differently. That is, different reasoners may build different representations of the same premises or reach different conclusions even though they understood the premises the same. The challenge, then, is to identify the individual traits that can reliably track these differences in reasoning.

Among all individual traits, the two undisputed stars of this research field are cognitive ability and disposition. Cognitive disposition (e.g., the *need for cognition* trait) is linked to the likelihood that one engages effortful mental processing, and cognitive ability (e.g., working memory span) is linked to the likelihood that this effortful mental processing is sustained until successful completion of the task. These two individual traits are central to dual-process models of reasoning because together they predict the likelihood of a Type-2 response to a reasoning task.

The past and present importance of dual-process models ensured that cognitive ability and disposition received the lion's share of attention, as individual predictors of reasoning performance. No other trait received comparable attention, with the possible exception of psychological disorders, as a general category.

We suggested looking at research in psycholinguistics for inspiration on how to investigate personality correlates of reasoning in nonclinical populations. In particular, we discussed research on the personality correlates of language use. We argued, for example, that if a personality trait correlated with the use of uncertain language, it may also correlate with the endorsement of uncertain premises and uncertain conclusions. Generally speaking, we suggested that different personality traits may predispose reasoners to endorse different premises and conclusions as a function of the congruence of these premises and conclusions with the personality traits.

We offered two illustrations of this general claim. First, we described published research that showed how personality predisposed reasoners to endorse conclusions derived from conditional rules that encapsulated their own "triggers" or personality signatures. Second, we reported original research showing that reasoners were predisposed to build causal models that fitted their personal BJW score. That is, reasoners with a strong *belief in a just world* were predisposed to construct causal models that minimized the probability of injustice.

So far, we have only considered what has to be an indirect link between language use and reasoning: we have assumed that an individual's linguistic style would be linked to that individual's reasoning style. However, we may expect a more direct link between language comprehension and reasoning, especially if we specifically target the comprehension of the words and phrases that serve as the building blocks of reasoning: connectives and quantifiers. The way reasoners interpret connectives (e.g., not, and, or, if) and quantifiers (e.g., some, most, possible, probable) has a large impact on the conclusions they draw (Bonnefon, 2014). Accordingly, any individual trait tracking the interpretation of connectives and quantifiers would also predict the conclusions that individuals derive from connected, quantified propositions.

We know of very little research investigating this link between personality and the pragmatic comprehension of connectives and quantifiers, and it has all addressed a single quantifier, the word "some." In their succinct review of individual differences in the interpretation of "some," Antoniou, Cummins, and Katsos (under review) only mentioned one hypothesis (reading *some* as *some and possibly all* might be linked to pedantry; Katsos & Bishop, 2011) and one result (reading *some* as *some but not all* is linked with trait Honesty; Feeney & Bonnefon, 2013). Much work is required, thus, before we obtain a clear understanding of how personality may affect reasoning through its effect on language comprehension. We believe this effort to be necessary however in order to draw a complete picture of personality effects on reasoning. We know a lot about the effect of cognitive ability and disposition and about the effect of various psychological disorders. We are now ready to start exploring personality correlates of reasoning beyond cognitive ability and disposition, in nonclinical samples.

References

Antoniou, K., Cummins, C., & Katsos, N. (under review). Why are some people under-informative? Evidence for a gricean two-stage model of implicature generation.

Bensi, L., & Giusberti, F. (2007). Trait anxiety and reasoning under uncertainty. *Personality and Individual Differences*, *43*, 827–838.

Bensi, L., Giusberti, F., Nori, R., & Gambetti, E. (2010). Individual differences and reasoning: A study on personality traits. *British Journal of Psychology*, *101*, 545–562.

Bonnefon, J. F. (2010). Deductive reasoning from if-then personality signatures. *Thinking & Reasoning*, *16*, 157–171.

Bonnefon, J. F. (2014). Politeness and reasoning: Face, connectives, and quantifiers. In T. M. Holtgraves (Ed.), *Oxford handbook of language and social psychology* (pp. 387–404). New York: Oxford University Press.

Bonnefon, J. F., Eid, M., Vautier, S., & Jmel, S. (2008). A mixed Rasch model of dual-process conditional reasoning. *Quarterly Journal of Experimental Psychology*, *61*, 809–824.

Bonnefon, J. F., & Vautier, S. (2008). Defective truth tables and falsifying cards: Two measurement models yield no evidence of an underlying fleshing-out propensity. *Thinking & Reasoning*, *14*, 231–243.

Bonnefon, J. F., Vautier, S., & Eid, M. (2007). Modelling individual differences in contrapositive reasoning with continuous latent state and trait variables. *Personality and Individual Differences, 42*, 1403–1413.

Dalbert, C. (1999). The world is more just for me than generally: About the Personal Belief in a Just World Scale's validity. *Social Justice Research, 12*, 79–98.

Demeure, V., Bonnefon, J. F., & Raufaste, E. (2009). Politeness and conditional reasoning: Interpersonal cues to the indirect suppression of deductive inferences. *Journal of Experimental Psychology: Learning, Memory, and Cognition, 35*, 260–266.

De Neys, W. (2006). Dual processing in reasoning: Two systems but one reasoner. *Psychological Science, 17*, 428–433.

De Neys, W., & Bonnefon, J. F. (2013). The whys and whens of individual differences in thinking biases. *Trends in Cognitive Sciences, 17*, 172–178.

Dudley, R. E. J., & Over, D. E. (2003). People with delusions jump to conclusions: A theoretical account of research findings on the reasoning of people with delusions. *Clinical Psychology & Psychotherapy, 10*, 263–274.

Evans, J. S. B. T. (2002). Logic and human reasoning: An assessment of the deduction paradigm. *Psychological Bulletin, 128*, 978–996.

Evans, J. S. B. T. (2007). On the resolution of conflict in dual-process theories of reasoning. *Thinking & Reasoning, 13*, 321–329.

Evans, J. S. B. T. (2008). Dual-processing accounts of reasoning. *Annual Review of Psychology, 59*, 255–278.

Evans, J. S. B. T., Barston, J. L., & Pollard, P. (1983). On the conflict between logic and belief in syllogistic reasoning. *Memory & Cognition, 11*, 295–306.

Evans, J. S. B. T., Handley, S. J., & Bacon, A. M. (2009). Reasoning under time pressure: A study of causal conditional inference. *Experimental Psychology, 56*, 77–83.

Evans, J. S. B. T., Handley, S. J., Neilens, H., & Over, D. E. (2010). The influence of cognitive ability and instructional set on causal conditional inference. *Quarterly Journal of Experimental Psychology, 63*, 892–909.

Feeney, A., & Bonnefon, J. F. (2013). Politeness and honesty contribute additively to the interpretation of scalar expressions. *Journal of Language and Social Psychology, 32*, 181–190.

Friesen, C., & Kammrath, L. K. (2011). What it pays to know about a close other: The value of contextualized "if-then" personality knowledge in close relationships. *Psychological Science, 22*, 567–571.

Fugard, A. J. B. (2009). *Exploring individual differences in deductive reasoning as a function of "autistic"-like traits*. Unpublished doctoral dissertation, University of Edinburgh, Scotland, UK.

Fumero, A., Santamaria, C., & Johnson-Laird, P. N. (2010). Ways of thinking: Personality affects reasoning. *Psicothema, 22,* 57–62.

Garety, P. A., & Freeman, D. (1999). Cognitive approaches to delusions: A critical review of theories and evidence. *British Journal of Clinical Psychology, 38,* 113–154.

Greene, J. (2013). *Moral tribes: Emotion, reason, and the gap between us and them.* New York: Penguin Press.

Hafer, C. L., & Bègue, L. (2005). Experimental research on just-world theory: Problems, developments, and future challenges. *Psychological Bulletin, 131,* 128–167.

Hall, D., Ouyang, B., Lonnquist, E., & Newcombe, J. (2011). Pragmatic communication is impaired in Parkinson disease. *International Journal of Neuroscience, 121,* 254–256.

Handley, S. J., Capon, A., Beveridge, M., Dennis, I., & Evans, J. S. B. T. (2004). Working memory, inhibitory control, and the development of children's reasoning. *Thinking & Reasoning, 10,* 175–195.

Hirayama, R., & Kusumi, T. (2004). Effect of critical thinking disposition on interpretation of controversial issues: Evaluating evidences and drawing conclusions. *Japanese Journal of Educational Psychology, 52,* 186–198.

Holtgraves, T., & McNamara, P. (2010). Pragmatic comprehension deficit in Parkinson's disease. *Journal of Clinical and Experimental Neuropsychology, 32,* 388–397.

Ireland, M. E., & Mehl, M. R. (2014). Natural language use as a marker of personality. In T. Holtgraves (Ed.), *Oxford handbook of language and social psychology* (pp. 201–218) New York: Oxford University Press.

Kahneman, D. (2011). *Thinking, fast and slow.* New York: Farrar, Straus & Giroux.

Kahneman, D., Slovic, P., & Tversky, A. (1982). *Judgment under uncertainty: Heuristics and biases.* New York: Cambridge University Press.

Kammrath, L. K. (2011). What we think we do (to each other): How the same relational behaviors mean different things to people with different personality profiles. *Journal of Personality and Social Psychology, 101,* 754–770.

Katsos, N., & Bishop, D. V. M. (2011). Pragmatic tolerance: Implications for the acquisition of informativeness and implicature. *Cognition, 120,* 67–81.

Kokis, J. V., MacPherson, R., Toplak, M. E., West, R. F., & Stanovich, K. E. (2002). Heuristic and analytic processing: Age trends and associations with cognitive ability and cognitive styles. *Journal of Experimental Child Psychology, 83,* 26–52.

Koudenburg, N., Postmes, T., & Gordijn, E. H. (2011). If they were to vote, they would vote for us. *Psychological Science, 22,* 1506–1510.

Lerner, M. J. (1980). *The belief in a just world: A fundamental delusion.* New York: Plenum Press.

Loukusa, S., & Moilanen, I. (2009). Pragmatic inference abilities in individuals with asperger syndrome or high functioning autism. A review. *Research in Autism Spectrum Disorders, 3,* 890–904.

MacPherson, R., & Stanovich, K. E. (2007). Cognitive ability, thinking dispositions, and instructional set as predictors of critical thinking. *Learning and Individual Differences, 17,* 115–127.

McKenzie, R., Evans, J. S. B. T., & Handley, S. J. (2010). Conditional reasoning in autism: Activation and integration of knowledge and belief. *Developmental Psychology, 46,* 391–403.

McKenzie, R., Evans, J. S. B. T., & Handley, S. J. (2011). Autism and performance on the suppression task: Reasoning, context and complexity. *Thinking & Reasoning, 17,* 182–196.

Mischel, W. (2004). Toward an integrative science of the person. *Annual Review of Psychology, 55,* 1–22.

Morsanyi, K., & Handley, S. J. (2012). Logic feels so good—I like it! Evidence for intuitive detection of logicality in syllogistic reasoning. *Journal of Experimental Psychology: Learning, Memory, and Cognition, 38,* 596–616.

Newstead, S. E., Handley, S. J., Harley, C., Wright, H., & Farelly, D. (2004). Individual differences in deductive reasoning. *Quarterly Journal of Experimental Psychology, 57A,* 33–60.

Noveck, I. A., & Reboul, A. (2008). Experimental pragmatics: A Gricean turn in the study of language. *Trends in Cognitive Sciences, 11,* 425–431.

Oaksford, M., & Chater, N. (1994). A rational analysis of the selection task as optimal data selection. *Psychological Review, 101,* 608–631.

Pijnacker, J., Geurts, B., van Lambalgen, M., Buitelaar, J., Kan, C., & Hagoort, P. (2009). Defeasible reasoning in high-functioning adults with autism: Evidence for impaired exception-handling. *Neuropsychologia, 47,* 644–651.

Politzer, G., & Bonnefon, J. F. (2006). Two varieties of conditionals and two kinds of defeaters help reveal two fundamental types of reasoning. *Mind & Language, 21,* 484–503.

Robbins, J. M., & Krueger, J. I. (2005). Social projection to ingroups and outgroups: A review and meta-analysis. *Personality and Social Psychology Review, 9,* 32–47.

Sa, W. C., West, R. F., & Stanovich, K. E. (1999). The domain specificity and generality of belief bias: Searching for a generalizable critical thinking skill. *Journal of Educational Psychology, 91,* 497–510.

Sloman, S. A., Barbey, A. K., & Hotaling, J. M. (2009). A causal model theory of the meaning of cause, enable, and prevent. *Cognitive Science, 33,* 21–50.

Stanovich, K. E. (2009). *Decision making and rationality in the modern world.* New York: Oxford University Press.

Stanovich, K. E. (2011). *Rationality and the reflective mind.* New York: Oxford University Press.

Stanovich, K. E., & West, R. F. (1998). Individual differences in rational thought. *Journal of Experimental Psychology. General, 127,* 161–188.

Stanovich, K. E., & West, R. F. (2000). Individual differences in reasoning: Implications for the rationality debate? *Behavioral and Brain Sciences, 23,* 645–665.

Stanovich, K. E., & West, R. F. (2008). On the relative independence of thinking biases and cognitive ability. *Journal of Personality and Social Psychology, 94,* 672–695.

Stanovich, K. E., West, R. F., & Toplak, M. E. (2011). Intelligence and rationality. In R. J. Sternberg & S. B. Kaufman (Eds.), *Cambridge handbook of intelligence* (pp. 784–826). New York: Cambridge University Press.

Stenning, K., & van Lambalgen, M. (2005). Semantic interpretation as computation in nonmonotonic logic: The real meaning of the suppression task. *Cognitive Science, 29,* 919–960.

Stenning, K., & van Lambalgen, M. (2008). *Human reasoning and cognitive science.* Cambridge, MA: MIT Press.

Stevenson, R. J., & Over, D. E. (1995). Deduction from uncertain premises. *Quarterly Journal of Experimental Psychology, 48A,* 613–643.

Stevenson, R. J., & Over, D. E. (2001). Reasoning from uncertain premises: Effects of expertise and conversational context. *Thinking & Reasoning, 7,* 367–390.

Trippas, D., Handley, S. J., & Verde, M. F. (2013). The SDT model of belief bias: Complexity, time and cognitive ability mediate the effects of believability. *Journal of Experimental Psychology: Learning, Memory, and Cognition, 39,* 1393–1402.

Vroling, M. S., & de Jong, P. J. (2010). Threat-confirming belief bias and symptoms of anxiety disorders. *Journal of Behavior Therapy and Experimental Psychiatry, 41,* 100–116.

Wason, P. C. (1968). Reasoning about a rule. *Quarterly Journal of Experimental Psychology, 20,* 271–281.

West, R. F., Toplak, M. E., & Stanovich, K. E. (2008). Heuristic and biases as measures of critical thinking: Associations with cognitive ability and thinking dispositions. *Journal of Educational Psychology, 100,* 930–941.

IV Implicit Thought: Its Role in Reasoning and in Creativity

12   Thinking beyond Boundaries

Tilmann Betsch, Johannes Ritter, Anna Lang, and Stefanie Lindow

Is Integration Really Effortful? Challenging an Undisputed Truism in Decision Research

Making decisions bears several challenges. One is that different pieces of information must be integrated. For instance, according to utility theory, an outcome's value must be weighted by its probability. Herbert Simon (1956) postulated that humans cannot meet the standards of utility theory because their cognitive capacity is limited. He suggested that humans apply simple rules that circumvent effortful processes such as the weighted integration of information. Building on this notion, research in JDM identified numerous strategies during the past decades that avoid weighting and integration (e.g., Gigerenzer & Gaissmaier, 2011; Payne, Bettman, & Johnson, 1993).

Notably, the quest for heuristics has not been pervasive across all domains of cognition. Outside the field of judgment and decision making (JDM), such as in perception, categorization, and speech processing, researchers comparatively rarely assume that integration is effortful and should thus be generally avoided through the use of simple heuristics. In this chapter we argue that integrating multiple pieces of information is *not* constrained by cognitive limitations. We aim at reattaching our field of JDM to other areas of cognition by reviewing evidence that convincingly documents weighting and integration as basic capacities of the mind. In doing so we propose a differential perspective on decision making that distinguishes processes that are bounded by limitations from those that are not.

Following Betsch and Haberstroh (2005), we distinguish between two broad categories of processes: preselectional and selectional processes. We claim that the assumptions of bounded rationality mainly apply to preselectional processes but are not necessarily involved in selectional processes.

Preselectional processes comprise searching and encoding information in the environment, retrieving information from memory, making inferences of decision criteria, and transferring information into a processable format. Often, preselectional processes require conscious and sequential consideration of information. For example, we search information in the Internet either by opening Web pages one after another or by purposely selecting an advice giver (e.g., an expert, a consumer magazine) to obtain information on choice alternatives. Memory search can also be an effortful process, especially in making judgments and decisions based on concrete samples of events retrieved from memory. Sometimes, relevant information is not directly accessible. For instance, in evaluating potential risks of technologies, beliefs must be inferred from a vast set of information that correlates with the criterion to different degrees. Such inferential thought processes can be effortful and time consuming.

If a decision situation requires sequential preselectional processes, effort will often be positively correlated with the number of informational and process units. In this case the predictions of bounded rationality apply; thus, search behavior is formed by cognitive limitations and the environmental structure (e.g., time and monetary costs of information acquisition). Humans certainly may apply different strategies or heuristics that help them to reduce information overload and shortcut the process of information acquisition. Lexicographic strategies, for example, are a useful means to sift through the problem space (Payne et al., 1993)—one begins with the most important attribute or outcome dimension and stops searching for information as soon as one option becomes dominant.

Selectional processes refer to cognitive operations that produce a judgment or decision based on encoded information. Here, information integration becomes an issue when working memory contains multiple pieces of information on a particular option. The application of a decision rule marks the end of the selectional process stage (thereafter, postselectional processes such as intention implementation begin). In utility theory selectional processes obey a weighted additive rule. Accordingly, outcome values are weighted by their probabilities (weights) and are summed up within each option to form expected values. The maximization principle guides the selection among the options (i.e., individuals presumably choose the option with the highest expected value).

We claim that the extensional integration of information, weighting procedures, and the selection of an option are virtually unbounded by processing limitations. All pieces of encoded information contained in working memory can be integrated without necessitating deliberation or conscious control. We propose that the integration of information is a fundamental, general, innate ability that applies to a vast array of cognitive tasks. It neither must be learned nor do its operations involve any considerable amount of cognitive effort because it operates in a parallel fashion (we specify this in the theoretical section below).

The Pervasiveness of Automatic Integration

In order to provide empirical evidence for our claim, we begin with a brief overview of research outside the field of JDM research. Specifically, we consider perception/

categorization and the understanding of speech. We then turn to the literature of our own field, judgment and decision making. Finally, we close with a review of pertinent findings from research with children.

Cheng, Shettleworth, Huttenlocher, and Rieser (2007) reviewed a large body of literature on the integration of spatial cues. They showed that spatial perception/categorization is often based on multiple cues of different modalities (e.g., visual, audio, haptic). Animals (e.g., snails, Gallistel, 1980) and humans are capable of performing weighted integration procedures "in a near-optimal fashion" (2007, p. 625) and without effortful and deliberative calculation.

As an example, consider the study by Alais and Burr (2004) on the utilization of visual and audio information in directional choice (right, left). Two Gaussian figures of different sizes (producing different degrees of overlap) were presented in succession on the computer screen (500 ms ISI). The angle between the two figures was varied so that deviation in direction became more or less apparent. Two speakers positioned right and left from the screen each emitted clicks. The interaural time difference between left and right clicks was varied. Participants in this study had to decide which of two subsequently presented stimuli was located left of the other. In bimodal trials with conflicting visual and audio information, participants' decisions were determined by a weighted sum of both cues. Subjective weights followed the level of cue precision (i.e., figure size and overlapping area). Weighting nearly perfectly followed the normative Bayesian model for integration. This research shows that individuals faced with a spatial choice task do not rely on a single source of information but rather on the integration of distinct information sources.

This capability also pertains to another modality. Ernst and Banks (2002) asked participants to judge the height of objects they could simultaneously touch and view. Results indicate that visual and haptic information are integrated in "a statistically optimal fashion" (p. 429). Specifically, estimates covaried with the predictions of a maximum-likelihood algorithm for weighted integration.

Perception and categorization regularly involve processing multiple cues. A fascinating example stems from the categorization of biological motion. Troje (2002) developed a weighted linear integration model to account for gender classification based on viewing body movements. He presented individuals with point-light displays of walking figures (Johansson, 1973). Although this paradigm abstracts from individuating information, a variety of structural and dynamic cues were found to systematically direct categorization, for example, arm swing amplitude and velocity, walking speed, amplitude of elliptical motion of shoulders and hips, lateral sway of upper body, elbow-body distance, and others. Viewers were intuitively capable of accurately identifying a walking pattern as male or female. Most notably, accuracy in gender classification drops to chance level when cues are *eliminated* (e.g., Kozlowski & Cutting, 1977). Hence, *more* information appears to *facilitate* the intuitive categorization process. Troje concludes that information about a walker's gender is not a matter of a single feature but rather a "complex process with a holistic character" (2002, p. 373) that corresponds to a weighted linear integration rule.

The literature on speech comprehension converges with the notion that individuals routinely integrate multiple pieces of information. Consider, for instance, the case of understanding irony. Adult recipients simultaneously utilize prosodic, mimic, gesture, and prior knowledge about the communicator to understand irony (see Gibbs & Colston, 2007, for overviews). Nevertheless, comprehension is a quick process and can be performed in less than 600 to 800 milliseconds (e.g., Schwoebel, Dews, Winner, & Srinivas, 2000). Interestingly, *increasing* the amount of information (e.g., by providing contextual cues) can result in a *faster* detection of irony (Amenta & Balconi, 2008). This finding is attributed to parallel processing based on neurophysiological data. Wang, Lee, Sigman, and Dapretto (2006) compared accuracy and speed in the identification of irony in average and clinical patients. In the latter, parallel activation of brain regions was impaired. Whereas average participants showed simultaneous activation of different brain regions and were quick to understand irony, clinical patients had to engage in time-consuming deliberative thought to grasp the ironic intention behind the message. These and other findings led researchers to postulate that essential parts of speech comprehension are driven by holistic processes operating automatically and in parallel (e.g., Long & Graesser, 1988).

Results of these studies converge in indicating that the brain is capable of quickly integrating information from different sources. In perceptual tasks the weight assigned to a certain cue seems to be sometimes a property of the stimulus itself. For example, if visual information is less reliable because it is perceptually noisy, it receives a lower weight than auditive information that is simultaneously present and more reliable (cf., Alais & Burr, 2004). The reliability of information—and accordingly the decision weight—is assumed to be encoded together with the location information as part of the perceptional process.

In decision tasks, in contrast, weights and values are often separated. For instance, in the gambling paradigm, weights and values are stated as separate symbols (i.e., numbers, graphical representations, etc.). Therefore, integration might be less likely in decision tasks compared to perception/categorization tasks because additional processing is required in the former to encode the weight information. Yet, what if the amount of additional processing necessary for encoding is reduced?

Glöckner and Betsch (2008a) presented adult participants with two versions of an information board in a probabilistic choice task. The hidden presentation format was similar to the standard Mouselab (Payne, Bettman, & Johnson, 1988) in which participant must uncover information (values, probabilities) by clicking on cells in the matrix with the computer mouse. In the open format all information was displayed simultaneously and could be inspected by the individuals at once. This manipulation had a tremendous effect. With the classic version individuals tended to employ simple strategies that reduced the effort of preselectional information search. Their decisions were informed by subsets of information. With the open board, however, the overwhelming majority of participants used all available information and employed weighted-additive-like integration procedures in an astoundingly narrow time frame. In another study the same authors showed that even an increase in information does not obstruct these integration capabilities (Glöckner & Betsch, 2012). Altogether, the results support the notion that integration is an automatic process.

Converging evidence stems from studies that present participants with classic two-option gambles and use eye-tracking technology. In a study by Fiedler and Glöckner (2012), participants' choices systematically reflected differences in expected value, although even mathematically skilled individuals would have had difficulty in deliberatively calculating the expected values. Moreover, eye-tracking data show a small proportion of transitions between gambles, indicating that information integration within gambles occurs as predicted by weighted-additive integration models.

Taken together, spontaneous integration can also be observed in decision-making tasks, especially if constraints for *preselectional* processes are reduced (e.g., by an open information board).

If it is true that integration is a basic and automated cognitive process, one should see evidence of its occurrence even in children. Therefore, it is worthwhile to acknowledge results from child research.

Integration abilities appear to be already developed a few hours after birth.

Streri, Coulon, and Guellaï (2012) reviewed evidence from studies on face–voice integration in infants. In one of these studies Coulon, Hemimou, and Streri (2013) measured newborns' facial imitations (mouth opening, lip spreading) elicited by visual (mouth opening, lip spreading) and audio (vowels /i/ vs. /a/) stimuli conveyed by an adult model. Imitations appeared more quickly when the model was audiovisual congruent as opposed to audiovisual incongruent or only visual. Streri and colleagues (2012) concluded that newborns focus on both cues and utilize both in behavior production.

Schlottmann (2001) studied evaluative judgments in four- to six-year-old children (in comparison to older children and adults). Specifically, participants judged how happy a puppet would feel if it won games differing in gain size (number of crayons) and probability. Probabilities were conveyed in a child-friendly visual presentation format. Each gamble was described by two potential outcomes and two associated probabilities. The probabilities were visualized by a glass tube containing a bicolored stripe. Each color occupied a varying number of segments in the tube corresponding to its relative probability. For the gamble, a marble was shaken in the tube, and the color of the segment it finally landed on determined the outcome, which consisted of a varying number of crayons. The results showed that children integrate probability and value in accordance with utility theory's weighted additive rule. Because the children in the sample presumably lack a formal understanding of the probability concept due to their age, Schlottmann concluded that four- to six-year-olds performed the weighting procedures intuitively (cf., Schlottmann & Wilkening, 2011, for a discussion).

Children are also capable of integrating more than two stimuli when forming judgments. Ebersbach (2009) investigated whether children are able to combine width, height, and length when estimating the volume of objects. Even five- to six-year-olds were able to integrate these dimensions in a multiplicative-like fashion. Performance did not dramatically improve with age, indicating that the integration of multidimensional information is a capability that is already well established in preschool children.

Again, consider the case of understanding irony. Climie and Pexman (2008) presented literal and ironic utterances of puppets to five- to nine-year-old children and adults. Puppets made either a literal or an ironic criticism or a literal or ironic compliment. Participants learned a priori to associate the puppet's intent with presented toys (i.e., being nice like the duck or mean like the shark) and pick up an object according to the speaker's intent (e.g., if the speaker made an ironic criticism, he was trying to be nice; thus, the duck should be picked). In order to correctly infer the speaker's intent, children had to take context information into account. Results show that children used multiple cues to infer the meaning of the puppet's utterances. They used not only contextual and prosodic information but also information about the puppet's personality (i.e., if it had been described as a serious or funny person).

Child research in the field of decision making completes the evidence. In a recent series of studies Betsch and colleagues (Betsch & Lang, 2013; Betsch, Lang, Lehmann, & Axmann, 2014) showed that even young children can employ weighting procedures in probabilistic decision environments similar to those used in adult research. Risky decisions in preschoolers (six-year-olds), elementary schoolers (nine-year-olds), and adults were studied with an information-board crossing either two (Betsch & Lang, 2013) or three cues (Betsch et al., 2013) with two options. Children learned cue-validities prior to decision making. In the more complex environment with three cues, presentation format was varied. Similar to what was done by Glöckner and Betsch (2008a), children were presented with either an open or a hidden information board. Results showed that increased complexity (three vs. two cues) did not decrease the rate of weighted integration. Most interestingly and converging with results from adult research, holistic information integration was most pronounced with an open presentation format.

In conclusion, there is considerable empirical evidence suggesting that weighted integration is pervasive across different tasks (perception, categorization, understanding of speech, judgment, and decision making) and modes (visual, audio, haptic) as well as that it can be performed in a narrow time frame, even by young children. Most interestingly, increasing the amount of information may result in faster processing.

Altogether, results indicate that integration is a fundamental and automated cognitive process capability that requires neither formal training nor deliberative thought.

When Integration Appears Effortful (Although It Is Not)

Three well-established findings in decision research seem to challenge the view that integration is effortless. First, decision time and the amount of information considered are often positively correlated (e.g., Payne et al., 1988). Second, the amount of information considered decreases with increasing constraints (e.g., Newell & Shanks, 2003; Newell, Weston, & Shanks, 2003; Payne et al., 1988). Third, individuals can make decisions using simple strategies that avoid weighted integration (e.g., lexicographic strategies, equal-weight strategies; Gigerenzer & Gaissmaier, 2011; Payne et al., 1993).

However, recall our distinction between preselectional and selectional processes. It is important to understand which step of decision making is bounded by constraints. Without separating these parts, one might mistake search effort for integration effort as well as encoding problems for weighting deficits.

Consider the everyday situation in which you search the Internet to gain access to information, perhaps regarding a new smartphone or tablet computer, in order to make a purchasing decision. It is impossible to consider all product-pertinent information available on the Internet. Therefore, you will likely apply a search strategy that focuses on what you consider the most relevant sources, perhaps the pages of a consumer test magazine. Despite the selective search, information acquisition is still effortful and time consuming. You must read narratives and consult tables and graphics. In the end you might experience some kind of fatigue in muddling through the vast information. Eventually, you may focus on the most important information (e.g., price) or a proxy for presumed quality (e.g., brand).

A decision researcher tracing your search would arrive at the conclusions that you used a simplifying strategy to reduce the complex task to a simpler one. Likely, you and your scientific monitor would agree that decision making is effortful. Note, however, that the entire episode tells us nothing about your integration capabilities. Because of the sequential and time-consuming nature of the search process, the information that you can retain in working memory is only a small subset of the plethora of information existing in the environment. In such a situation input formation is a bottleneck and hampers the brain's potential, that is, the handling and integration of multiple pieces of information to arrive at a decision. This example illustrates the problem: when analyzing evidence, we must take care not to mistake search effort for integration effort. On the methodological level we must design environments that can reveal the potentials of the processing system. If constraints of input formation are removed, weighted integration becomes the default (Betsch & Glöckner, 2010; Betsch et al., 2013; Glöckner & Betsch, 2008a).

Another reason processing capabilities are misjudged relates to encoding problems during input formation. Encoding requires prior knowledge. Individuals cannot encode a stimulus unless they have a corresponding coding system (Bruner, 1957) with appropriate categories and schemas.

Assume we presented four-year-old children with a gambling task in the same format as used in adult research. Presumably, we would find that choices are unsystematic because the children could not properly encode the meaning of the stated information—for example, they would likely not understand what the statement "with a probability of 0.78" means. Arguably, any conclusions about processing capabilities (such as deficits in weighting by probabilities) are unjustified in this case.

However, decision researchers still did not consider the problem of encoding and mental representation until the 1990s (Maule, 2005). Gigerenzer and Hoffrage (1995) proposed that the presentation format of probabilistic information must map to processing formats. Otherwise, judgmental performance will decrease and will be misjudged as being fallacious. *General evaluability theory* (Hsee & Zhang, 2010) assumes that given values and probabilities require reference information to be evaluated properly. Mode of presentation, knowledge, and nature are sources of such reference anchors. These approaches and others highlight the preconditions for weighting processes to be performed. If probabilistic information or the importance of attributes cannot be encoded properly (e.g., due to presentation mode, lacking knowledge, or a mismatch between natural and presentation formats), it is likely that the individual will apply strategies that ignore weights. However, it would be false to conclude that unweighted decision making in such cases evidences weighting deficits of the computational system.

How Does the Mind Intuitively Integrate Information?

A person educated in algebra is capable of consciously performing integration operations. Accordingly, one can instruct an adult decision maker to apply a weighted additive rule for decision making, that is, to weigh all outcome values by their corresponding probabilities, sum up products within options to form an expected value, and then choose the option with the highest expected value. In a task with two or three options, binary outcomes (0,1), and three or six cues (with different probabilities), university undergraduates require 20 seconds on average to perform this task (Glöckner & Betsch, 2008a) when provided with unconstrained access to information (open information board).

Without this instruction, participants from the same subject pools made their decisions in less than 4 seconds (with six cues appearing in randomized order across trials; cf. Glöckner & Betsch, 2008a, p. 1067). Their choice pattern, however, converged with the predictions of a weighted additive rule. This finding suggests that intuitive integration rests on operations that are distinct from conscious weighting and adding. In various areas of cognition these operations are described by connectionist models (e.g., perception/reading: McClelland & Rumelhart, 1981; Rumelhart & McClelland, 1982; explanatory coherence: Thagard, 1989; attribution: Read & Marcus-Newhall, 1993; stereotyping: Kunda & Thagard, 1996; decision making: Betsch, 2005; Glöckner & Betsch, 2008b; Holyoak & Simon, 1999; Thagard & Millgram, 1995).

Glöckner and Betsch (2008b; see also Betsch & Glöckner, 2010) embedded a connectionist model of integration into a component approach to decision making. They assumed that intuition and analyses are not different modes of thinking but rather *component* processes responsible for guiding different subtasks of decision making. Selectional processes such as information integration and formation of a choice preference do not necessitate controlled processes such as deliberative analyses. Rather, they are assumed to function intuitively, that is, automatically and autonomously. Preselectional processes, however, often require cognitive control, such as directed information search, understanding their implications (e.g., assessing how given consequences affect personal goals), anticipating future events, making inferences to generate information, and so forth.

It is a basic property of the brain that different nets of neurons are activated simultaneously and process information in parallel. Connectionist models mimic nature by assuming that cognition rests on parallel processing. Applying the network metaphor to model selectional processes, options, outcomes, and goals are represented by nodes in a working network. The nodes are interconnected and pass energy in parallel through the network. By passing energy, the activation of the nodes can change. The amount of change reflects the properties of the network structure that are represented by associations among the nodes. The associations can activate or deactivate other nodes such as synaptic linkages between neurons. Some connections pass energy on a high level, others on a low level—that is, the connections differ in strength. These properties of the connections are formally modeled by weights that are assigned to the linkages. A weight can reflect a variety of relational properties between entities, such as semantic, functional, or logic relations. In the case of decision making, a weight between two elements, say a cue and an option, can reflect probabilistic relations.

Glöckner and Betsch (2008b) applied a parallel-constraint satisfaction (PCS) rule for integration and decision (McClelland & Rumelhart, 1981; Read, Vanman, & Miller, 1997; Rumelhart & McClelland, 1982). The mathematical properties are described elsewhere in detail (e.g., Glöckner & Betsch, 2012; Glöckner, Betsch, & Schindler, 2010). For the present purpose it is sufficient to mention the rule's basic properties. First, it is holistic in that it considers all information contained in the network. In other words the assumption is that automatic selectional processes use all information represented in a working network at the time that a decision is made. Second, the updating process works in parallel and is constrained not by cognitive limitations but rather by the network structure that makes it more or less easy to find a solution. The third assumption is that the process aims at finding the most coherent activation pattern that minimizes overall energy by simultaneously satisfying the constraints imposed by the network structure. Fourth, after the network reaches a stable state of activation, the option with the highest expected value is chosen (see Betsch, 2005, for a variant of this rule). Although the PCS rule processes information in a nonlinear fashion, its outcome on the choice level converges with the predictions of the weighted additive rule of information integration and choice.

In an advanced PCS framework Glöckner and Betsch (2008b) outlined how analytic and intuitive processes can be integrated in a component model. Preselectional processes beyond spontaneously encoding salient information were described as processes of deliberative construction. Controlled processes can also interfere at the selectional stage (e.g., inhibition of information, deliberate change of weights). However, the initiation of deliberate control and construction can be modeled as a choice that follows the same automatic processes as the decision among the options. The authors put forward a two-network model to capture the instantiation of deliberate processes. They also came up with first predictions when these processes set in. An important endogenous factor is the coherence that can be achieved in the primary network devoted to the option decision. The more difficult it is to achieve an acceptable level of coherence, the greater the likelihood that a second network is formed that selects among different deliberative operations at the level of preselectional activity and/or deliberative supervision of selectional process (i.e., weight change).

Implications for Decision Research

We showed that weighted integration processes are fundamental cognitive operations that are not task specific and apply to a variety of domains such as perception, categorization, speech comprehension, estimation, evaluative judgment, and decision making. Integration does not appear to require the deliberate application of a rule. Rather, it simply takes place in the background while the conscious mind focuses on those aspects of the task that do require control such as actively searching information in the environment.

We decision researchers, however, still neglect these potentials of the mind. Raised in the spirit of the bounded rationality program, we are prone to overemphasize the role of conscious cognition. With reference to the assumed functioning of the weighted additive rule (e.g., assessing weights, using a linear weighted integration rule), Shah and Oppenheimer (2008, p. 207) conclude:

Clearly, such an algorithm requires great mental effort; however, people do not have unlimited processing capacity. People must operate within the constraints imposed by both their cognitive resources and the task environment—a concept known as bounded rationality (Simon, 1955, 1956). As the demands on limited cognitive resources increase, people may employ methods or strategies that reduce the effort they expend on computation. We will therefore refer to heuristics as methods that use principles.

This reasoning implies that heuristics are necessary because integration and other operations are effortful. If, however, automatic processes can solve these operations with virtually no noticeable effort, this reasoning is challenged. It has been the aim of this chapter to show that evidence across domains exists that challenges one of the basic assumptions of the bounded rationality approach. Note, however, that it would clearly be false to conclude that all processes involved in decision making are cognitively unlimited. Decision making often involves and necessitates controlled processes such as a thorough search for information. These processes are indeed effortful. They must be performed in a serial fashion and are bound by time, cognitive costs, and other constraints. Memory span also bottlenecks JDM. However, these processes pertain primarily to the preselectional phase in which the input to the decision itself is formed. Selectional processes (integration of information after encoding and the application of a choice rule) occur automatically by default. Under certain conditions, however, they may require supervision and controlled meta-processing, causing decision making to become difficult and effortful—for example, when individuals attempt to suppress information that they wish to exclude from a decision or attempt to overcome habits (see Betsch, 2005; Betsch & Held, 2012; Glöckner & Betsch, 2008b, for discussions).

Our conclusions have implications for further research on the methodological and theoretical level. Researchers must develop tools that allow one to more precisely analyze the allocation of cognitive effort. How does the decision environment determine how many resources are used for information acquisition as opposed to the decision process itself? This question cannot be answered satisfactorily at present because process-tracing methods regularly focus on detecting strategies that *combine* information search and decision processes. A resource allocation view would motivate researchers to design tools that more precisely track the distribution of cognitive effort between preselectional and selectional processes.

On the theoretical level, the field of JDM is still dominated by a compound rather than component view of processes. Although there are attempts to separate the building blocks of strategies and heuristics (e.g., Gigerenzer et al., 1999; Payne et al., 1993), researchers still consider strategies to be amalgams of preselectional and selectional processes. Accordingly, a combination of a certain search process with a certain decision rule is considered to be a distinct entity. Only this theoretical assumption justifies the use of learning procedures for strategies in our research as well as the assumption that specific strategies can be reinforced by positive feedback (Bröder, Glöckner, Betsch, Link, & Ettlin, 2013 ; Bröder & Schiffer, 2006; Rieskamp & Otto, 2006). The theoretical component approach put forward in this chapter suggests that rules for preselectional and selectional process are independent and do not merge to compound strategies. According to the PCS model, there is only one all-purpose rule guiding selectional processes (integration and choice), but there are a variety of strategies in the preselectional stage that guide input formation. As decision researchers, we should acknowledge the distinct properties of the two processes—both their potentials and their boundaries.

References

Alais, D., & Burr, D. (2004). The ventriloquist effect results from near-optimal cross-modal integration. *Current Biology*, *14*, 257–262.

Amenta, S., & Balconi, M. (2008). Understanding irony: An ERP analysis on the elaboration of acoustic ironic statements. *Neuropsychological Trends*, *3*, 7–27.

Betsch, T. (2005). Preference theory—an affect-based approach to recurrent decision making. In T. Betsch & S. Haberstroh (Eds.), *The routines of decision making* (pp. 39–65). Mahwah, NJ: Lawrence Erlbaum Associates.

Betsch, T., & Glöckner, A. (2010). Intuition in judgment and decision making: Extensive thinking without effort. *Psychological Inquiry*, *21*, 1–16.

Betsch, T., & Haberstroh, S. (2005). Research on the routines of decision making: Advances and prospects. In T. Betsch & S. Haberstroh (Eds.), *The routines of decision making* (pp. 359–376). Mahwah, NJ: Lawrence Erlbaum Associates.

Betsch, T., & Held, C. (2012). Run and jump modes of analysis. *Mind & Society*, *11*, 69–80.

Betsch, T., & Lang, A. (2013). Utilization of probabilistic cues in the presence of irrelevant information: A comparison of risky choice in children and adults. *Journal of Experimental Child Psychology*, *115*, 108–125.

Betsch, T., Lang, A., Lehmann, A., & Axmann, J. M. (2014). Utilizing probabilities as decision weights in closed and open information boards: A comparison of children and adults. *Acta Psychologica*, *153*, 74–86.

Bröder, A., Glöckner, A., Betsch, T., Link, D., & Ettlin, F. (2013). Do people learn option or strategy routines in multi-attribute decisions? The answer depends on subtle factors. *Acta Psychologica*, *143*, 200–209.

Bröder, A., & Schiffer, S. (2006). Adaptive flexibility and maladaptive routines in selecting fast and frugal decision strategies. *Journal of Experimental Psychology: Learning, Memory, and Cognition*, *32*(4), 904–918.

Bruner, J. S. (1957). Going beyond the information given. In H. E. Gulber, K. R. Hammond, & R. Jessor (Eds.), *Contemporary approaches to cognition* (pp. 41–69). Cambridge, MA: Cambridge University Press.

Cheng, K., Shettleworth, S. J., Huttenlocher, J., & Rieser, J. J. (2007). Bayesian integration of spatial information. *Psychological Bulletin, 133*, 625–637.

Climie, E., & Pexman, P. (2008). Eye gaze provides a window on children's understanding of verbal irony. *Journal of Cognition and Development, 9*, 257–285.

Coulon, M., Hemimou, C., & Streri, A. (2013). Effects of seeing and hearing vowels on neonatal facial imitation. *Infancy, 18*, 782–796.

Ebersbach, M. (2009). Achieving a new dimension: Children integrate three stimulus dimensions in volume estimations. *Developmental Psychology, 45*, 877–883.

Ernst, M. O., & Banks, M. S. (2002). Humans integrate visual and haptic information in a statistically optimal fashion. *Nature, 415*, 429–433.

Fiedler, S., & Glöckner, A. (2012). The dynamics of decision making in risky choice. An eye-tracking analysis. *Frontiers in Psychology, 3*, 335.

Gallistel, C. R. (1980). *The organization of action: A new synthesis.* Hillsdale, NJ: Lawrence Erlbaum Associates.

Gibbs, R., & Colston, H. (Eds.). (2007). *Irony in language and thought: A cognitive science reader.* New York: Lawrence Erlbaum Associates.

Gigerenzer, G., & Gaissmaier, W. (2011). Heuristic decision making. *Annual Review of Psychology, 62*, 451–482.

Gigerenzer, G., & Hoffrage, U. (1995). How to improve Bayesian reasoning without instruction: Frequency formats. *Psychological Review, 102*, 684–704.

Gigerenzer, G., Todd, P. M., & the ABC Research Group. (1999). *Simple heuristics that make us smart.* Oxford: Oxford University Press.

Glöckner, A., & Betsch, T. (2008a). Multiple-reason decision making based on automatic processing. *Journal of Experimental Psychology: Learning, Memory, and Cognition, 34*, 1055–1075.

Glöckner, A., & Betsch, T. (2008b). Modelling option and strategy choices with connectionist networks: Towards an integrative model of automatic and deliberate decision making. *Judgment and Decision Making, 3*, 215–228.

Glöckner, A., & Betsch, T. (2012). Decisions beyond boundaries: When more information is processed faster than less. *Acta Psychologica, 139*, 532–542.

Glöckner, A., Betsch, T., & Schindler, N. (2010). Coherence shifts in probabilistic inference tasks. *Journal of Behavioral Decision Making, 23*, 439–462.

Holyoak, K. J., & Simon, D. (1999). Bidirectional reasoning in decision making by constraint satisfaction. *Journal of Experimental Psychology. General, 128*, 3–31.

Hsee, C. K., & Zhang, J. (2010). General evaluability theory. *Perspectives on Psychological Science, 5*, 343–355.

Johansson, G. (1973). Visual perception and a model for its analysis. *Perception & Psychophysics, 14*, 201–211.

Kozlowski, L. T., & Cutting, J. E. (1977). Recognizing the sex of a walker from a dynamic point-light display. *Perception & Psychophysics, 21*, 575–580.

Kunda, Z., & Thagard, P. (1996). Forming impressions from stereotypes, traits, and behaviors: A parallel-constraint-satisfaction theory. *Psychological Review, 103*, 284–308.

Long, D. L., & Graesser, A. C. (1988). Wit and humour in discourse processes. *Discourse Processes, 11*, 35–60.

Maule, J. (2005). Re-framing decision framing. *Presidential address at the 20th Biennial Conference on Subjective Probability, Utility and Decision Making, 22–24 August, Stockholm, Sweden.*

McClelland, J. L., & Rumelhart, D. E. (1981). An interactive activation model of context effects in letter perception: I. An account of basic findings. *Psychological Review, 88*, 375–407.

Newell, B. R., & Shanks, D. R. (2003). Take-the-best or look at the rest? Factors influencing "one-reason" decision making. *Journal of Experimental Psychology: Learning, Memory, and Cognition, 29*, 53–65.

Newell, B. R., Weston, N. J., & Shanks, D. R. (2003). Empirical tests of a fast-and-frugal heuristic: Not everyone "takes-the-best." *Organizational Behavior and Human Decision Processes, 91*, 82–96.

Payne, J. W., Bettman, J. R., & Johnson, E. J. (1988). Adaptive strategy selection in decision making. *Journal of Experimental Psychology: Learning, Memory, and Cognition, 14*, 534–552.

Payne, J. W., Bettman, J. R., & Johnson, E. J. (1993). *The adaptive decision maker.* Cambridge: Cambridge University Press.

Read, S. J., & Marcus-Newhall, A. (1993). Explanatory coherence in social explanations: A parallel distributed processing account. *Journal of Personality and Social Psychology, 65*, 429–447.

Read, S. J., Vanman, E. J., & Miller, L. C. (1997). Connectionism, parallel constraint satisfaction processes, and Gestalt principles: (Re)introducing cognitive dynamics to social psychology. *Personality and Social Psychology Review, 1*, 26–53.

Rieskamp, J., & Otto, P. E. (2006). SSL: A theory of how people learn to select strategies. *Journal of Experimental Psychology. General, 135*, 207–236.

Rumelhart, D. E., & McClelland, J. L. (1982). An interactive activation model of context effects in letter perception: Part 2. The context enhancement effect and some tests and extensions of the model. *Psychological Review, 89*, 60–94.

Schlottmann, A. (2001). Children's probability intuitions: Understanding the expected value of complex gambles. *Child Development, 72*, 103–122.

Schlottmann, A., & Wilkening, F. (2011). Judgment and decision making in young children. In M. Dhami, A. Schlottmann, & M. Waldmann (Eds.), *Judgment and decision making as a skill: Learning, development and evolution* (pp. 55–83). Cambridge: Cambridge University Press.

Schwoebel, J., Dews, S., Winner, E., & Srinivas, K. (2000). Obligatory processing of the literal meaning of ironic utterances: Further evidence. *Metaphor and Symbol, 15,* 47–61.

Shah, A. K., & Oppenheimer, D. M. (2008). Heuristics made easy: An effort-reduction framework. *Psychological Bulletin, 134,* 207–222.

Simon, H. A. (1955). A behavioral model of rational choice. *Quarterly Journal of Economics, 69,* 99–118.

Simon, H. A. (1956). Rational choice and the structure of the environment. *Psychological Review, 63,* 129–138.

Streri, A., Coulon, M., & Guellaï, B. (2012). The foundations of social cognition: Studies on face/voice integration in newborn infants. *International Journal of Behavioral Development, 37,* 79–83.

Thagard, P. (1989). Explanatory coherence. *Behavioral and Brain Sciences, 12,* 435–502.

Thagard, P., & Millgram, E. (1995). Inference to the best plan: A coherence theory of decision. In A. Ram & D. B. Leake (Eds.), *Goal-driven learning* (pp. 439–454). Cambridge, MA: MIT Press.

Troje, N. F. (2002). Decomposing biological motion: A framework for analysis and synthesis of human gait patterns. *Journal of Vision (Charlottesville, Va.), 2,* 371–387.

Wang, A. T., Lee, S. S., Sigman, M., & Dapretto, M. (2006). Neural basis of irony comprehension in children with autism: The role of prosody and context. *Brain, 129,* 932–943.

13 Implicit and Explicit Processes: Their Relation, Interaction, and Competition

Ron Sun

The distinction between "intuitive" and "reflective" thinking has been arguably one of the most important distinctions in cognitive science. There are currently many dual-process theories (two-system views) out there. They seem to have captured popular imagination nowadays. However, although the distinction itself is important, these terms involved (e.g., "intuitive thinking" and "reflective thinking") have been somewhat loaded and thus ambiguous. Not much finer-grained analysis has been done, especially not in a precise, mechanistic, process-based way. In this chapter, toward developing more fine-grained and more comprehensive dual-process theories, I adopt the less loaded and somewhat more precise terms of *implicit* and *explicit processes* (Reber, 1989; Sun, 2002) and present a more nuanced view of these processes centered on a computational *cognitive architecture* (Sun, 2003, 2015).

Given that there has already been an overwhelming amount of research on explicit processes ("reflective thinking"), it is important, in studying the human mind, to emphasize implicit processes (including intuition), given the apparent significance of implicit processes in human psychology (Sun, 1994, 2002). I have argued before that we need to treat implicit processes as an integral part of human thinking, reasoning, and decision making, not as an add-on or an auxiliary (see Sun, 1994, 2002). Therefore, in this chapter I explore implicit processes in a variety of domains as well as their interaction with explicit processes. A theoretical model is presented that addresses, in a mechanistic, process-based sense, various relevant issues. Issues addressed include different types of implicit processes, their relations to explicit processes, and their relative speeds, that is, fast versus slow implicit processes in relation to explicit thinking. The notions of instinct, intuition, and creativity are important in this endeavor and are briefly addressed in this chapter also.

Some Background

There are many two-system views (dual-process theories) currently available (e.g., as reviewed by Evans and Frankish, 2009). One such two-system view was proposed early on by Sun (1994, 1995). For example, in Sun (1994), the two systems were characterized as follows: "It is assumed in this work that cognitive processes are carried out in two distinct 'levels' with qualitatively different mechanisms. Each level encodes a fairly complete set of knowledge for its processing, and the coverage of the two sets of knowledge encoded by the two levels overlaps substantially" (Sun, 1994, p. 44).

That is, the two "levels" (i.e., the two modules or two components) encode somewhat similar or overlapping contents. But they encode their contents in different ways: symbolic and subsymbolic representation are used, respectively. Symbolic representation is used by explicit processes at one "level," and subsymbolic representation is used by implicit processes at the other. Consequently, different mechanisms are involved at these two "levels." Therefore, one "level" is explicit and the other implicit (Sun, 2002). It was further hypothesized by Sun (1994) that these two different levels can potentially work together synergistically, complementing and supplementing each other (see also Sun, 1995 and Sun et al., 2005). This is, at least in part, the reason why evolution led to these two separate levels.

However, some more recent two-system views are somewhat different and their claims seem more contentious. For instance, a more recent two-system view was proposed by Kahneman (2003). The gist of his ideas was as follows: There are two styles of processing: intuition and reasoning. Intuition (or System 1) is based on associative reasoning, fast and automatic, involving strong emotional bonds, based on formed habits, and difficult to change or manipulate. Reasoning (or System 2) is slower, more volatile, and subject to conscious judgments and attitudes.

Evans (2003) espoused a similar view. According to him, System 1 is "rapid, parallel and automatic in nature: only its final product is posted in consciousness"; its "domain-specific nature of the learning" was also noted. System 2 is "slow and sequential in nature and makes use of the central working memory system"; it "permits abstract hypothetical thinking that cannot be achieved by System 1." Moreover, in terms of the relationship between the two systems, Evans argued for a default-interventionist view: System 1 is the default system, and System 2 may intervene when feasible and called for (see Evans, 2003, for more details).

However, some of these claims above seem somewhat simplistic. For one thing, intuition (System 1) can be slow, not necessarily faster than explicit processes (System 2) (Bowers et al., 1990; Helie & Sun, 2010). For another, intuition may sometimes be subject to conscious control and manipulation; that is, it may not be entirely "automatic" (Berry, 1991; Curran & Keele, 1993; Stadler, 1995). Furthermore, decisions made implicitly can be subject to conscious "judgment" (Gathercole, 2003; Libet, 1985). In terms of the relationship between the two systems, implicit and explicit processes may be parallel and mutually interactive in complex ways, not necessarily in a default-interventionist way (Sun, 2002).

It therefore seems necessary for us to come up with a more nuanced and more detailed characterization of the two systems (i.e., the two types of processes). To do so, it is important that we ask some key questions that bear on these issues. For instance, for either type of processes, there can be the following relevant questions in any given situation:

- How deep is its processing (in terms of precision, certainty, and other criteria)?
- How much information is involved (i.e., how broad is its processing)?
- How incomplete, inconsistent, or uncertain is the information available for processing?
- How many processing cycles are needed considering the factors above and beyond?

And there are many other similar or related questions (see, for example, Evans & Frankish, 2009, Evans & Stanovich, 2013, and Kruglanski & Gigerenzer, 2011). Answers to these questions determine characteristics of the two systems in specific circumstances (including, e.g., their relative speeds). They may lead to better characterizations of the two systems and useful interpretations of related notions, but doing so has to start with some basic theoretical frameworks.

A Theoretical Framework

In order to sort out these issues and answer these questions in a more tangible way, below I present a theoretical framework that can potentially provide some clarity to these issues and questions. The framework, I should note here, is based on the CLARION cognitive architecture (Sun, 2002, 2003, 2015) viewed at a theoretical level and used as a conceptual tool for theoretical interpretations and explanations (Sun, 2009b).

The theoretical framework consists of a number of basic principles. The first principle of this framework is the division between procedural (action-centered) and declarative (non-action-centered) processes, which is rather uncontroversial (Anderson & Lebiere, 1998; Sun, 2012, 2015; Tulving, 1985). The next two principles concern the division between implicit and explicit processes, but not just one simple division as in many other dual-process theories. They are unique to this theoretical framework and thus may require some justifications (which have been done in, e.g., Sun, 2012, 2015). The second principle is the division between implicit and explicit procedural processes (Sun et al., 2005). The third principle is the division between implicit and explicit declarative processes (Helie & Sun, 2010). Therefore, in this framework, there is a four-way division: implicit and explicit procedural processes and implicit and explicit declarative processes. These different processes may run in parallel and interact with each other in complex ways (as described in detail, e.g., in Sun, 2002, 2003, 2015).

The divisions above between implicit and explicit processes may be related to some existing computational paradigms, for example, symbolic-localist versus connectionist distributed representation (Sun, 1994, 1995). As has been extensively argued before (e.g.,

Sun, 1994, 2002), the (relatively) consciously inaccessible nature of implicit knowledge may be captured by distributed connectionist representation because distributed representational units are subsymbolic and generally not individually meaningful. This characteristic of distributed representation, which renders the representational form less accessible computationally, accords well with the relative inaccessibility of implicit knowledge in a phenomenological sense. In contrast, explicit knowledge may be captured by symbolic or localist representation, in which each unit is more easily interpretable and has a clearer conceptual meaning.

A Sketch of a Cognitive Architecture

Now that the principles have been outlined, I will go on to sketch an overall picture of the CLARION cognitive architecture itself (without getting too much into technical details however), which is centered on these principles.

CLARION is a generic computational cognitive architecture—that is, a comprehensive, domain-generic model of psychological processes of a wide variety, specified computationally. It has been described in detail and justified on the basis of psychological data (Sun, 2002, 2003, 2015).

CLARION consists of a number of subsystems. Its subsystems include the Action-Centered Subsystem (the ACS), the Non-Action-Centered Subsystem (the NACS), the Motivational Subsystem (the MS), and the Meta-cognitive Subsystem (the MCS). Each of these subsystems consists of two "levels" of representations, mechanisms, and processes as theoretically posited earlier (Sun, 2002). Generally speaking, in each subsystem, among the two "levels," the "top level" encodes explicit knowledge (using symbolic-localist representation), and the "bottom level" encodes implicit knowledge (using distributed representation; Rumelhart et al., 1986). See figure 13.1 for a sketch of the CLARION cognitive architecture.

Among these subsystems, the ACS is responsible for procedural processes, that is, for controlling actions (regardless of whether the actions are for external physical movements or for internal mental operations along the line of executive control) utilizing procedural knowledge. Among procedural processes, implicit procedural processes are captured by MLP networks (i.e., backpropagation networks) at the bottom level of the ACS within the cognitive architecture. Explicit procedural processes, on the other hand, are captured by explicit "action rules" at the top level of the ACS.

The NACS is responsible for declarative processes, that is, for maintaining and utilizing declarative (non-action-centered) knowledge for information and inferences. Among these processes, implicit declarative processes are captured by associative memory networks (Hopfield-type networks or MLP networks) at the bottom level of the NACS. Explicit declarative processes are captured by explicit "associative rules" at the top level. Figure 13.2 highlights the resulting four-way division as mentioned earlier.

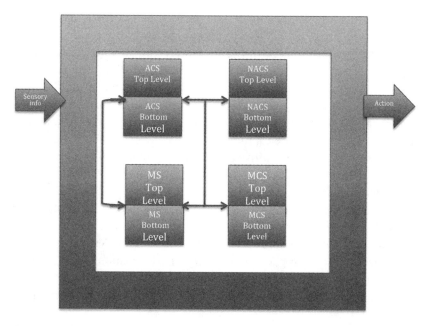

Figure 13.1
The four subsystems of CLARION. ACS stands for the Action-Centered Subsystem. NACS stands for the Non-Action-Centered Subsystem. MS stands for the Motivational Subsystem. MCS stands for the Meta-cognitive Subsystem.

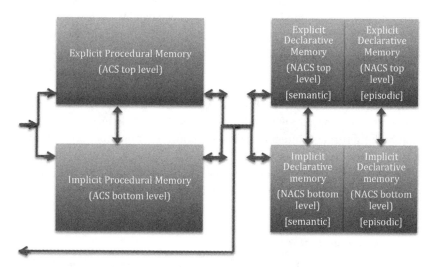

Figure 13.2
The different memory stores within the ACS and the NACS.

The MS is responsible for motivational dynamics, that is, for providing underlying motivations for perception, action, and cognition (in terms of providing impetus and feedback). Implicit motivational processes are captured by MLP networks for *drive* (basic motive) activation at the bottom level of the MS (Sun, 2009a). Explicit motivational processes are centered on explicit *goal* representation at the top level.

The MCS is responsible for meta-cognitive functions; that is, its responsibility is to monitor, direct, and modify the operations of the other subsystems dynamically. Implicit meta-cognitive processes are captured by MLP networks at the bottom level of the MCS, whereas explicit meta-cognitive processes are captured by explicit rules at the top level.

The two levels within each subsystem (the ACS, the NACS, the MS, and the MCS) interact. The interaction between the two levels includes bottom-up and top-down activation flows. *Bottom-up activation* is the "explicitation" of implicit information through activation of nodes at the top level by corresponding nodes at the bottom level. *Top-down activation* is the "implicitation" of explicit information through activation of nodes at the bottom level by corresponding nodes at the top level. For example, there may be an inhibitory role for explicit processes in suppressing implicit information (Gathercole, 2003).

The interaction between the two levels also includes bottom-up and top-down learning. *Bottom-up learning* means implicit learning first and explicit learning on that basis (Sun et al., 2001). This may be viewed as online knowledge extraction from neural networks: implicit knowledge may be learned implicitly (through trial-and-error learning, that is, through reinforcement learning within the neural networks at the bottom level) and then may be explicated to become explicit knowledge (at the top level). *Top-down learning* means explicit learning first and implicit learning on that basis. Explicit knowledge may be explicitly learned and may then be assimilated into implicit processes through a gradual process (e.g., through reinforcement learning at the bottom level, guided by explicit rules).

The interaction between the two levels also includes the integration of the results from the two levels. For example, within the ACS, the two levels may cooperate in action decision making through integration of the action recommendations from the two levels of the ACS in reaching the final action decisions. The proportion of implicit versus explicit processing in the integration may be determined by the MCS taking into consideration a number of factors (Sun, 2003, 2015). See figure 13.1 for the sketch of CLARION.

Interpreting Folk Psychological Notions

Based on the framework outlined above (i.e., the theoretical principles and the CLARION cognitive architecture that embodies them), we may reinterpret some common folk psychological notions, to give them some clarity and precision (hopefully).

For instance, the notion of *instinct* may be reinterpreted and made more specific by appealing to the CLARION framework. Instinct, according to its common, colloquial usage, involves mostly implicit processes and is mostly concerned with action. Within CLARION, instinct may be roughly equated with the following chain of activation: *stimulus → drive → goal → action*. This chain goes from stimuli received to the MS, the MCS, and eventually the ACS. That is, stimuli activate drives (especially those representing essential, innate motives; Sun, 2009a); drive activations lead to goal setting in a (mostly) direct, implicit (and often innate) way; based on the goal set and the stimuli, actions are selected in a (mostly) implicit way to achieve the goal. Instinct is mostly implicit, but it may become more explicit, especially with regard to the part of "goal → action" (Sun et al., 2001).

For another instance, the notion of *intuition* can also be made more specific using the CLARION framework. Intuition, according to CLARION, is roughly the following chain: *stimulus → drive → goal → implicit thinking*. This chain goes from stimuli received to the MS, the MCS, the ACS, and the NACS. As such, intuition mostly involves implicit declarative processes within the NACS (at its bottom level, directed by the ACS), including common functionalities within the NACS such as associative memory retrieval, soft constraint satisfaction, and partial pattern completion. Intuition is often complementary to explicit reasoning, and the two types are used often in conjunction with each other (Helie & Sun, 2010).

Some other folk psychological notions may be reinterpreted and made more precise in a similar manner. For example, the notion of *creativity* can be captured within the CLARION framework. Creativity may be achieved through complex, multiphased implicit-explicit interaction, that is, through the interplay between intuition and explicit reasoning, according to Helie and Sun's (2010) theory of creative problem solving—a theory derived from the CLARION cognitive architecture. More specifically, creative problem solving involves (1) the explicit phase—processing given information (mostly) explicitly through reasoning using explicit declarative knowledge (mostly at the top level of the NACS); (2) the implicit phase—developing intuition using (mostly) implicit declarative knowledge (mostly at the bottom level of the NACS); and (3) finally, the explicit phase again when implicit information emerges into explicit processes—verifying and validating the result using (mostly) explicit declarative knowledge (mostly at the top level of the NACS). See Helie and Sun (2010) for further details. This theory has been successful in accounting for a variety of empirical data related to creativity (e.g., in relation to incubation and insight generation).

What about the competition among these different types of processes, especially in terms of their relative time courses, as alluded to earlier? An issue was raised earlier concerning fast versus slow processes with regard to different two-system views (e.g., Kahneman, 2011; Evans, 2003). The twin distinctions in CLARION, procedural versus declarative and implicit versus explicit, definitely have significant implications for identifying slow versus fast processes. We may question the conventional wisdom on a number of fronts in this regard, instead of simply assuming the seemingly obvious as in some existing two-system views. For instance:

• In terms of the division between procedural and declarative processes, can fast procedural versus slow declarative processes be posited?
• In terms of the division between implicit and explicit procedural processes, can fast implicit versus slow explicit procedural processes be posited?
• In terms of the division between implicit and explicit declarative processes, can fast implicit versus slow explicit declarative processes be likewise posited?
• What about relative speeds if we consider the four-way division together?

And so on. The conjectures as implied by the questions above may be true sometimes to some extent, but they may not be exactly accurate and may well be misleading. That is, the whole picture is not that simple; it is substantially more complex according to the framework discussed thus far.

In this regard we may view existing models and simulations of these types of processes as a form of theoretical interpretation (in particular, concerning their time courses; Sun, 2009b). In that case, we have the following potential answers:

• Fast procedural versus slow declarative processes (as hypothesized earlier): Fortunately, this conjecture of the speed difference is generally true if we examine many existing models and simulations viewing them as theoretical interpretations (Sun, 2003, 2015; see also Anderson & Lebiere, 1998).
• Fast implicit versus slow explicit procedural processes: This conjecture of the speed difference is again generally true, based on theoretical interpretations through modeling and simulation (Sun et al., 2001, 2005).
• Fast implicit versus slow explicit declarative processes: This conjecture, however, is generally not true. Intuition (implicit declarative processes) may (or may not) take a long time compared with explicit declarative processes. See, for example, Helie and Sun (2010) and Bowers et al. (1990) for some possible interpretations that contradicted this conjecture.

The upshot is that we need to be careful in evaluating sweeping generalizations. We may need to characterize different types of processes in a more fine-grained fashion than the conventional wisdom would have it. Characteristics of different processes (e.g., their relative time courses) may also vary in relation to contexts such as task demands.

Many empirical and simulation studies have indeed been conducted within the CLARION framework that shed light on these issues and substantiate the postulates and interpretations made above. See, for example, Sun et al. (2005, 2009), Sun and Zhang (2006), Helie and Sun (2010), and so on. Below some examples of empirical and simulation studies are described.

Some Empirical and Simulation Studies

We can look into many studies that address the questions and issues raised. I focus on the issue of relative speeds of implicit and explicit processes as an illustration of what a cognitive architecture can provide. (For other issues touched on earlier, the reader may consult the references cited earlier.)

Below two studies are sketched. To summarize the main findings with these two studies, it was confirmed that:

• Automatization may lead to faster response times; that is, the more implicit, the faster, in the context of a task involving mainly procedural processes (which is somewhat consistent with the conventional wisdom on this issue).
• Incubation may have positive effects on performance; that is, implicit processes may be slow, and the more time they have, the better the outcomes may be, in the context of a memory task involving declarative processes (which contradicts the conventional wisdom).

Let us look into some details below.

Alphabetic Arithmetic

First, let us look into an alphabetic arithmetic task, which captures the process of automatization (turning explicit processes into implicit processes). In relation to relative speeds of implicit and explicit processes, it produced unsurprising results, confirming the general folklore about implicit and explicit processes (as cited earlier).

The task of alphabetic arithmetic was as follows (Rabinowitz & Goldberg, 1995): Subjects were asked to solve alphabetic arithmetic problems of the following forms: *letter*$_1$ + *number* = *letter*$_2$, or *letter*$_1$ − *number* = *letter*$_2$, where *letter*$_2$ is *number* positions up or down from *letter*$_1$, depending on whether + or − was used. They were given *letter*$_1$ and *number* and asked to produce *letter*$_2$.

In experiment 1 of Rabinowitz and Goldberg (1995), during the training phase, one group of subjects, the consistent group, received 36 blocks of training in which each block consisted of the same 12 addition problems. Another group, the varied group, received six blocks of training in which each block consisted of the same 72 addition problems. Although both groups received 432 trials, the consistent group practiced 36 times on each problem, but the varied group only 6 times.

In the transfer phase of this experiment, each group received 12 new addition problems, repeated three times. The findings were that, at the end of training, the consistent group performed far better than the varied group. However, during the transfer phase, the consistent group performed worse than the varied group. The varied group showed perfect transfer, whereas the consistent group showed considerable slowdown (see figure 13.3).

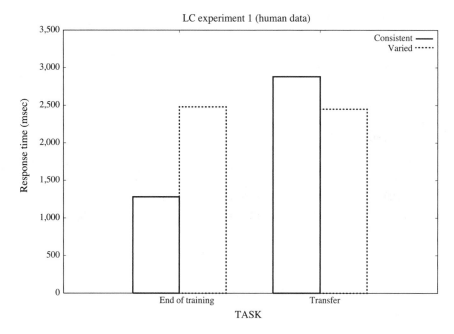

Figure 13.3
The results of experiment 1 of Rabinowitz and Goldberg (1995).

In experiment 2, the training phase was identical to that of experiment 1. However, during the transfer phase, both groups received 12 subtraction (not addition) problems, which were the reverse of the original addition problems used for training (for both groups), repeated three times. The findings were that, in contrast to experiment 1, during transfer the consistent group actually performed better than the varied group. Both groups performed worse than their corresponding performance at the end of training (see figure 13.4).

To investigate possible mechanistic underpinnings of the data pattern, simulations were carried out with CLARION. The simulation, as described in Sun et al. (2009), was based on top-down learning (automatization): that is, a set of rules for capturing prior knowledge concerning letter counting were coded at the top level of the ACS to begin with. Then, on the basis of these rules, performance was carried out, and in the meantime implicit learning at the bottom level of the ACS took place. The NACS was also involved for storing and retrieving experienced instances.

The simulation results were as follows. First, at the end of the training phase of experiment 1, the simulation matched the response time difference between the consistent and the varied group. See the simulation data in figure 13.5. CLARION provided the following explanations for this difference. The simulated consistent group had a

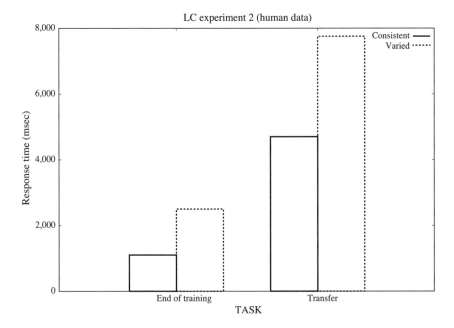

Figure 13.4
The results of experiment 2 of Rabinowitz and Goldberg (1995).

lower response time because it had more practices on a smaller number of instances, which led to the better-performing bottom level of the ACS (implicit processes) as well as better-performing instance retrieval from the NACS. Because these two components had lower response times than other components, a lower overall response time resulted for the simulated consistent group.

CLARION also matched the transfer performance difference between the two groups in experiment 1, as shown in figure 13.5. The simulated consistent group relied more on the bottom level of the ACS (implicit processes) and on the NACS during training, and therefore its counting rules at the top level of the ACS (explicit processes) were less practiced. As a result, it took more time to apply the counting rules, which it had to apply, during transfer. This was because it had to deal with a different set of problems during transfer, and implicit processes (at the bottom level of the ACS) were specific to training instances and thus not applicable to transfer problems. The performance of the simulated varied group hardly changed, compared with its performance at the end of training, because it relied mostly on the counting rules at the top level of the ACS (explicit processes) during training, which was equally applicable to both training problems and transfer problems. As a result, it performed better than the simulated consistent group during transfer.

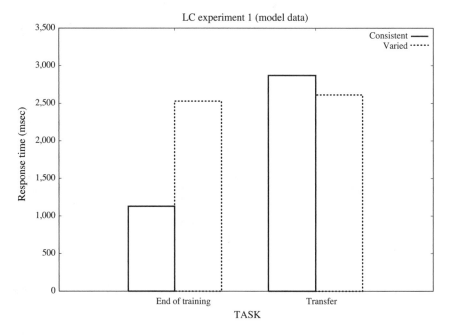

LC experiment 1 (model data)

Figure 13.5
The CLARION simulation of experiment 1.

As indicated by figure 13.6, the CLARION simulation also accurately captured the human data of experiment 2. During transfer in experiment 2, due to the change in the task setting (counting down as opposed to counting up), the practiced rule for counting up was no longer useful. Therefore, both simulated groups had to use a new counting rule (for counting down instead of counting up). Similarly, both simulated groups had to use a new instance retrieval rule (for reversed retrieval). Both simulated groups performed worse than at the end of training for that reason.

The simulation also captured the fact that the varied group performed worse than the consistent group during transfer (figure 13.6). In CLARION, this difference was explained by the fact that the simulated consistent group had more practice with these instances involved during the transfer phase of experiment 2 because of the reverse relationship between the training and the transfer problems. Therefore, the simulated consistent group performed better than the simulated varied group during the transfer phase.

Other possible simulations based on other possible interpretations of the human data were also tested using CLARION for the sake of exploring alternatives. No other simulation produced better results; that is, we found no better alternative explanations of the data. Therefore, the explanations above seemed reasonable (including implicit procedural processes being faster than explicit procedural processes).

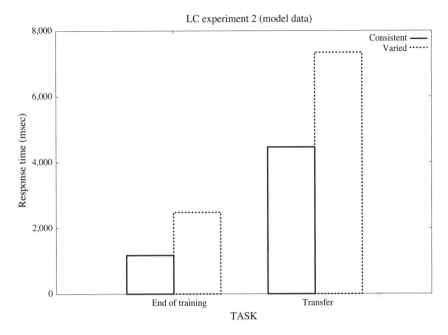

Figure 13.6
The CLARION simulation of experiment 2.

Incubation

Next, we examine a demonstration of slow intuition, as described by Helie and Sun (2010). This result regarding relative time courses of implicit and explicit declarative processes is contrary to the conventional wisdom that intuition is faster than explicit reasoning (as discussed earlier).

As described by Helie and Sun (2010), Smith and Vela's (1991) task explored the effect of incubation on reminiscence score (defined as the number of new words recalled during the second free recall phase in a two-phased free recall). In this experiment subjects had 5 minutes to memorize 50 line drawings. After that, they took part in the first free recall test, which lasted 1, 2, or 4 minutes. Then, they had a 0-, 1-, 5-, or 10-minute incubation phase. After the incubation phase the subjects took part in a second free recall test, the length of which was the same as the first. The results of the experiment showed that there was no significant effect of test duration, but incubation interval had a significant effect on reminiscence score (see figure 13.7).

To simulate this task, mainly the NACS of CLARION was used. According to Helie and Sun (2010), parallel memory searches were conducted in explicit and implicit memory (at the top level and the bottom level of the NACS) during the free recall tests. In general, explicit memory search requires attentional resources, whereas implicit

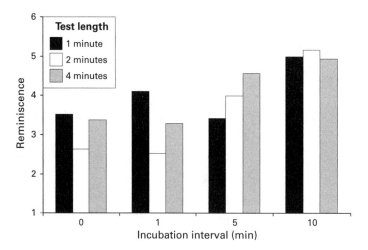

Figure 13.7
Reminiscence effect found in Smith and Vela's (1991) experiment. The black bars represent 1-minute tests, the white bars represent 2-minute tests, and the gray bars represent 4-minute tests.

memory search is mostly automatic. Implicit memory search within the NACS often requires many cycles of settling in an attractor neural network and therefore may be slow. Both forms of memory search were present in this task. However, the incubation phase was different: mostly implicit processes (the bottom level of the NACS) were deployed during the incubation phase (due to the absence of conscious attention to the task; Helie & Sun, 2010). Words were retrieved from implicit memory during that period. These words were then output at the beginning of the second test, increasing the number of words recalled during the second test (but not the first test).

The results were as shown in figure 13.8. As predicted, mean reminiscence scores were positively affected by the incubation length, similar to the human data. Overall, CLARION was successful in capturing the effects of incubation (slow intuition).

Concluding Remarks

The distinction between implicit and explicit processes (i.e., between "intuitive" and "reflective" thinking) has arguably been one of the most important distinctions in cognitive science. It is therefore necessary to further explore implicit processes ("intuitive" thinking) and their interaction with explicit processes ("reflective" thinking). The CLARION cognitive architecture addresses, in a mechanistic, process sense, such important issues, especially the relation, interaction, and competition between implicit and explicit processes.

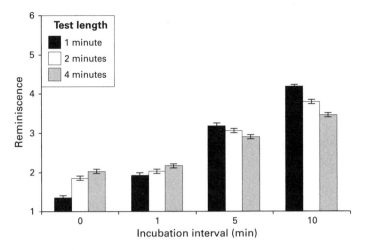

Figure 13.8

Simulated reminiscence effect. Error bars indicate standard errors. The black bars represent 1-minute tests, the white bars represent 2-minute tests, and the gray bars represent 4-minute tests.

Whereas dual-process theories are prevalent, some issues involved in these theories (e.g., the issue of relative speeds of implicit and explicit processes) are more complex than they might appear at first glance—there is often no simple answer to related questions. Moreover, the two orthogonal distinctions are important in this regard and need to be taken into consideration: procedural versus declarative, and implicit versus explicit. Thinking in terms of a cognitive architecture may go a long way in helping us disentangle these issues. Of course, in using a cognitive architecture to explore these issues, theoretical interpretations and speculations inevitably have been involved. Therefore, further empirical and theoretical work is needed for the sake of necessary validation.

In particular, with the use of the CLARION cognitive architecture, it has been shown that, often, the more implicit a process is, the faster it is (e.g., as shown by the alphabetic arithmetic task above). But this is not always the case. Implicit intuition can be slow, especially the emergence of explicit insight from implicit intuition (see, e.g., Bowers et al., 1990; Helie & Sun, 2010). Slow intuition often cannot be sped up by forcing explicitness (e.g., Schooler et al., 1993; see Helie & Sun, 2010, for a detailed, mechanistic interpretation).

Going back to the theoretical framework interpreting folk psychological notions described earlier (as derived from CLARION), we may posit the following:

• Instinct (implicit procedural processes) is usually fast.
• Explicit action decision making (explicit procedural processes) may be slower than instinct, as one would expect.

• Intuition (implicit declarative processes) may be fast or slow depending on circumstances.
• Explicit reasoning (explicit declarative processes) is usually slow; it may be faster or slower than intuition depending on circumstances.

Nevertheless, it may be difficult to come up with a strict hierarchy in terms of speed—many contextual factors are involved.

Likewise, many other issues related to implicit and explicit processes should be equally carefully treated. With many more detailed and more careful explorations, a better understanding of implicit and explicit processes shall be achieved.

Acknowledgments

The work reviewed here has been supported in part by the ONR grants N00014–08–1–0068 and N00014–13–1-0342. Thanks are due to my students and collaborators, past and present, whose work has been reviewed here. The CLARION software, along with simulation examples, may be found at www.clarioncognitivearchitecture.com (courtesy of Nick Wilson and Mike Lynch).

References

Anderson, J. R., & Lebiere, C. (1998). *The atomic components of thought.* Mahwah, NJ: Lawrence Erlbaum Associates.

Berry, D. (1991). The role of action in implicit learning. *Quarterly Journal of Experimental Psychology, 43A,* 881–906.

Bowers, K., Regehr, G., Balthazard, C., & Parker, J. (1990). Intuition in the context of discovery. *Cognitive Psychology, 22,* 72–110.

Curran, T., & Keele, S. (1993). Attention and structure in sequence learning. *Journal of Experimental Psychology: Learning, Memory, and Cognition, 19,* 189–202.

Evans, J. S. B. T. (2003). In two minds: Dual-process accounts of reasoning. *Trends in Cognitive Sciences, 7*(10), 454–459.

Evans, J. S. B. T., & Frankish, K. (Eds.). (2009). *In Two Minds: Dual Processes and Beyond.* Oxford: Oxford University Press.

Evans, J. S. B. T., & Stanovich, K. E. (2013). Dual-process theories of higher cognition: Advancing the debate. *Perspectives on Psychological Science, 8,* 223–241, 263–271.

Gathercole, S. (2003). *Short-term and working memory.* London: Taylor and Francis.

Helie, S., & Sun, R. (2010). Incubation, insight, and creative problem solving: A unified theory and a connectionist model. *Psychological Review, 117,* 994–1024.

Kahneman, D. (2003). A perspective on judgment and choice: Mapping bounded rationality. *American Psychologist, 58*(9), 697–720.

Libet, B. (1985). Unconscious cerebral initiative and the role of conscious will in voluntary action. *Behavioral and Brain Sciences, 8,* 529–566.

Kruglanski, A. W., & Gigerenzer, G. (2011). Intuitive and deliberate judgments are based on common principles. *Psychological Review, 118,* 97–109.

Rabinowitz, M., & Goldberg, N. (1995). Evaluating the structure-process hypothesis. In F. Weinert & W. Schneider (Eds.), *Memory performance and competencies* (pp. 225–242). Hillsdale, NJ: Lawrence Erlbaum Associates.

Reber, A. S. (1989). Implicit learning and tacit knowledge. *Journal of Experimental Psychology, 118,* 219–235.

Rumelhart, D., McClelland, J., & the PDP Research Group. (1986). *Parallel distributed processing: Explorations in the microstructures of cognition.* Cambridge, MA: MIT Press.

Schooler, J. W., Ohlsson, S., & Brooks, K. (1993). Thoughts beyond words: When language overshadows insight. *Journal of Experimental Psychology. General, 122,* 166–183.

Smith, S. M., & Vela, E. (1991). Incubated reminiscence effects. *Memory & Cognition, 19,* 168–176.

Stadler, M. A. (1995). Role of attention in implicit learning. *Journal of Experimental Psychology: Learning, Memory, and Cognition, 21,* 674–685.

Sun, R. (1994). *Integrating rules and connectionism for robust commonsense reasoning.* New York: John Wiley & Sons.

Sun, R. (1995). Robust reasoning: Integrating rule-based and similarity-based reasoning. *Artificial Intelligence, 75*(2), 241–296.

Sun, R. (2002). *Duality of the mind: A bottom-up approach toward cognition.* Mahwah, NJ: Lawrence Erlbaum Associates.

Sun, R. (2003). *A Tutorial on CLARION 5.0.* Technical Report, Cognitive Science Department, Rensselaer Polytechnic Institute. http://www.cogsci.rpi.edu/rsun/sun.tutorial.pdf

Sun, R. (2009a). Motivational representations within a computational cognitive architecture. *Cognitive Computation, 1*(1), 91–103.

Sun, R. (2009b). Theoretical status of computational cognitive modeling. *Cognitive Systems Research, 10*(2), 124–140.

Sun, R. (2012). Memory systems within a cognitive architecture. *New Ideas in Psychology, 30,* 227–240.

Sun, R. (2015). *Anatomy of the mind.* New York: Oxford University Press.

Sun, R., Merrill, E., & Peterson, T. (2001). From implicit skills to explicit knowledge: A bottom-up model of skill learning. *Cognitive Science, 25,* 203–244.

Sun, R., Slusarz, P., & Terry, C. (2005). The interaction of the explicit and the implicit in skill learning: A dual-process approach. *Psychological Review, 112*, 159–192.

Sun, R., & Zhang, X. (2006). Accounting for a variety of reasoning data within a cognitive architecture. *Journal of Experimental & Theoretical Artificial Intelligence, 18*(2), 169–191.

Sun, R., Zhang, X., & Mathews, R. (2009). Capturing human data in a letter counting task: Accessibility and action-centeredness in representing cognitive skills. *Neural Networks, 22*, 15–29.

Tulving, E. (1985). How many memory systems are there? *American Psychologist, 40*, 385–398.

14 Meta-reasoning: Monitoring and Control of Reasoning, Decision Making, and Problem Solving

Valerie A. Thompson, Nicole H. Therriault, and Ian R. Newman

The goal of this chapter is to summarize what is currently known in the nascent field of meta-reasoning (cf. Ackerman & Thompson, 2014). Meta-reasoning refers to the processes that monitor and control reasoning, problem solving, and decision making (Ackerman & Thompson, 2014). One of the primary goals of this chapter is to demonstrate what the reasoning community has to gain from the study of meta-reasoning and what we could we learn from that pursuit that we would not otherwise have learned by sticking to the study of reasoning proper.

Meta-reasoning, like other meta-processes, refers to second-order processes that monitor and control object-level cognition. In other words the study of meta-reasoning is not necessarily concerned with the processes by which conclusions or decisions are reached per se but with the processes that determine how satisfied one is with a conclusion or decision as well as the processes that determine whether, for how long, and how one allocates resources to a problem. Take, for example, a well-known theory of inference, the *mental models theory* (Johnson-Laird & Byrne, 2002). This is a theory of how people represent information in a problem space and how those representations afford inferences about that information. A key assumption of the theory is that people often form an incomplete representation of the problem space, which can be, but which is not always, fleshed out. This object-level theory specifies the processes by which conclusions are reached; the meta-level theory is concerned with how those processes are monitored and controlled. From a meta-level perspective, one of the more important questions would be why and when reasoners are content with their initial representations and why and when they decide to expend the necessary effort to supplement it? Understanding meta-level processes is critical to all theories of reasoning that suppose that analytic or reflective thinking are a core component of human rationality (Thompson, 2009).

In addition, understanding meta-reasoning processes has other, practical implications. For example, the confidence with which a conclusion is held determines the commitment one has to a course of action arising from that conclusion (Adidam & Bingi, 2011); confidence in a conclusion is also contagious, in that a person's confidence is a sign for others to share that confidence (Fox & Walters, 1986; Vidmar, Coleman, & Newman, 2010). Thus, confident eyewitnesses have been known to persuade both investigating officers and jurors despite ample evidence to show that the relationship between confidence and accuracy is very low. As argued below, research on meta-reasoning, like research on other meta-cognitive processes, shows that people's understanding of their reasoning processes is faulty, which makes them prone to misassessing their performance and misdirecting their cognitive resources (Bjork, Dunlosky, & Kornell, 2013).

The twin elements of meta-reasoning encompass (a) the ongoing monitoring of object-level reasoning and (b) the control of subsequent processes (Ackerman & Thompson, 2014). For example, a very basic control function is the termination of reasoning, either because one is satisfied with the current answer, because one believes that an answer cannot be achieved, or because although an answer might be possible, pursuing it is not deemed to be worth the effort. Alternatively, control processes may shift resources to a different strategy, they may allocate additional resources to the problem, reject the current answer, and so forth. In all cases control functions are sensitive to monitoring judgments, whose purpose it is to determine how well things are going. The goal of this chapter is to summarize the extant work on meta-reasoning, including what we know about both monitoring and control processes.

Meta-reasoning and the Control of Reasoning Outcomes

In essence, control processes are about allocating working memory (WM) resources, an issue of particular importance to dual-process theories of reasoning (e.g., Evans & Stanovich, 2013; Thompson, Prowse Turner, & Pennycook, 2011). These theories assume that reasoning and decision making are mediated by two qualitatively different processes (Evans, 2014; Kahneman, 2003; Sloman, 1996; Stanovich, 2011). Type 1 processes are autonomous, meaning they are fast and often form a default response (Evans & Stanovich, 2013). Type 2 (analytic) processes require WM and are generally slower (Evans & Stanovich, 2013). These processes may or may not be engaged in a given context and, even when engaged, may not be employed to maximum effectiveness (Shynkaruk & Thompson, 2006; Stanovich, 2009; Thompson & Johnson, 2014). From a meta-reasoning perspective, the key issue is to understand when and why reasoners are content with the default answer and when (and how) analytic processes are engaged (Thompson, 2009).

On the surface, meta-level processes appear to be most relevant for the default interventionist version of dual process theories (Evans, 2007b), because this group of theories posit a sequence of events (i.e., a Type 1 response that always occurs and precedes Type 2 interventions). Nonetheless, meta-level considerations are equally relevant for parallel theories, which assume that Type 1 and Type 2 processes are initiated in parallel

(Sloman, 1996), and single-process theories (Kruglanski & Gigerenzer, 2011) that do not propose two types of processing. This is because, regardless of the architecture that is assumed, people will give more and deeper analysis to some problems than others, some people are more inclined toward such engagement, and circumstances vary in the degree to which they facilitate it. Thus, regardless of the type of reasoning mechanism that is proposed, it is important to understand how and when analytic processes are engaged: that is the job of the monitoring processes described below. For example: What is the signal that a given problem requires a lot of thought whereas another does not? As above, in the mental models view, when are representations fleshed out and when are they not? Many theories of reasoning posit that the availability of counterexamples plays a crucial role in reasoning outcomes (e.g., Cummins, Lubart, Alksnis, & Rist, 1991; Markovits, 1986; Thompson, 2000), but when do reasoners initiate such a search? What determines the confidence with which a conclusion is held? How do people evaluate (a) whether a problem is solvable and (b) that a solution is worth the effort required to achieve it? Meta-reasoning asks about the cues that initiate or terminate analytic processes, and that decide which of several possible strategies are appropriate in the current context.

Meta-reasoning and the Monitoring of Reasoning Processes

It is assumed that meta-reasoning processes are engaged to monitor reasoning and decision-making activities in an analogous manner to meta-memory and other meta-cognitive processes (Ackerman & Thompson, 2014; Thompson, 2009; Thompson et al., 2011). In some instances these processes may be engaged explicitly, as, for example, when one is proofreading a manuscript. In most cases, however, these processes are assumed to run in the background and to respond to cues present in the environment, or to expereintial cues associated with the perceived ease or difficulty of processing (e.g., Koriat, 2007). For example, did you catch the typo in the last sentence? The chances are good that you were not reading this paragraph with the goal of detecting spelling mistakes, but, nonetheless, your monitoring processes alerted you to the error.

Monitoring is assumed to be continuous, engaged from the beginning of a reasoning attempt and continuing after its completion. The research into these monitoring processes is in its infancy; below, we summarize what we know and have grounds to speculate about.

Judgments of Solvability

The principle of cognitive miserliness (see Stanovich, 2009, for a review) suggests that reasoners should be reluctant to invest cognitive resources in tasks that they perceive to have low probabilities of success (Ackerman & Thompson, 2014). Consequently, we have speculated that the first monitoring judgment to be made concerns the probability that the problem will be solved (Thompson, 2009). In extreme cases this initial judgment of solvability (JOS) may suggest to the reasoner that it is not worthwhile to attempt a solution at all (Ackerman & Thompson, 2014).

Although there is no current research specifically pertaining to JOSs, we know that analogous judgments in meta-memory play an important control function in determining both choice of processing strategy and allocation of study time. Thus, problems that are accompanied by low JOS are expected to lead to "giving up" when motivation is low or when there is time pressure and to result in a high investment when motivation is high (Ackerman & Thompson, 2014). As an everyday example, imagine ordering some furniture that is delivered in boxes rather than assembled. An initial judgment of solvability might lead one to think that it is not worth the effort to attempt assembly; alternatively, one might judge that the challenge would be enjoyable and so worth the time required.

In addition to determining whether a solution is attempted, JOSs may determine the probability that one persists with a problem. This conclusion is made by analogy to meta-memory, where it has been shown that participants allocate more study time to those items they think will be difficult to learn (e.g., Son & Metcalfe, 2000). This effect is context dependent, so that under time pressure, participants tend to allocate their time to easier items (Son & Metcalfe, 2000; Thiede & Dunlosky, 1999). Thus, when the time is limited, people skip the items they judge to be time consuming and strategically invest effort in easier ones.

Because JOSs are made prior to an attempt to solve a problem, it is probable that they would be based on the reasoner's beliefs about the task at hand, on their past experience and success at solving similar problems, and on surface-level cues to difficulty (Ackerman & Thompson, 2014). For example, a person who believes herself to be good at solving problems, doing logic, or thinking probabilistically should have correspondingly higher JOS for problems in those domains than someone who does not hold such beliefs (Prowse Turner & Thompson, 2009; Shynkaruk & Thompson, 2006). This confidence should then translate into a higher probability of attempting to solve the problem and more perseverance once the attempt has been made.

Other cues to difficulty might include familiarity with the material (Markovits, Thompson, & Brisson, 2015) and ease of representation (Stupple, Ball, & Ellis, 2013). Fluency of processing could also give rise to a strong JOS. Topolinski (2014) summarized a large body of research on the Remote Associate Test, in which people are asked to find a fourth word that is associated with a triad of words (Mednick & Mednick, 1967). He argued that solvable items are processed more fluently than nonsolvable ones, and this gives rise to a positive affective response allowing participants to make very fast JOSs. Consequently, participants are able to make JOSs in less time than it takes to actually solve the problem.

Without such cues, however, it seems likely that the accuracy of JOSs will be low (Ackerman & Thompson, 2014). Payne and Duggan (2011) found that participants could not tell that a water-jar problem was unsolvable until after a lengthy attempt to solve it, and sometimes not even then. However, the ability to calibrate JOSs may increase with experience. Novick and Sherman (2003) asked participants to judge the solvability of anagrams. Accuracy increased with experience, possibly because experienced participants learned reliable cues that were not available to their less experienced counterparts.

Ongoing Confidence

Many reasoning and problem-solving processes require a substantial amount of time to solve. Ackerman (2014) posited that these ongoing problem-solving processes are continuously monitored and that this monitoring controls the allocation of analytic resources to that problem. In her study participants provided confidence ratings while solving CRT and RAT problems. Confidence increased over the problem-solving interval (as per Metcalfe & Wiebe, 1987). She also observed that initial confidence judgments, which were made 5–10 seconds after the problem was presented, predicted the amount of additional time spent solving the problem, similar to the relationship observed between Feelings of Rightness (FOR) and analytic thinking (below; see Thompson et al., 2011, 2013). Finally, she observed that a by-product of spending additional time solving a problem was that participants became more and more willing to give less confidently held answers. That is, early in the process, answers would be given only if the participants were fairly confident in them; however, as time progressed, participants were willing to compromise on this standard and provided answers with less degrees of confidence.

Feelings of Rightness

FOR is a special-case judgment that applies to situations in which an answer comes quickly to mind, as in the examples below:

A ball and a bat together cost $1.10. The bat cost $1.00 more than the ball. How much does the ball cost? ___ cents

(Frederick, 2005)

If today is Thursday, then tomorrow is Friday. Today is Thursday; therefore, …

Linda is 31 years old, single, outspoken, and very bright. She majored in philosophy. As a student, she was deeply concerned with issues of discrimination and social justice and also participated in antinuclear demonstrations. What is the probability that:

Linda is a bank teller____

Linda is a bank teller who is active in the feminist movement___

(Adapted from Tversky & Kahneman, 1983)

In all cases an answer comes easily to mind. In some cases the answer that comes to mind is consistent with rules of probability or logic, as in the second example; in others it is generated by other processes, such as representativeness (Kahneman & Tversky, 1973; Tversky & Kahneman, 1983). Other situations where answers come easily to mind include insight in problem solving (Topolinski & Reber, 2010), impulsive choices, and a range of other decisions that are based on fast-acting heuristic processes (Thompson et al., 2011, 2013). Meta-reasoning theory (Thompson, 2009; Thompson et al., 2011) assumes that these initial answers have two components: the first is the answer itself, and the second is the FOR that accompanies the answer, which can vary in strength. That is, whereas answers to the three questions above might be generated with a strong FOR, in other instances, the initial response may be generated with less confidence, as when, for example, there are two conflicting bases of response (Thompson et al., 2011):

In a study 1000 people were tested. Among the participants there were 5 engineers and 995 lawyers. Jack is a randomly chosen participant of this study. Jack is 36 years old. He is not married and is somewhat introverted. He likes to spend his free time reading science fiction and writing computer programs.
   What is the probability that Jack is an engineer? ____

(De Neys & Glumicic, 2008)

Data gathered in our laboratory have established a solid link between an initial FOR and subsequent analytic engagement. Specifically, reasoners spend more time rethinking their initial answer and are more likely to change that answer when it is accompanied by a weak than a strong FOR (Thompson et al., 2011, 2013; Thompson, Evans, & Campbell, 2013; Thompson & Johnson, 2014). That is, a weak FOR is a signal that something is not quite right with the initial answer and that additional thought may be warranted. This relationship has been established with a variety of classic reasoning tasks such as conditional and syllogistic reasoning (Thompson et al., 2011), the Wason selection task (Thompson et al., 2013), and base rate neglect tasks (Thompson et al., 2011, 2013). Given that the strength of the FOR mediates analytic engagement, the next logical question is where does this feeling come from? As argued below, the FOR, as well as a range of other meta-reasoning judgments, is likely to be multiply determined and based largely on the experience associated with deriving a solution.

Final Judgments of Confidence (FJC) refer to the final degree of satisfaction one has in an answer. Again, there is relatively little research on FJC in a reasoning context, despite the existence of a large literature on confidence in other cognitive tasks. We know, for example, that FJCs are higher than FORs even when an initially correct answer has been changed to an incorrect one (Shynkaruk & Thompson, 2006). Indeed, the calibration of FJCs with accuracy is modest at best (Prowse Turner & Thompson, 2009; Shynkaruk & Thompson, 2006), for reasons outlined below. Although we have never actually reported the correlations in all of our studies, FJCs and FORs are highly correlated, suggesting that the same variables underlie both types of judgments.

Monitoring as Inference

It is common to see meta-cognition defined as "cognition about cognition" or "knowing about knowing." Describing it in these terms strongly implies that meta-cognition involves explicit knowledge and beliefs that we hold about our cognitive processes. In contrast, leading meta-memory theorists (e.g., Koriat, 1997, 2007; Schwartz, Benjamin, & Bjork, 1997; Bjork, Dunlosky, & Kornell, 2013; Metcalfe, Schwartz, & Joaquim, 1993) have argued that, counterintuitively, people seldom form explicit judgments of the strength and coherence of their memories; instead, these judgments are made inferentially on the basis of a variety of cues such as characteristics of the learning task and the rememberer's subjective experiences of learning and recall (Miele & Molden, 2010). In other words meta-cognitive judgments such as the FOR and JOS are not likely to be based on the quality or accuracy of the outcome but, instead, to be made inferentially based on the qualities of the experience associated with solving the problem. Consequently, the reliability of monitoring, measured by FOR, JOS, and similar measures, will depend on the reliability of the cues on which it is based (Koriat, 2007). In some cases the cues may be reliable, giving rise to accurate meta-cognitive judgments. In others they may be misleading.

Take, as a case in point, *answer fluency*, which refers to the ease or speed with which an answer comes to mind. Answer fluency has been studied extensively in the domain of meta-memory: fluency of retrieval gives rise to a sense of familiarity or knowing; thus, items that come to mind easily are likely to be perceived as correct, easily learnable, likely to be recalled in a subsequent test, and so on (e.g., Benjamin et al., 1998; Jacoby et al., 1989; Kelley & Jacoby, 1993, 1996; Matvey, Dunlosky, & Guttentag, 2001; Whittlesea & Leboe, 2003). Fluency of processing is a powerful cue to subjective confidence, such that fluency of retrieval begets confidence in the answer (Costermans, Lories, & Ansay, 1992; Kelley & Lindsay, 1993; Robinson, Johnson, & Herndon, 1997).

Because ease of retrieval is a valid cue to difficulty in many contexts, relying on this cue will often produce accurate monitoring judgments (Koriat, Ackerman, Lockl, & Schneider, 2009). However, in contexts where ease of retrieval is not diagnostic of difficulty, it can mislead people into misplaced confidence. For example, when one is answering questions, answers that are retrieved with a lot of effort are more likely to be recalled on a later test than answers that come to mind easily (Gardiner, Craik, & Bleasdale, 1973), presumably because the difficulty of retrieval established a strong memory trace for subsequent recall (Schwartz et al., 1997). However, because monitoring judgments are based on cues such as fluency, participants wrongly predict that they would be more likely to remember the easy-to-retrieve items than the difficult-to-retrieve ones (Schwartz et al., 1997). Indeed, fluent processing can produce an illusion that an item has been previously experienced, regardless of whether it has or has not (e.g., Jacoby et al., 1989; Whittlesea, Jacoby, & Girard, 1990).

Below, I document some of the cues that underlie meta-reasoning judgments. As is the case for meta-memory, the monitoring of conclusions, decisions, and problem solutions is posited to be cue driven, and these cues are derived from the environment, from the experience of solving the problems, from beliefs held by the reasoner, and so on. To the extent that the cues in question are correlated with accuracy or other qualities of good reasoning, so too will monitoring be. The review is uneven because research in the area is in its infancy, and, as such, early work has focused on obvious possibilities such as fluency. Most of the monitoring judgments that are referred to below are FORs and FJCs; exceptions are noted.

Answer Fluency

By analogy to meta-memory, I have defined *answer fluency* as the ease with which an initial conclusion comes to mind (Thompson, 2009). The study of answer fluency, therefore, is most relevant to those types of problems that elicit an immediate response, either because they are syntactically simple (If p, then q; p, therefore ??) or because a fast-acting heuristic process produces an answer (as in the ball and bat and Linda examples above). Consequently, the most straightforward prediction about the origins of the FOR accompanying an initial answer is that it is determined by the speed or fluency with which that answer is produced.

Several pieces of data support the role of answer fluency in meta-reasoning judgments. On a variety of different reasoning tasks Thompson and colleagues have demonstrated that fluently generated answers (as measured by speed of response) produce a stronger FOR than slower answers. This is true for tasks as diverse as conditional reasoning and base rate neglect (Thompson et al., 2011, 2013) and extend also to the Wason selection task (Thompson, Evans, & Campbell, 2012) and a syllogistic reasoning task (Thompson et al., 2011). Indeed, on these latter two tasks, there is evidence to suggest that responses based on heuristic strategies, such as matching on the Wason task and the *min* heuristic (Chater & Oaksford, 1999) on the syllogistic task, are faster than their nonheuristic counterparts. This relative fluency results in a stronger FOR and a correspondingly lower level of analytic engagement for the heuristic responses. Consequently, Thompson et al. (2012) suggest that one reason that so many so-called reasoning biases are held with such confidence is that they are fluently produced by heuristic processes, which, in turn, create a sense of confidence in the answer.

Indeed, Ackerman and Zalmanov (2012) showed that fluency is a potent cue to confidence, regardless of its relationship with accuracy. In one experiment participants solved two versions of problems whose modal response is incorrect, such as Frederick's (2005) Cognitive Reflection Test:

If it takes 5 machines 5 minutes to make 5 widgets, how long would it take 100 machines to make 100 widgets? ____ minutes

For half the participants, items were presented in the standard format, as illustrated above, for the others, participants were presented with a list of options to choose from. The traditional format produced more incorrect answers than the multiple-choice format, but the relationship between fluency and confidence was the same in both conditions. In a second experiment participants solved items from the remote associate test (Mednick & Mednick, 1967), in which they were asked to find the common associate for three common words ("nut" is the solution to pea, chest, shell). The fluency-confidence relationship was found to be identical for correct and incorrect solutions. Thus, regardless of whether the answer is correct, fluency is a powerful cue to confidence.

The next logical step in this research program would be to manipulate fluency experimentally. In all of the preceding studies relative fluency is a naturally occurring product either of individual items, that is, some answers are made more fluently than others, or of experimental conditions (such as matching vs. nonmatching cards in the Wason task). Ideally, one would like to have an experimental manipulation that directly affected the speed of responding and then to demonstrate that fast answers were provided with stronger FORs than slower ones.

To date, however, such attempts have not been successful. Thompson and Therriault have made several attempts to prime conclusions to problems. This procedure derives from the extensive literature on semantic priming, whereby prior presentation of a word such as "doctor" facilitates recognition of a related word such as "nurse" (Meyer & Schvaneveldt, 1971). Our goal was to extend this manipulation to a reasoning context and to prime the conclusions to reasoning problems by preceding them with either semantically related or unrelated words, for example:

Semantically related primes: carrot, vegetable, soil, rabbit, lettuce

Semantically unrelated primes: castle, moat, palace, dungeon, chamber

Reasoning problem: All vegetables need soil. A carrot is a vegetable. A carrot needs soil.

Participants were to give the first word that came to mind while reading the word lists and were then immediately presented with the reasoning problem, for which they made a judgment of logical validity. It was anticipated that the semantically related words would prime the conclusion and produce faster responses, which, in turn, would produce higher FORs. This manipulation failed, however, as both RTs and FORs were identical in the related and unrelated conditions.

Thinking that the word lists might have primed the reading of the conclusion but not necessarily the production of the answer itself (which was a yes/no response), we modified the procedure so that the priming manipulation related more directly to the production of the answer. For this, we changed the task from an evaluation task to a generation one, where participants were provided the first two premises and asked to provide a verbal conclusion (e.g., to say aloud that "a carrot needs soil"). Again, however, there were no differences in the time required to generate the answer (answer fluency) or FORs for the related and unrelated primes.

Clearly, priming the answer to a reasoning problem will not be as straightforward as priming words. Nonetheless, in order for the concept of answer fluency to have much traction, it needs to move from a correlational to an experimental variable; otherwise, it will be difficult to conclude that the relationship between answer fluency and FOR is a causal one and not one brought about by a third, as yet unknown variable. One possible approach might be to examine relative, rather than absolute, fluency (Whittlesea & Williams, 2000; Hansen, Dechêne, & Wänke, 2008; see Hansen & Wänke, 2012, for review). That is, the key variable may be whether the current answer was less or more fluent than the one before rather than how quickly the answer came to mind in absolute terms. Thus, a promising line of inquiry might be to embed a relatively disfluent problem in a series of fluent ones and vice versa; the change in fluency should produce a corresponding shift in FOR.

Perceptual Fluency

Whereas answer fluency concerns the ease with which an answer pops into mind, *perceptual fluency* concerns the ease with which a problem is perceived. For example, problems can be made difficult to read (i.e., disfluent) by changing the font in which they are presented or altering the contrast between the target and background. Many studies have found that fluently processed items are perceived more positively than those requiring more effort (see Alter & Oppenheimer, 2009; Topolinski & Reber, 2010, for recent reviews) and that this positive feeling carries over to seemingly unrelated judgments; for example, fluently processed statements are judged true more often than their disfluent counterparts (Reber & Schwarz, 1999).

On this basis Alter, Oppenheimer, Epley, and Eyre (2007) proposed that whereas problems presented in a fluent manner would create a sense of subjective ease, their disfluent counterparts would result in a sense of meta-cognitive unease. This meta-cognitive unease was postulated to act as a cue that additional analytic thinking is required. Indeed, evidence from the domain of meta-memory suggests that meta-memory judgments are sensitive to perceptual fluency, such as making a text or word easy or difficult to read (e.g., Kornell, Rhodes, Castel, & Tauber, 2011; Rhodes & Castel, 2008). Consistent with their hypotheses, Alter et al. found that presenting reasoning problems (i.e., the widget and ball and bat examples above as well as three-term syllogisms) in a difficult-to-read font increased the probability of correct answers; similarly, when people were cued to interpret their efforts as disfluent by furrowing their brow, they showed reduced levels of base-rate neglect.

However, recent work by Thompson and colleagues (Thompson et al., 2013) challenges these findings. In seven experiments, using both identical tasks and conceptually similar tasks to those of Alter et al. (2007), they failed to find that problems presented in a perceptually disfluent manner (e.g., presented in a difficult-to-read font) were solved more accurately than their fluent counterparts, nor that disfluency created a sense of "meta-cognitive unease" as measured by either FOR ratings or FJCs. They did find that in four out of five experiments where response time was measured, reasoners took slightly longer to produce answers in the disfluent conditions, although this additional time did not result in more correct responses. Thus, although there was some evidence that disfluent perceptions were associated with analytic thinking, there was little evidence that this resulted in more correct answers.

Thompson et al. (2013) also noted that Alter et al.'s (2007) participants were all drawn from elite Ivy League universities (Harvard and Princeton) and speculated that the relationship between perceptual fluency and accuracy may be observed only among the intellectually elite. However, even among participants from an elite Israeli university (the Technion-Israel Institute of Technology), problems presented in a fluent font were solved just as often as problems presented in a disfluent font. However, in two studies, Thompson et al. (2013) demonstrated that the font manipulation interacted with IQ: persons of high IQ were better at solving disfluent than fluent problems, and, if anything, the effect was reversed at the lower end of the IQ scale. Thus, although perceptual disfluency might trigger a sense of unease that produces longer thinking time, this thinking time results in correct answers only among the most cognitively able (although even this relationship is suspect; see Meyer et al., 2015, for a comprehensive test of the perceptual fluency hypothesis). Also, there was little evidence to suggest that disfluency affected meta-cognitive judgments, possibly because the effect of fluency may have been diffuse, creating a global sense of unease that was not captured by item-specific measures such as FOR (Thompson et al., 2013). It is also possible that the effect of perceptual fluency is subtle relative to a very salient alternative, such as answer fluency, so that it does not manifest itself in FOR judgments (Ackerman & Thompson, 2014).

Alternatively, as was the case with answer fluency, it is possible that absolute levels of perceptual fluency do not influence meta-cognitive judgments but, rather, that these judgments would be sensitive to relative changes in perceptual fluency. For example, statements that are easy to read are more likely to be judged true than those that are difficult to read, but only when there is a change of difficulty (Hansen, Dechêne, & Wänke, 2008). That is, difficult-to-read statements in a string of difficult-to-read statements are as likely to be judged to be true as easy-to-read statements in a string of easy-to-read statements. It is the shift or change of fluency that seems to act as the basis of a cue. Thus, the fact that Thompson et al. (2013) manipulated fluency as a between-subjects variable may have reduced the probability of observing an effect on reasoning performance, although we note that Alter et al.'s (2007) manipulation of perceptual fluency was also between subjects.

Regardless, preliminary data from our lab does not support this hypothesis. As part of the conditional reasoning experiment reported earlier, Thompson and Therriault also manipulated perceptual fluency within subjects. The manipulation was the same as in Alter et al. (2007) and Thompson et al. (2013) and involved presenting problems in a difficult- and easy-to-read font. The font manipulation was blocked, so that four fluent or disfluent problems were presented in a row. There was no difference in FORs, RT, or accuracy for the two font conditions, nor were there differences when we compared the last trial of a block to the first trial of the next block (i.e., when fluency changed). We also increased the number of trials per block to six (as per Hansen et al., 2008) but still observed null effects. Thus, to the best of our knowledge, we can find no evidence to suggest that perceptual fluency per se affects either reasoning outcomes or monitoring judgments.

Monitoring for Conflict and Coherence

There is a substantial body of evidence to suggest that meta-reasoning judgments are sensitive to coherence and response conflict. These are related but somewhat different constructs in that response conflict is a subset of incoherence. That is, information can be incoherent because it leads to irreconcilable responses but also because it does not cohere around any response at all.

Investigations of coherence have focused on a variant of Mednick & Mednick's (1967) Remote Association task. In this paradigm, participants are presented with both coherent and incoherent word triads and were asked to judge whether they are coherent. Coherent triads are those in which the words are associated with a fourth word (e.g., playing, credit, and report are associated with card), whereas in incoherent trials there is no solution (e.g., house, lion, butter). Bolte and Goschke (2005) found that participants' judgments of coherence were above chance, even when they were unable to retrieve the answer and were required to make their judgments in a very short period of time (i.e., 1.5 seconds after the presentation of the triad). Although not commonly referred to in such terms, judgments of coherence are very similar to meta-cognitive judgments. Specifically, they resemble JOSs described above in that they are made prior to a solution attempt and concern the solvability of the problem at hand.

Topolinski (2011) argues that coherence judgments measure a broader ability to detect incoherence and inconsistency in the world. In his view intuitions of coherence are mediated by priming (i.e., coherent triads all prime their remote associate), which gives rise to fluent processing, which, in turn, results in a positive affective experience (see Thompson & Morsanyi, 2012, for review). As evidence for this *fluency-affect intuition model*, lexical decision times for coherent triads are faster than those for incoherent triads (Toplinski & Strack, 2009a); similarly, coherent triads are read faster than incoherent triads (Topolinski & Strack, 2009b). Manipulating the fluency of processing increases judgments of coherence, for example, when they are primed (Topolinksi & Strack, 2009c). As evidence for the affective component, coherent triads activate the facial muscles associated with smiling (Topolinski, Likowski, Weyers, & Strack, 2009), and are "liked" better than incoherent trials (Topolinski & Strack, 2009b).

Response conflict has also been shown to affect meta-reasoning judgments. In many classic reasoning problems participants are asked to make inferences about two types of trials: trials that are congruent, in which the answers based on logic or probability are the same as those based on Type 1 outputs, such as representativeness or beliefs, and incongruent ones, such as when base rate information contradicts a stereotype (as in the lawyers and engineers problem above). People tend to perform more poorly on conflict than nonconflict problems in terms of both accuracy (Evans et al., 1983) and RT (Bonner & Newell, 2010; Thompson et al., 2011). From a meta-reasoning perspective, they are also less confident (De Neys, Rossi, & Houdé, 2013; Thompson et al., 2011; Thompson & Johnson, 2014) in answers given to conflict than nonconflict problems, which suggests that monitoring processes respond to conflict. Finally, consistent with Topolinski's hypothesis about the role of affect in monitoring, skin conductance responses indicated that conflict can produce a mild state of arousal (De Neys, Moyens, & Vansteenwegen, 2010).

Finally, Koriat's most recent theory of confidence also brings coherence to the fore (Koriat, 2012). He argues that in situations where people must decide between two alternatives, the information gathered about those alternatives forms the basis of a decision, with confidence in the decision being a by-product of the relative number of pros and cons favoring the chosen option. In this view fluency is a by-product of consistency, such that choices with a high degree of consistency are made fluently relative to less consistent ones.

In sum, the evidence is mounting that reasoners monitor for coherence and consistency and that violations of either produce disfluent processing and lower judgments of confidence or rightness. Notably, conflict lowers FORs, even if it does not affect fluency of processing (Thompson & Johnson, 2014), and FORs are sensitive to conflict, even when answers must be provided under time pressure (Thompson et al., 2011, 2013). Thus, the conflict must arise quickly, implying that at least some (simple) probability judgments and logical evaluations are made by fast, Type 1 processes (Handley et al., 2011; Pennycook et al., 2013).

Diagrams, Illustrations, and Illusions of Comprehension

Diagrams and illustrations are often perceived to be helpful in a variety of problem-solving contexts. However, the presence of a diagram can create a false sense of confidence in one's understanding of the problem. Ackerman, Leiser, and Shpigelman (2013) presented college students with challenging problems (e.g., combinatorial problems). After attempting to solve the problems, participants were immediately given an explanation of how to solve the problem. Some of the explanations were accompanied by a noninformative illustration (e.g., a combinatorial problem involving drawing socks from a drawer was accompanied by an illustration depicting pairs of socks), whereas others were presented just as text. Participants judged their comprehension of the illustrated explanations to be higher than the text-alone versions; however, their performance on the transfer problems was, in fact, lower than the text-alone versions. The authors speculated that the illustrations focused attention on surface details, rather than on the deep structure of the problem.

A similar argument was made by McCabe and Castel (2008) regarding the persuasive impact of fMRI images. Participants were asked to rate the soundness of the arguments presented in fictitious scientific articles that were accompanied by a brain image, an informationally equivalent bar graph, or by no illustration at all. In all cases the illustrations were redundant with the text in that they did not present any new data. In three studies texts that were accompanied by brain images were rated to be more persuasive than any of the control conditions, possibly because they appeared to provide a tangible, physical explanation for an invisible, abstract phenomenon (McCabe & Castel, 2008).

Familiarity

There is also preliminary evidence that monitoring judgments of reasoning outcomes are sensitive to the familiarity of the content of the problem as they are in the domain of meta-memory (Costermans, Lories, & Ansay, 1992; Koriat & Levy-Sadot, 2001; Shanks & Serra, 2014). For example, Shynkaruk and Thompson (2006) found that confidence in syllogistic conclusions was higher when they were either believable or unbelievable than when they were neutral (e.g., Some Abens are not Welps). In other words when reasoners could make judgments on the basis of prior knowledge, they appeared to be more confident in the conclusions than when such knowledge was absent. Unfortunately, the believable conditions differed from the others on a number of other dependent variables, including the probability of a correct response and response times, both of which might have contributed to differences in confidence. Recently, however, Markovits and colleagues (Markovits, Thompson, & Brisson, 2015) have found that in a conditional reasoning task, familiar conditionals (e.g., if a finger is cut, then the finger will bleed) invited stronger degrees of confidence in their conclusions than their less familiar counterparts (e.g., If someone glebs, then they are brandup), even when controlling for accuracy of response. Although we cannot rule out differences in fluency as a potential explanation for these findings, they nonetheless constitute preliminary evidence that familiarity with the items being reasoned about per se increases confidence in the answers.

Accuracy of Monitoring

As the preceding discussion suggests, judgments of confidence, comprehension, solvability, and so on are based on cues such as fluency, coherence, familiarity, and the presence of diagrams. To the extent that such cues are correlated with accuracy or good reasoning outcomes, one would expect good concordance between those judgments and accuracy or other indices of good reasoning outcomes. For example, as discussed earlier, fluency is often diagnostic of difficulty, such that problems that take a long time to solve are also likely to be difficult and therefore error prone. Similarly, incoherent information tends to make decisions difficult, and diagrams can be used to summarize or illustrate complex and abstract relationships and facilitate problem solving. These cues, however, are imperfect and, at times, misleading, meaning that people may have high confidence in poor reasoning outcomes, or misallocate time and WM resources based on poorly calibrated JOS or FORs. That is, fast-acting heuristics (belief, representativeness, availability, etc.) may give rise to a fluent response, which in turn engenders a strong FOR, with the result that reasoners have a strong sense of confidence in an answer that might, instead, benefit from additional analysis (Thompson et al., 2011).

Thus, it is not surprising that the relationship between various monitoring judgments and accuracy on logical reasoning tasks is poor. In the case of complex tasks, such as syllogistic reasoning, the relationship is close to zero (Shynkaruk & Thompson, 2006), even after training in the task (Prowse Turner & Thompson, 2009). On simpler tasks, such as conditional reasoning, we have observed correlations as high as 0.3 with both realistic conditionals and abstract ones (Markovits et al., 2015; Thompson et al., 2011). In problem-solving tasks Ackerman and Zalmanov (2012) also observed that confidence was higher for correct than incorrect problem solutions, even after controlling for the fluency of response. This last finding suggests that there are cues, other than fluency, that people rely on to discriminate accurate and inaccurate responses, although it not clear what they might be (Ackerman & Zalmanov, 2012).

Implicit versus Explicit Monitoring

Koriat (2007) distinguished between implicit and explicit monitoring cues. Implicit cues are likely experienced as an affective response, carry little cognitive content, and have origins that are not accessible to conscious processes (Koriat & Levy-Sadot, 1999). The FOR (Thompson, 2009) and Topolinski's coherence judgments are posited to fall into this category. In contrast, explicit cues are beliefs about cognitive processes that are accessed and consciously applied to making judgments. It is widely assumed that most monitoring cues are implicit (Mueller, Dunlosky, Tauber, & Rhodes, 2014); indeed, people often appear to have difficulty accessing explicit beliefs when they are relevant. For example, people (rightly) believe that forgetting increases as a function of time; however, such beliefs are not adequately reflected in their judgments of future memory except when they are explicitly reminded to use them (Koriat, Bjork, Sheffer, & Bar, 2004).

There is preliminary evidence to suggest that explicit beliefs may also play a role in meta-reasoning judgments. For example, in our first study of meta-reasoning (Shynkaruk & Thompson, 2006) and in every subsequent study we have done, we have noted that reasoners' final judgments of confidence are higher than their initial FORs; moreover, this happens regardless of whether they change a right to a wrong answer or vice versa (Shynkaruk & Thompson, 2006). We have speculated that this increase in confidence represents a belief that judgments that are deliberated on are better than those that are made quickly. Interestingly, as with beliefs about forgetting, these beliefs appear to affect judgments only when cued. In our first study of FOR we included three groups: a group that gave both an intuitive and final answer, a group that gave only intuitive answers, and a group that gave only final answers. Confidence levels for the latter two groups were identical, regardless of the fact that the latter took about 40% more time to generate their answers (the data, but not this particular analysis, were reported by Thompson et al., 2011). Only when the two responses were given in tandem, that is, in the two-response group, was there a difference between initial and final confidence. Thus, although it appears that people believe that answers that are given slowly are better than those given quickly, this belief is reflected in their judgments only when they are explicitly cued to do so.

Another source of explicit beliefs that underlie monitoring judgments might be a general assessment of one's skills or competence as a reasoner, as measured, for example, by questionnaires such as the Rational Experiential Inventory (REI; Pacini & Epstein, 1999). This questionnaire measures the self-reported tendency to rely on analytic approaches to solving problems or to rely on past experiences and intuitions. Prowse Turner & Thompson (2009) found that rationality scores from the REI were positively correlated with confidence in solving syllogistic reasoning problems, although the correlation with accuracy was close to zero. In other words those who self-report a reliance on analytic thinking were more confident on the reasoning test, although not necessarily more accurate, than those who report a preference for more intuitive styles.

However, although it is possible that some monitoring is based on explicit beliefs, it is widely assumed that most cues, such as answer fluency, give rise to feelings and attendant judgments whose basis is opaque. Koriat (2007, p. 315) refers to these judgments as "sheer subjective feelings, which [although they] lie at the heart of consciousness, may themselves be the product of unconscious processes." Indeed, as described above, Topolinski (2011) has argued affect is a crucial component of meta-cognitive judgments, which may be caused by cognitive processes such as priming, but whose content is essentially affective rather than cognitive. Moreover, the WM demands that would be required for continuous, explicit monitoring of our cognitive processes would be prohibitive, making the lower-cost, parallel, implicit processes more likely. In other words in order to be engaged in a manner that does not take capacity away from the object-level processes that they are monitoring, these processes need to operate quietly in the background, which, in turn, means they are most likely to be implicit. Take, for example, the goal of proofreading a manuscript. One may set about this task with explicit intent, but if one had to attend to each word individually, and consider issues of spelling, grammar, and semantics for each, it would take days. Instead, the processes that detect errors, like the processes by which we read and understand words, grammar, and syntax, are likely automated.

This blurring of the line between implicit and explicit processes has profound implications for our notion of rationality. As argued above, one outcome of these monitoring processes is the initiation and control of analytic processes. On the view that monitoring processes are implicit, it follows that analytic thinking is engaged in response to an implicit cue. For example, analytic thinking might be engaged in response to a weak FOR without the need for an explicit evaluation of that feeling (Thompson, 2009). Thus, analytic thinking, which forms the cornerstone of many theories of rationality, may be initiated unintentionally, prompted by a vague sense of unease about the answer at hand (in response, for example, to disfluent processing). Conversely, one might have a strong sense of confidence in a conclusion that precludes the need for further analysis, again without awareness of the origin of that feeling. Thus, analytic processes may share an important characteristic with implicit ones, namely, that they are initiated autonomously (Evans & Stanovich, 2013) in response to environmental cues (Thompson, 2013). Unlike implicit processes, however, analytic processes do not necessarily run to completion, as the reasoner may exercise discretion regarding how and for how long they are implemented.

An important caveat to this discussion is that although monitoring is widely assumed to be implicit, the evidence in favor of that hypothesis is slim to nonexistent. That is, although many researchers assume that cues such as fluency have a direct and unconscious influence on monitoring judgments, little research has been done to substantiate this claim. Thus, it is possible that the relationship between fluency and a variety of meta-cognitive judgments is mediated by reasoners' beliefs about fluency, namely that easily accessed answers are likely to be correct. The first step in testing such a hypothesis would be to demonstrate that reasoners hold accurate explicit theories about the basis of their monitoring judgments, and there is some evidence to support this. For example, there is evidence to suggest that participants believe that items that come quickly to mind will be easier to remember (Matvey, Dunlosky, & Guttentag,

2001) and that items that are easier to read will be easier to remember (Mueller et al., 2014). Moreover, the effect of perceptual fluency on meta-memory judgments may vary according to one's theory of intelligence (Miele & Molden, 2010), a finding that would not be anticipated if the sole basis for the relationship were unavailable to introspection. Note, however, that all of these studies have examined the effect of perceptual fluency (i.e., manipulating font size, legibility, etc.), which is a visible and striking effect, so that it is possible that people induce these theories in the context of the task; it seems less probable that more subtle effects, such as those of answer fluency, might be mediated by explicit beliefs.

A recent study with Ian Newman provides some support for this assertion. Participants ($N = 42$) completed a conditional reasoning task in which they were asked to judge the validity of inferences derived from conditional statements, for example:
If a person is cut, she will bleed.
A person is cut. Does it follow that she is bleeding?
A person is not bleeding. Does it follow that she was not cut?
A person is bleeding. Does it follow that she was cut?
A person is not cut. Does it follow that she is not bleeding?
Some of the statements described familiar, concrete relations, as above, whereas others employed nonsense words (e.g., if a person is pik, then he will nomp). Each participant received both the familiar and unfamiliar statements; for each statement, they were asked to evaluate all four inferences and provide a FOR about their answers. After the test was complete, participants were asked to describe, if they could, the bases of their FORs. A total of 52 responses were recorded, almost all of which described salient features of the task: 46% mentioned familiarity as a contributor (interestingly, four of those indicated that the familiar materials seemed harder); 11% indicated that negations were difficult; and 6% mentioned order (i.e., that the order of information in the inference was the same as in the conditional). Almost no one mentioned one of the most powerful determinants of those FORs, that is, the speed or fluency with which they are generated; moreover, of the two who did make mention of speed, one described the relationship in the wrong direction (longer thinking equals more confidence) and the other vaguely referred to "speed of responding." These preliminary data suggest that people's meta-cognitive beliefs (a) may be strongly cued by the task demands and (b) may not be about fluency.

Finally, the fact that participants have explicit theories about the origins of their monitoring judgments does not mean that the judgments themselves require access to those theories; indeed, as described above, participants often need to be strongly guided or cued to take advantage of those theories (Koriat et al., 2004; Thompson et al., 2011). Thus, although the jury is still awaiting clear evidence on the matter, it seems reasonable to keep the provisional assumption that the source of many meta-cognitive judgments is implicit.

Conclusions

One of the goals of this chapter has been to argue that researchers have much to gain by expanding the scope of their enquiry to include meta-reasoning processes. Meta-reasoning refers to the processes by which resources are allocated or terminated, strategies are selected, and by which confidence in an answer or conclusion is determined. Consequently, understanding meta-reasoning is central to understanding the outcomes of object-level reasoning processes.

The second goal has been to discuss some of the research that has investigated meta-reasoning processes. I have argued that meta-reasoning, like other meta-cognitive processes, are like a ghost in the machine, operating in the background, exercising implicit control over explicit reasoning processes. Thus, explicit, analytic thinking may be initiated or terminated in response to implicit monitoring cues such as fluency or familiarity. As a consequence, implementation of reasoning processes is imperfect, such that answers or decisions may be confidently held when they should, instead, have been reconsidered.

Research into meta-reasoning is in its infancy, and there are many important, yet unanswered questions. We have identified a handful of cues underlying meta-cognitive judgments, but clearly, we have only scratched the surface. It is not clear, for example, why participants can be more confident in right answers than wrong ones, even after controlling for powerful cues to confidence, such as fluency. The construct of fluency itself requires additional research and new paradigms that will allow us to manipulate, rather than measure, the fluency of processing. There has been some initial work on the feeling of rightness, but none so far on the feeling of wrongness. However, a sense that something is out of place may act as a powerful meta-cognitive cue in real-world settings (Klein, 1996). Finally, the "new paradigm" of reasoning (Elqayam & Over, 2013) also presents a challenge to meta-cognitive analysis in that participants are asked to reason about and draw conclusions that have degrees of belief rather than to make binary judgments of validity. These degrees of belief are often referred to as confidence, blurring the distinction between object and meta-level processes.

References

Ackerman, R. (2014). The diminishing criterion model for metacognitive regulation of time investment. *Journal of Experimental Psychology. General, 143*, 1349–1368.

Ackerman, R., Leiser, D., & Shpigelman, M. (2013). Is comprehension of problem solutions resistant to misleading heuristic cues? *Acta Psychologica, 143*(1), 105–112.

Ackerman, R., & Thompson, V. (2014). Meta-reasoning: What can we learn from meta-memory? In A. Feeney and V. A. Thompson (Eds.), *Reasoning as Memory* (pp. 164–182). New York: Psychology Press.

Ackerman, R., & Zalmanov, H. (2012). The persistence of the fluency—confidence association in problem solving. *Psychonomic Bulletin & Review, 19*(6), 1189–1192.

Adidam, P. T., & Bingi, R. P. (2011). The importance of decision confidence to strategy outcomes. *Journal of Business Research, 16*, 25–50.

Alter, A. L., & Oppenheimer, D. M. (2009). Uniting the tribes of fluency to form a metacognitive nation. *Personality and Social Psychology Review, 13*(3), 219–235.

Alter, A. L., Oppenheimer, D. M., Epley, N., & Eyre, R. N. (2007). Overcoming intuition: Metacognitive difficulty activates analytic reasoning. *Journal of Experimental Psychology. General, 136*(4), 569–576.

Benjamin, A. S., Bjork, R. A., & Schwartz, B. L. (1998). The mismeasure of memory: When retrieval fluency is misleading as a metamnemonic index. *Journal of Experimental Psychology. General, 127*, 55–68.

Bjork, R. A., Dunlosky, J., & Kornell, N. (2013). Self-regulated learning: Beliefs, techniques, and illusions. *Annual Review of Psychology, 64*, 417–444.

Bolte, A., & Goschke, T. (2005). On the speed of intuition: Intuitive judgments of semantic coherence under different response deadlines. *Memory & Cognition, 33*(7), 1248–1255.

Bonner, C., & Newell, B. R. (2010). In conflict with ourselves? An investigation of heuristic and analytic processes in decision making. *Memory & Cognition, 38*(2), 186–196.

Chater, N., & Oaksford, M. (1999). The probability heuristics model of syllogistic reasoning. *Cognitive Psychology, 38*, 191–258.

Costermans, J., Lories, G., & Ansay, C. (1992). Confidence level and feeling of knowing in question answering: The weight of inferential processes. *Journal of Experimental Psychology: Learning, Memory, and Cognition, 18*(1), 142–150.

Cummins, D. D., Lubart, T., Alksnis, O., & Rist, R. (1991). Conditional reasoning and causation. *Memory & Cognition, 19*, 274–282.

De Neys, W., & Glumicic, T. (2008). Conflict monitoring in dual process theories of reasoning. *Cognition, 106*, 1248–1299. doi:10.1016/j.cognition.2007.06.002.

De Neys, W., Moyens, E., & Vansteenwegen, D. (2010). Feeling we're biased: Autonomic arousal and reasoning conflict. *Cognitive, Affective & Behavioral Neuroscience, 10*(2), 208–216. doi:10.3758/CABN.10.2.208.

De Neys, W., Rossi, S., & Houdé, O. (2013). Bats, balls, and substitution sensitivity: Cognitive misers are no happy fools. *Psychonomic Bulletin & Review, 20*(2), 269–273.

Elqayam, S., & Over, D. E. (2013). New paradigm psychology of reasoning: An introduction to the special issue edited by Elqayam, Bonnefon, and Over. *Thinking & Reasoning, 19*(3–4), 249–265.

Evans, J. S. B. T. (2007a). *Hypothetical thinking: Dual processes in reasoning and judgement.* Hove, UK: Psychology Press.

Evans, J. S. B. T. (2007b). On the resolution of conflict in dual-process theories of reasoning. *Thinking & Reasoning, 13*, 321–329. doi:10.1080/13546780601008825.

Evans, J. S. B. T. (2014). Two minds rationality. *Thinking & Reasoning, 20*, 129–146.

Evans, J. S. B. T., Barston, J., & Pollard, P. (1983). On the conflict between logic and belief in syllogistic reasoning. *Memory & Cognition, 11*(3), 295–306. doi:10.3758/BF03196976.

Evans, J. S. B. T., & Stanovich, K. E. (2013). Dual-process theories of higher cognition: Advancing the debate. *Perspectives on Psychological Science, 8*(3), 223–241.

Fox, S. G., & Walters, H. A. (1986). The impact of general versus specific expert testimony and eyewitness confidence upon mock juror judgment. *Law and Human Behavior, 10*(3), 215.

Frederick, S. (2005). Cognitive reflection and decision making. *Journal of Economic Perspectives, 19*(4), 25–42.

Gardiner, F. M., Craik, F. I., & Bleasdale, F. A. (1973). Retrieval difficulty and subsequent recall. *Memory & Cognition, 1*(3), 213–216.

Handley, S. J., Newstead, S. E., & Trippas, D. (2011). Logic, beliefs, and instruction: A test of the default interventionist account of belief bias. *Journal of Experimental Psychology, 37*(1), 28–43. doi:10.1037(73)90004-2.

Hansen, J., Dechêne, A., & Wänke, M. (2008). Discrepant fluency increases subjective truth. *Journal of Experimental Social Psychology, 44*(3), 687–691.

Hansen, J., & Wänke, M. (2012). Fluency in context: Discrepancy makes processing experiences informative. In C. Unkelbach & R. Greifeneder (Eds.), *The experience of thinking: How feelings from mental processes influence cognition and behaviour* (pp. 70–84). Hove, UK: Psychology Press.

Jacoby, L. L., Kelley, C. M., & Dywan, J. (1989). Memory attributions. In H. L. Roediger III & F. I. M. Craik (Eds.), *Varieties of memory and consciousness: Essays in honour of Endel Tulving* (pp. 391–422). Hillsdale, NJ: Lawrence Erlbaum Associates.

Johnson-Laird, P. N., & Byrne, R. M. (2002). Conditionals: A theory of meaning, pragmatics, and inference. *Psychological Review, 109*(4), 646.

Kahneman, D. (2003). A perspective on judgment and choice: Mapping bounded rationality. *American Psychologist, 58*(9), 697–720.

Kahneman, D., & Tversky, A. (1973). On the psychology of prediction. *Psychological Review, 80*(4), 237–251. doi:10.1037/h0034747.

Kelley, C. M., & Jacoby, L. L. (1993). The construction of subjective experience: Memory attributions. In M. Davies & G. W. Humphreys (Eds.), *Consciousness: Psychological and philosophical essays (Readings in mind and language)* (pp. 74–89). Malden, MA: Blackwell.

Kelley, C. M., & Jacoby, L. L. (1996). Memory attributions: Remembering, knowing, and feeling of knowing. In L. M. Reder (Ed.), *Implicit memory and metacognition* (pp. 287–308). Hillsdale, NJ: Lawrence Erlbaum Associates.

Kelley, C. M., & Lindsay, D. S. (1993). Remembering mistaken for knowing: Ease of retrieval as a basis for confidence in answers to general knowledge questions. *Journal of Memory and Language*, *32*, 1–24.

Klein, P. C. (1996). Pricing Black-Scholes options with correlated credit risk. *Journal of Banking & Finance*, *20*, 1211–1229.

Koriat, A. (1997). Monitoring one's own knowledge during study: A cue-utilization approach to judgments of learning. *Journal of Experimental Psychology. General*, *126*, 349–370.

Koriat, A. (2007). Metacognition and consciousness. In P. D. Zelazo, M. Moscovitch, & E. Thompson (Eds.), *The Cambridge handbook of consciousness* (pp. 289–325). Cambridge: Cambridge University Press.

Koriat, A. (2012). The self-consistency model of subjective confidence. *Psychological Review*, *119*(1), 80. doi: 10.1037.

Koriat, A., Ackerman, R., Lockl, K., & Schneider, W. (2009). The memorizing-effort heuristic in judgment of memory: A developmental perspective. *Journal of Experimental Child Psychology*, *102*, 265–279. doi:10.1016/j.jecp.2008.10.005.

Koriat, A., Bjork, R. A., Sheffer, L., & Bar, S. (2004). Predicting one's own forgetting: The role of experience-based and theory-based processes. *Journal of Experimental Psychology. General*, *133*(4), 643–656.

Koriat, A., & Levy-Sadot, R. (1999). Processes underlying metacognitive judgments: Information-based and experience-based monitoring of one's own knowledge. In S. Chaiken & Y. Trope (Eds.), *Dual process theories in social psychology* (pp. 483–502). New York: Guilford Publications.

Koriat, A., & Levy-Sadot, R. (2001). The combined contributions of the cue-familiarity and accessibility heuristics to feelings of knowing. *Journal of Experimental Psychology: Learning, Memory, and Cognition*, *27*(1), 34–53.

Kornell, N., Rhodes, M. G., Castel, A. D., & Tauber, S. K. (2011). The ease-of-processing heuristic and the stability bias. *Psychological Science*, *22*(6), 787–794.

Kruglanski, A. W., & Gigerenzer, G. (2011). Intuitive and deliberate judgments are based on common principles. *Psychological Review*, *118*(1), 97–109.

Markovits, H. (1986). Familiarity effects in conditional reasoning. *Journal of Educational Psychology*, *78*(6), 492–494.

Markovits, H., Thompson, V. A., & Brisson, J. (2015). Metacognition and abstract reasoning. *Memory & Cognition*, *43*, 681–693.

Matvey, G., Dunlosky, J., & Guttentag, R. (2001). Fluency of retrieval at study affects judgements of learning (JOLs): An analytic or nonanalytic basis for JOLs? *Memory & Cognition*, *29*, 222–233. doi:10.3758/BF03194916.

McCabe, D. P., & Castel, A. D. (2008). Seeing is believing: The effect of brain images on judgments of scientific reasoning. *Cognition*, *107*(1), 343–352.

Mednick, S. A., & Mednick, M. T. (1967). *Examiner's manual, remote associates test.* Boston: Houghton Mifflin.

Metcalfe, J., Schwartz, B. L., & Joaquim, S. G. (1993). The cue familiarity heuristic in metacognition. *Journal of Experimental Psychology: Learning, Memory, and Cognition, 19,* 851–861.

Metcalfe, J., & Wiebe, D. (1987). Metacognition in insight and non-insight problem solving. *Memory & Cognition, 15,* 238–246.

Meyer, A., Frederick, S., Burnham, T. C., Guevara Pinto, J. D., Boyer, T. W., Ball, L. J., et al. (2015). Disfluent fonts don't help people solve math problems. *Journal of Experimental Psychology. General, 144*(2), e16–e30.

Meyer, D. E., & Schvaneveldt, R. W. (1971). Facilitation in recognizing pairs of words: Evidence of a dependence between retrieval operations. *Journal of Experimental Psychology, 90*(2), 227.

Miele, D. B., & Molden, D. C. (2010). Naive theories of intelligence and the role of processing fluency in perceived comprehension. *Journal of Experimental Psychology. General, 139*(3), 535.

Mueller, M. L., Dunlosky, J., Tauber, S. K., & Rhodes, M. G. (2014). The font-size effect on judgments of learning: Does it exemplify fluency effects or reflect people's beliefs about memory? *Journal of Memory and Language, 70,* 1–12.

Novick, L. R., & Sherman, S. J. (2003). On the nature of insight solutions: Evidence from skill differences in anagram solution. *Quarterly Journal of Experimental Psychology, 56*(2), 351–382.

Pacini, R., & Epstein, S. (1999). The relation of rational and experiential information processing styles to personality, basic beliefs, and the ratio-bias phenomenon. *Journal of Personality and Social Psychology, 76*(6), 972–987.

Payne, S. J., & Duggan, G. B. (2011). Giving up problem solving. *Memory & Cognition, 39*(5), 902–913.

Pennycook, G., Trippas, D., Handley, S., & Thompson, V. A. (2013). Base rates: Both neglect and intuitive. *Journal of Experimental Psychology: Learning, Memory, and Cognition, 40,* 544–554.

Prowse Turner, J. A., & Thompson, V. A. (2009). The role of training, alternative models, and logical necessity in determining confidence in syllogistic reasoning. *Thinking & Reasoning, 15*(1), 69–100.

Reber, R., & Schwarz, N. (1999). Effects of perceptual fluency on judgments of truth. *Consciousness and Cognition, 8,* 338–342.

Rhodes, M. G., & Castel, A. D. (2008). Memory predictions are influenced by perceptual information: Evidence for metacognitive illusions. *Journal of Experimental Psychology. General, 137*(4), 615–625.

Robinson, M. D., Johnson, J. T., & Herndon, F. (1997). Reaction time and assessments of cognitive effort as predictors of eyewitness memory accuracy and confidence. *Journal of Applied Psychology, 82,* 416–425.

Schwartz, B. L., Benjamin, A. S., & Bjork, R. A. (1997). The inferential and experiential basis of metamemory. *Current Directions in Psychological Science, 6*, 132–137.

Shanks, L. L., & Serra, M. J. (2014). Domain familiarity as a cue for judgments of learning. *Psychonomic Bulletin & Review, 21*, 445–453.

Shynkaruk, J. M., & Thompson, V. A. (2006). Confidence and accuracy in deductive reasoning. *Memory & Cognition, 34*(3), 619–632.

Sloman, S. A. (1996). The empirical case for two systems of reasoning. *Psychological Bulletin, 119*, 3–22. doi:apa.org/journals/bul/119/1/3.

Son, L. K., & Metcalfe, J. (2000). Metacognitive and control strategies in study-time allocation. *Journal of Experimental Psychology: Learning, Memory, and Cognition, 26*(1), 204–221.

Stanovich, K. E. (2009). Distinguishing the reflective, algorithmic, and autonomous minds: Is it time for a tri-process theory? In J. Evans & K. Frankish (Eds.), *In two minds: Dual processes and beyond* (pp. 55–88). Oxford: Oxford University Press.

Stanovich, K. E. (2011). *Rationality and the reflective mind.* New York: Oxford University Press.

Stupple, E. J. N., Ball, L. J., & Ellis, D. (2013). Matching bias in syllogistic reasoning: Evidence for a dual-process account from response times and confidence ratings. *Thinking & Reasoning, 19*(1), 54–77.

Thiede, K. W., & Dunlosky, J. (1999). Toward a general model of self-regulated study: An analysis of selection of items for study and self-paced study time. *Journal of Experimental Psychology: Learning, Memory, and Cognition, 25*, 1024–1037.

Thompson, V. A. (2000). The task-specific nature of domain-general reasoning. *Cognition, 76*(3), 209–268.

Thompson, V. A. (2009). Dual-process theories: A metacognitive perspective. In J. Evans & K. Frankish (Eds.), *In two minds: Dual processes and beyond* (pp. 171–195). Oxford: Oxford University Press.

Thompson, V. A. (2013). Why it matters: The implications of autonomous processes for dual-process theories—Commentary on Evans & Stanovich (2013). *Perspectives on Psychological Science, 8*, 253–256. doi:10.1177/1745691613483476.

Thompson, V. A., Evans, J. S. B. T., & Campbell, J. I. D. (2013). Matching bias on selection task: It's fast and it feels good. *Thinking & Reasoning, 13*, 431–452.

Thompson, V. A., & Johnson, S. C. (2014). Conflict, metacognition, and analytic thinking. *Thinking & Reasoning, 20*, 215–244.

Thompson, V., & Morsanyi, K. (2012). Analytic thinking: Do you feel like it? *Mind & Society, 11*(1), 93–105.

Thompson, V. A., Prowse Turner, J., & Pennycook, G. (2011). Intuition, reason, and metacognition. *Cognitive Psychology, 63*(3), 107–140.

Thompson, V. A., Turner, J. A. P., Pennycook, G., Ball, L. J., Brack, H., Ophir, Y., et al. (2013). The role of answer fluency and perceptual fluency as metacognitive cues for initiating analytic thinking. *Cognition, 128*, 237–251.

Topolinski, S. (2011). A process model of intuition. *European Review of Social Psychology, 22*(1), 274–315. doi:10.1080/10463283.2011.640078.

Topolinski, S. (2014). Intuition: Introducing affect into cognition. In A. Feeney & V. A. Thompson (Eds.), *Reasoning as memory* (pp. 146–163). New York: Psychology Press.

Topolinski, S., Likowski, K. U., Weyers, P., & Strack, F. (2009). The face of fluency: Semantic coherence automatically elicits a specific pattern of facial reactions. *Cognition and Emotion, 23*(2), 260–271.

Topolinski, S., & Reber, R. (2010). Gaining insight into the "aha" experience. *Current Directions in Psychological Science, 19*(6), 402–405.

Topolinski, S., & Strack, F. (2009a). The analysis of intuition: Processing fluency and affect in judgements of semantic coherence. *Cognition and Emotion, 23*(8), 1465–1503.

Topolinski, S., & Strack, F. (2009b). Scanning the "fringe" of consciousness: What is felt and what is not felt in intuitions about semantic coherence. *Consciousness and Cognition, 18*, 608–618. doi:10.1016/j.concog.2008.06.002.

Topolinski, S., & Strack, F. (2009c). The architecture of intuition: Fluency and affect determine intuitive judgments of semantic and visual coherence and judgments of grammaticality in artificial grammar learning. *Journal of Experimental Psychology. General, 138*, 39–63. doi:10.1037/a0014678.

Tversky, A., & Kahneman, D. (1983). Extension versus intuitive reasoning: The conjunction fallacy in probability judgment. *Psychological Review, 90*(4), 293–315. doi:10.1037/0033-295X.90.4.293.

Vidmar, N., Coleman, J. E., & Newman, T. A. (2010). Rethinking reliance on eyewitness confidence. *Judicature, 94*(1), 16–19.

Whittlesea, B. W. A., Jacoby, L. L., & Girard, K. (1990). Illusions of immediate memory: Evidence of an attributional basis for feelings of familiarity and perceptual quality. *Journal of Memory and Language, 29*(6), 716–732.

Whittlesea, B. W. A., & Leboe, J. P. (2003). Two fluency heuristics (and how to tell them apart). *Journal of Memory and Language, 49*, 62–79.

Whittlesea, B. W. A., & Williams, L. D. (2000). The source of feelings of familiarity: The discrepancy-attribution hypothesis. *Journal of Experimental Psychology: Learning, Memory, and Cognition, 26*, 547–565.

15 Incubation in Creative Thinking

Ken Gilhooly

Creative problems are generally defined as problems that require the production of new approaches and solutions, where, by "new," I mean novel to the solver and not necessarily historically novel (Boden, 2004). Explaining how such personally novel solutions are reached is still a major challenge for the psychology of thinking. In analyses of creative problem solving it has often been claimed that setting creative problems aside for a period can lead to novel solution ideas occurring, either spontaneously while one is attending to other matters or very rapidly when a previously intractable problem is revisited.

Wallas (1926, p. 80) built on Poincaré's (1913) analysis of mathematical creation, and in his well-known stage analysis of creative problem solving, Wallas labeled the stage in which a problem is set aside and not consciously addressed as *incubation*. (Poincaré himself did not use the term "incubation" in his 1913 paper, although he reported examples of incubation periods from his own experience.) Incubation is proposed as a useful stage after conscious *preparation* and preceding *illumination* (or *inspiration*) and *verification*, and the incubation stage is the focus of the present discussion. Clues to processes underlying creative thinking would be obtained if we could understand why incubation during creative problem solving can be beneficial. Personal accounts by eminent creative thinkers in a range of domains have offered support for the existence of this phenomenon (Poincaré, 1913; Ghiselin, 1952; Csikszentmihalyi, 1996). However, since such accounts are often given long after the events occurred, the reliability of such reports is questionable. For example, frequently cited accounts by Coleridge, Mozart, and Kekulé have proven to be false (Weisberg, 2006, pp 73–78). After Wallas (1926) and Poincaré (1913), who relied on their own introspections and on personal reports by others (e.g., Wallas drew on daydream reports by Varendonck, 1921), a substantial body of experimental research on incubation effects has accumulated using both (a) *insight* problems, in which there is a single solution, but the solver has to develop a new way of representing or structuring the task to reach that solution, and (b) *divergent* problems, in which there is no single correct solution but as many novel and useful ideas as possible are sought. The prototypical divergent task, which has often been used in such studies, is the Alternative Uses Task, in which participants are asked to generate as many uses different from the normal use to one or more familiar objects, such as a brick (Guilford, 1971; Guilford, Christensen, Merrifield, & Wilson, 1978; Gilhooly, Fioratou, Anthony, & Wynn, 2007).

Early work on incubation used a laboratory paradigm, which can be labeled the *delayed incubation paradigm,* in which participants in the incubation condition work on the target problem for an experimenter-determined time (preparation time) and are then given an *interpolated activity* away from the target task for a fixed time (incubation period) and finally return to the target problem for a postincubation work period. Performance of the incubation group is contrasted with that of a control group who work continuously on the target task for a time equal to the sum of the preparation and postincubation conscious working time of the incubation group. A recently developed variant (*immediate incubation paradigm*) employs an interpolated task for a fixed period *immediately* after instructions on the target problem and *before* any conscious work has been undertaken on the target problem, followed by uninterrupted work on the target problem (Dijksterhuis & Meurs, 2006).

Delayed and Immediate Incubation effects

There is now considerable evidence from laboratory studies for the efficacy of delayed incubation—that setting a problem aside after a period of work is beneficial (see Dodds, Ward, & Smith, 2012, for a qualitative review). A recent meta-analysis by Sio and Ormerod (2009) of 117 studies identified a positive effect of delayed incubation, where the overall average effect size was in the low-medium band (mean $d = 0.32$) over a range of insight and divergent tasks. For divergent tasks considered separately, the mean d was larger at 0.65, which may be considered to be in the high-medium band of effect sizes. Overall, the basic existence of delayed incubation effects can now be regarded as well established, particularly in the case of divergent problem solving.

Regarding the efficacy of immediate incubation, Dijksterhuis and Nordgren (2006) reported studies in which better decisions and more creative solutions were found when immediate incubation breaks were given after the decision problems or divergent tasks were presented. In the realm of decision problems, Nordgren, Bos, and Dijksterhuis (2011) found that delayed incubation produced better decisions than immediate incubation, and both were better than no incubation.

However, the beneficial effects of immediate incubation on decision making have not always been easy to reproduce, and a number of unsuccessful replication attempts have been reported (e.g., Acker, 2008; Newell, Wong, Cheung, & Rakow, 2009; Payne, Samper, Bettman, & Luce, 2008; Rey, Goldstein, & Perruchet, 2009). Despite such reports, a meta-analysis by Strick, Dijksterhuis, Bos, Sjoerdsma, van Baaren, and Nordgren (2011), which integrated data from 92 decision studies, found a significant beneficial aggregate effect size of $g = 0.224$ for immediate incubation. Their results also pointed to a number of moderating factors that can amplify or reduce the beneficial effects of an immediate incubation period after presentation of information about options and before the decision response stage. For example, beneficial effects were greater, with more options, shorter presentation times, shorter incubation times, and induction of a configural mindset versus a feature-based mindset.

The present chapter focuses on creative thinking, which was studied by Dijksterhuis and Meurs (2006) using a divergent task. It was found that, in their experiment 3, participants produced responses of higher-rated average creativity when the instructions to list things one can do with a brick were followed immediately by a 3-minute distractor task (immediate incubation) before generating uses, compared to participants who began generating uses right away. It may be noted that the instructions did not ask for unusual uses, which is the norm in divergent thinking tasks, and so it is not clear whether participants had a goal of being creative. They may have been reporting infrequent uses that they happened to know rather than generating uses novel to them. Raters tend to score infrequent responses as creative, although such uses may have been previously known and therefore could reflect memory retrieval rather than generation of subjectively novel responses (Quellmalz, 1985). However, Gilhooly, Georgiou, Garrison, Reston, and Sirota (2012), using more standard instructions with a stress on unusual uses, found a stronger beneficial effect of immediate incubation than of delayed incubation, with both incubation effects being superior to controls (no incubation) in terms of fluency and novelty of responses. Thus, the benefit of immediate incubation was also found when the task involved novelty as well as fluency.

Zhong, Dijksterhuis, and Galinsky (2008), using the immediate incubation paradigm with the Remote Associates Task (RAT) in which participants have to retrieve an associate common to three given words (e.g., *cottage, blue, mouse*? Answer: *cheese*), found that, although immediate incubation did not facilitate the actual solution, it appeared to activate solution words on unsolved trials, as indicated by lexical decision measures, compared to unsolved trials without immediate incubation.

Overall, it seems from meta-analyses (Sio & Ormerod, 2009; Strick et al., 2011) and from recent studies (Gilhooly et al., 2012, 2013) that incubation periods, whether delayed or immediate, have beneficial effects. But how might these effects arise? We now outline the main theories regarding incubation effects.

Theories of Incubation Effects

1. *Intermittent conscious work* This theory suggests that although incubation is intended to be a period without conscious work on the target task, nevertheless, participants may carry out intermittent conscious work (Seifert, Meyer, Davidson, Patalano, & Yaniv, 1995, p. 82; Weisberg, 2006, pp. 443–445). Any conscious work during the supposed

incubation period would reduce the time required when the target problem was read-dressed but would be expected to impair performance on the interpolated task. This approach is parsimonious and essentially explains incubation away as not involving any special processes.

2. *"Fresh look"* This view (e.g., Simon, 1966; Segal, 2004; see also, Dijksterhuis & Meurs, 2006) suggested an important role for automatic passive reduction in idea strength or activation during the incubation period. The proposal is that misleading strategies, mistaken assumptions, and related "mental sets" weaken through forgetting during the incubation period, and thus, a fresh start or "set shifting" is facilitated when the problem is resumed. On this account, incubation works by allowing weakening of misleading approaches to the task during a break after a period of work (delayed incubation), thus allowing a fresh start. Segal (2004) proposed a variant of the *fresh look* view in which simply switching attention away from the main task allowed a new start, with no forgetting or unconscious work proposed. The fresh look view would not expect a beneficial effect of immediate incubation because with immediate incubation, there is no time for sets or fixations to develop, and so forgetting of misleading approaches cannot occur.

3. *Unconscious work* This approach proposes that incubation effects occur through active but unconscious processing of the problem materials (as against the passive forgetting processes envisaged in the fresh look approach.) The term "unconscious work" seems to have first been used in the context of problem solving by Poincaré (1913, p. 393). Other phrases referring to the same notion include "nonconscious idea generation" (Snyder et al., 2004) and "unconscious thought" (Dijksterhuis & Nordgren, 2006), but I generally use the phrase "unconscious work" in this chapter. The question naturally arises of what form unconscious work might take? Is it really possible that *unconscious* work could be just like *conscious* work but carried out without conscious awareness? Or is it better thought of as an automatic spreading activation along associative links as against a rule- or strategy-governed activity? Wallas (1926) proposed the idea of spreading "associative chains," which anticipated the modern idea of spreading activation. Poincaré (1913) argued for quite specific mechanisms of automatic idea generation and selection tailored to his domain of interest, which was mathematical creation. Both Poincaré and Wallas argued that the suddenness of illumination or inspiration coupled with the feeling of confidence in the sudden insight arose from prolonged unconscious work. Wallas's analysis is often labeled as a four-stage theory, incorporating preparation, incubation, illumination, and verification, but he also proposed a substage of illumination, which he dubbed *intimation* (Wallas, 1926, p. 97). This substage is often overlooked in discussions of Wallas's analysis, although Wallas considered it to be important, practically and theoretically. Intimation is the moment in the illumination period when the solver becomes aware that a flash of success is imminent. Theoretically, Wallas saw intimation as reflecting increasing activation of

a successful association train about to become conscious. Thus, intimation was consistent with the view that incubation involved unconscious work. Practically, Wallas felt it was important that the solver recognize the intimation feeling and desist from distracting activities to allow the solution to come. Overall, unconscious work has long been favored as a possible explanation of incubation effects. We consider the question of what form unconscious work might take more fully in the Theoretical Discussion section.

The mechanisms outlined above are not mutually exclusive. A delayed incubation condition could conceivably evoke all three, with the person engaging in some intermittent conscious work when attention wanders from the interpolated incubation task, and with some beneficial forgetting and unconscious work also taking place when the person is attending to the interpolated incubation task. However, an immediate incubation effect would not be consistent with a fresh look explanation in that there is not time in the immediate paradigm for sets or misleading directions to be established, but the immediate paradigm could allow some intermittent conscious work and/or some unconscious work.

Theories of Incubation: Empirical Evidence

What does previous research suggest regarding the possible mechanisms of incubation?

Intermittent Work

As a check against the possibility of intermittent conscious work, performance on the interpolated task during the incubation period should be compared with performance of a control group working on the same interpolated task without being in an incubation condition. A deficit in the interpolated task on the part of the incubation group would be consistent with the hypothesis of some conscious work on the target task during incubation. Although this seems a rather basic methodological check, surprisingly it does not appear to have been carried out (Dodds et al., 2012; Sio & Ormerod, 2009) until recently. Gilhooly et al. (2012, 2013) incorporated suitable checks for intermittent conscious work on the target divergent thinking task during the incubation period. In two experiments there were no signs of any impairment to the interpolated incubation period tasks (which were mental rotations and anagram solving) relative to controls as a result of the tasks being carried out during incubation periods as against being carried out as stand-alone tasks. These studies also found positive incubation effects despite lack of evidence for intermittent conscious work. Indeed, the trends in the data were the opposite of those that would be predicted by the intermittent work hypothesis. If anything, mental rotation and anagrams were somewhat facilitated by being carried out as distractor tasks during incubation.

Baird, Smallwood, Mrazek, Kam, Franklin, and Schooler (2012) found in a thought-monitoring study that frequency of target task-related intermittent thoughts during incubation was not related to quality of performance after the incubation period. So, it seems that even if intermittent thoughts about the target task occurred, they were ineffective and could not explain the incubation effect. Overall, from Gilhooly et al. (2012, 2013) and Baird et al. (2012), it seems safe to rule out the intermittent conscious work explanation of incubation effects.

Fresh Look

Segal (2004) addressed the fresh look hypothesis directly. The study involved a spatial insight problem (figure 15.1) in which a square has a parallelogram superimposed on it, and the task is to find the sum of the areas of the two shapes. The problem is greatly facilitated by an insight that the shapes can be restructured as two equal right-angled triangles that, if slid in a certain way, form a rectangle whose area is readily calculated. Participants engaged in this target task until they felt they were experiencing an *impasse,* that is, a state in which they could generate any more task-relevant actions (or had not solved after 20 minutes). After impasse or 20 minutes without solution, some of the participants were given 4 or 12 minutes on either a demanding verbal task (crossword) or an undemanding task (browsing through newspapers) and then returned to the main task for up to 6 more minutes. The remaining participants formed a control group who continued to work for up to 6 minutes after the impasse or 20 minutes without solving. Results indicated significant benefits for the incubation break versus no break but no effects for short versus long break or for the demandingness of the activity during the incubation interval. Segal argued that these results were consistent with a version of the fresh look view, that simply removing attention from the target task

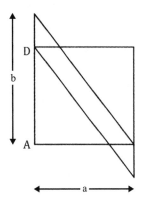

Figure 15.1
Find the sum of the areas of the square and the parallelogram.

was sufficient and that it was not important what was done in the incubation period or how long that period was.

Unconscious Work

In contrast to Segal (2004), Dijksterhuis and Meurs (2006) argued that in the immediate incubation paradigm, the fresh look approach may be ruled out, as there is no period of initial work in which misleading fixations and sets could be developed. Thus, if immediate incubation is shown to be effective, the unconscious work hypothesis must remain in contention for immediate incubation effects at least and would also be a candidate explanation for delayed incubation. Dijksterhuis and Meurs (2006) took the beneficial effects of the immediate incubation paradigm on a divergent task in their experiment 3 as support for the role of unconscious work in incubation. However, as already mentioned, the task in this study did not clearly meet the usual criteria for a creative task, and the scoring did not distinguish infrequent from genuinely novel responses. Hence, this study did not unequivocally address creative thinking as against free recall of possibly rare but previously experienced events from episodic and semantic memory.

Gilhooly et al. (2012), using explicit instructions for novelty, did find that both delayed and immediate incubation were effective in the Alternative Uses task and that immediate incubation produced more facilitation than delayed incubation. The results supported a role for unconscious work in creative divergent thinking, particularly in the case of immediate incubation.

Snyder, Mitchell, Ellwood, Yates, and Pallier (2004) also found evidence consistent with unconscious work from a study using the delayed incubation paradigm but with a surprise return to the target task. Even although the return to the main task was unexpected, beneficial effects were found, suggesting that automatic continuation of unconscious work could have occurred when the task was set aside. It should be noted, however, that Snyder et al., used a task that simply required production of uses for a piece of paper as against generation of novel uses, and so their task did not necessarily involve creative thinking as against recall.

It is notable that both Segal (2004) and Dijksterhuis and Meurs (2006) used interpolated tasks during their incubation periods that were different in character from the target tasks. Segal's target task was spatial, whereas the interpolated tasks were verbal; Dijksterhuis and Meurs's target task was verbal, but the interpolated task was spatial. The similarity relationship between target and interpolated tasks could be important in that the main competing hypotheses suggest different effects of similarity. If unconscious work is the main process, then interpolated tasks similar to the target task should interfere with any unconscious work using the same mental resources and so lead to weaker (or even reversed) incubation effects when compared with effects of dissimilar interpolated tasks. On the other hand, a forgetting account would suggest that interpolated tasks similar to the target task would cause greater interference, which would lead to more forgetting and enhanced incubation benefits.

Helie, Sun, and Xiong (2008) explored the effects of different interpolated tasks on the *reminiscence* paradigm in free recall. This paradigm is analogous to incubation in that an initial free recall is followed by a gap filled with interpolated tasks, and then the same free recall is attempted a second time. The reminiscence score was the number of new items recalled on the second test. Helie et al. found that more executively demanding interpolated tasks reduced reminiscence scores for free recall of pictures when a surprise free recall was required after the interpolated task. In this study participants studied booklets of pictures for a set period, freely recalled the items, and then did various different interpolated activities before being retested with free recall of the pictures. The results were consistent with Helie and Sun's (2010) Explicit-Implicit Interaction model, which can be applied to creative problem solving and allows for unconscious implicit processes in parallel with conscious explicit processes. However, the target task of Helie et al. (2008) was free recall rather than creative thinking, and so, although suggestive, it does not speak directly to creative thinking, which is the focus of the present chapter.

Ellwood, Pallier, Snyder, and Gallate (2009) found a beneficial effect on a number of responses postincubation of a dissimilar interpolated task in a delayed incubation experiment. However, this study used a fluency-of-uses task rather than a novel-uses task. Also, as Ellwood et al. pointed out, although their findings are consistent with an explanation in terms of unconscious work, an explanation in terms of selective relief of fatigue could also be invoked to account for the effects of similarity between incubation and target tasks. On this view, for example, a spatial delayed incubation task very different from a main verbal task could allow more recovery from specific fatigue of verbal processes than would an intervening verbal task. Gilhooly et al. (2013) included tests of the effects of incubation–target task similarity in an immediate incubation paradigm, where fatigue could be ruled out.

The Gilhooly et al. (2013) study on effects of varying incubation activities (verbal–anagram solving vs. spatial–mental rotations) used either a clearly creative verbal divergent task (alternate uses) or a clearly spatial divergent task (mental synthesis), and both divergent tasks were scored for novelty as well as fluency, unlike Ellwood et al. (2009) or Helie et al. (2008). Significant incubation effects were found together with interactions, in that spatial incubation benefitted verbal divergent thinking, and verbal incubation activity benefitted spatial divergent thinking but not vice versa. The results supported a role for unconscious work during incubation periods in creative thinking tasks and did not support the hypotheses that incubation effects are due to selective forgetting or attention shifting. The selective forgetting account predicted the opposite pattern of facilitation, and the attention shifting view predicted no differential effects of different types of interpolated tasks.

Theoretical Discussion

It seems from the above review of recent research that some form of the unconscious work hypothesis offers viable mechanisms, given the benefits of immediate incubation, in which sets are unlikely to have been developed. Furthermore, Gilhooly et al. (2012, 2013) found no evidence for intermittent work when suitable control conditions were included. This leaves unconscious work as the most likely explanation for the benefits of immediate incubation. We found that delayed incubation was beneficial but less so than immediate incubation in a divergent thinking task. It may be suggested that in delayed incubation, sets could build up and then be overcome by selective forgetting; however, Gilhooly et al.'s (2013) pattern of results did not fit the pattern of interference/facilitation for interpolated tasks similar to and different from the target tasks predicted by a forgetting account. Again, unconscious work seems a viable mechanism for delayed incubation as well as for immediate incubation.

However, the question still arises of what form unconscious work might take? Is it really possible that unconscious work could be just like conscious work but carried out without conscious awareness? Conscious work is generally rule or strategy governed. Could unconscious work also be rule governed? Poincaré (1913, p. 390) considered the possibility of a "subliminal self" that worked in the same way as the conscious self, but without consciousness, and might even be a superior "self" because it could find solutions that evaded the conscious mind. Or is incubation better thought of as involving automatic spreading of activation along associative links? The latter reflects Poincaré's final position, as he did reject the notion of a superior but unconscious subliminal self.

To explore the idea that unconscious work might be a subliminal version of conscious work, let us consider the nature of conscious processing in the alternate uses task. This was addressed in a think-aloud study of a brick uses task by Gilhooly et al., (2007) in which it was found that participants used strategies, such as scanning the target object's properties ("It's heavy") and using the retrieved properties to cue uses ("Heavy objects can hold down things like sheets, rugs, tarpaulin and so on, so a heavy brick could do those things too"). However, it seems unlikely that unconscious work could simply duplicate this form of conscious work but without awareness. It is generally accepted in cognitive science (a) that mental contents vary in activation levels, (b) that above some high activation level mental contents become available to consciousness, (c) that we are conscious of only a limited number of highly activated mental elements at any one time (that is, the contents of working memory), and (d) that strategy- or rule-based processing, as found in Gilhooly et al.'s (2007) think-aloud study of alternative use generation, requires such highly activated (conscious) material as inputs and generates highly activated (conscious) outputs. That is, the kind of processing that is involved in conscious work requires the highly activated contents of working memory, of which we are necessarily aware, given that material is in consciousness if and only if it is above a high activation threshold. Thus, it seems logically impossible that unconscious processes could duplicate conscious processes in every respect and stay unconscious. For example, using rules and working memory to multiply two three-digit numbers (e.g., $364 \times 279 = ?$) seems impossible without having highly activated representations in working memory of the numbers, the goal, and intermediate results, and such representations are necessarily conscious. Unconscious multiplication of even moderately large numbers, not previously practiced, seems impossible. (With practice, of course, it would be possible to store many three-digit multiplication results in long-term memory that could then be directly retrieved—a type of unconscious process—but this is not the same as mental multiplication.) Poincaré (1913, p. 394) made a very similar point when he wrote "It never happens that the unconscious work gives us the result of a somewhat long calculation *all made*, where we only have to apply fixed rules." Overall, then, the idea that unconscious work or thought could be just the same as conscious work minus awareness of any mental content can be discounted. What, then, might unconscious work consist of? Many theorists, such as Poincaré (1913), Campbell (1960), and Simonton (1995), have argued that unconscious work in incubation involves a quasi-random generation of associations between mental elements to produce novel combinations of ideas, some of which may be useful. Processes such as parallel spreading activation through a semantic network could serve to form remote and unusual associations (Jung-Beeman, Bowden, Haberman, Frymaire, Arambel-Liu, Greenblatt, et al., 2004) without requiring activation levels to rise above the threshold of consciousness. In Helie and Sun's (2010) Explicit-Implicit Interaction model, incubation is regarded as involving unconscious implicit associative processes that demand little attentional capacity in contrast with conscious explicit rule-governed, attentionally demanding processes. According to Dijksterhuis and Nordgren's (2006) unconscious thought theory, unconscious thought, or work, has the following characteristics. It is parallel, bottom-up, inexact, and, importantly for the present discussion, divergent, whereas conscious thought is serial, exact, and generally convergent. Thus, there is broad agreement among a number of theorists that unconscious thinking, or work, in the form of implicit associative processes based on spreading activation (similar to Wallas's [1926] concept of "associative trains"), is a possible explanation of incubation effects.

How might the suddenness of inspiration be explained? Both Poincaré and Wallas saw this feature of creative thinking as indicative of prolonged unconscious work that found a solution and delivered it to consciousness. However, here Poincaré identified a problem for the unconscious work account. How did the good idea become selected for promotion to consciousness? He proposed that selection was based on the mathematician's special sensibility to beauty in mathematics and that the subliminal self possessed this sensibility. Poincaré was focused on mathematical creation, so his theory as stated in the 1913 paper is narrow in scope; generalization to other fields, such as poetry, music, physics, and so on would presumably require specific sensibilities to be proposed for those fields. An alternative possibility that has general applicability is that when a problem is set aside, the goal representation remains somewhat active, although below the threshold for consciousness. The goal helps control the activation flow to relevant paths, and when a solution complex becomes active, the solution and the goal mutually activate each other more and more in a positive feedback loop, leading both to become conscious. The rising activation ("rising train of association," as Wallas put it) is experienced as intimation. This account has the benefit of automaticity and is parsimonious in not requiring special sensibilities to be invoked.

Concluding Comments

Overall, it can be concluded that the field, although still acknowledging the pioneering work of Poincaré and Wallas, has made considerable progress. The existence of incubation as a beneficial stage in creative thinking has been established through a large number of empirical studies (Sio & Ormerod, 2009), so the field does not depend on potentially unreliable introspective accounts. New paradigms such as immediate incubation have been established and have helped justify a role for unconscious work. Theoretical ideas have been sharpened and refined and could soon lead to computer simulations (Helie & Sun, 2010).

References

Acker, F. (2008). New findings on unconscious versus conscious thought in decision making: Additional empirical data and meta-analysis. *Judgment and Decision Making, 3,* 292–303.

Baird, B., Smallwood, J., Mrazek, M. D., Kam, J. W. Y., Franklin, M. S., & Schooler, J. W. (2012). Inspired by distraction: Mind wandering facilitates creative incubation. *Psychological Science, 23,* 1117–1122.

Boden, M. (2004). *Creative mind: Myths and mechanisms* (2nd ed.). London: Routledge.

Campbell, D. T. (1960). Blind variation and selective retention in creative thought as in other knowledge processes. *Psychological Review, 67,* 380–400.

Csikszentmihalyi, M. (1996). *Creativity: Flow and the psychology of discovery and invention.* New York: HarperCollins.

Dijksterhuis, A., & Meurs, T. (2006). Where creativity resides: The generative power of unconscious thought. *Consciousness and Cognition, 15,* 135–146.

Dijksterhuis, A., & Nordgren, L. F. (2006). A theory of unconscious thought. *Perspectives on Psychological Science, 1,* 95–109.

Dodds, R. A., Ward, T. B., & Smith, S. M. (2012). A review of the experimental literature on incubation in problem solving and creativity. In M. A. Runco (Ed.), *Creativity research handbook* (Vol. 3, pp. 251–284). Cresskill, NJ: Hampton Press.

Ellwood, S., Pallier, P., Snyder, A., & Gallate, J. (2009). The incubation effect: Hatching a solution? *Creativity Research Journal, 21,* 6–14.

Ghiselin, B. (1952). *The creative process: A symposium.* New York: Mentor.

Gilhooly, K. J., Fioratou, E., Anthony, S. H., & Wynn, V. (2007). Divergent thinking: Strategies and executive involvement in generating novel uses for familiar objects. *British Journal of Psychology, 98,* 611–625.

Gilhooly, K. J., Georgiou, G. J., & Devery, U. (2013). Incubation and creativity: Do something different. *Thinking & Reasoning, 19,* 137–149.

Gilhooly, K. J., Georgiou, G. J., Garrison, J., Reston, J., & Sirota, M. (2012). Don't wait to incubate: Immediate versus delayed incubation in divergent thinking. *Memory & Cognition, 40,* 966–975.

Guilford, J. P. (1971). *The nature of human intelligence.* New York: McGraw-Hill.

Guilford, J. P., Christensen, P. R., Merrifield, P. R., & Wilson, R. C. (1978). *Alternate Uses: Manual of instructions and interpretations.* Orange, CA: Sheridan Psychological Services.

Helie, S., & Sun, R. (2010). Incubation, insight, and creative problem solving: A unified theory and a connectionist model. *Psychological Review, 117,* 994–1024.

Helie, S., Sun, R., & Xiong, L. (2008). Mixed effects of distractor tasks on incubation. In B. C. Love, K. McRae, & V. M. Sloutsky (Eds.), *Proceedings of the 30th annual meeting of the Cognitive Science Society* (pp. 1251–1256). Austin, TX: Cognitive Science Society.

Jung-Beeman, M., Bowden, E. M., Haberman, J., Frymiare, J. L., Arambel-Liu, S., Greenblatt, R., et al. (2004). Neural activity when people solve verbal problems with insight. *Public Library of Science—Biology, 2,* 500–510.

Newell, B. R., Wong, K. Y., Cheung, J. C. H., & Rakow, T. (2009). Think, blink or sleep on it? The impact of modes of thought on complex decision making. *Quarterly Journal of Experimental Psychology, 62,* 707–732.

Nordgren, L. F., Bos, M. W., & Dijksterhuis, A. (2011). The best of both worlds: Integrating conscious and unconscious thought best solves complex decisions. *Journal of Experimental Social Psychology, 47,* 509–511.

Payne, J., Samper, A., Bettman, J. R., & Luce, M. F. (2008). Boundary conditions on unconscious thought in complex decision making. *Psychological Science, 19,* 1118–1123.

Poincaré, H. (1913). *The foundations of science.* New York: Science House.

Quellmalz, E. (1985). Test review of Alternate Uses. In J. V. Mitchell Jr. (Ed.), *The ninth mental measurements yearbook.* [Electronic version]. Retrieved from Buros Institute's *Test Reviews Online* website: http://buros.org/

Rey, A., Goldstein, R. M., & Perruchet, P. (2009). Does unconscious thought improve complex decision making? *Psychological Research, 73*, 372–379.

Segal, E. (2004). Incubation in insight problem solving. *Creativity Research Journal, 16*, 141–148.

Seifert, C. M., Meyer, D. E., Davidson, N., Patalano, A. L., & Yaniv, I. (1995). Demystification of cognitive insight: Opportunistic assimilation and the prepared-mind perspective (pp. 65–124). In R. J. Sternberg & J. E. Davidson (Eds.), *The nature of insight*. Cambridge, MA: MIT Press.

Simon, H. A. (1966). Scientific discovery and the psychology of problem solving. In R. Colodny (Ed.), *Mind and cosmos* (pp. 22–40). Pittsburgh, PA: University of Pittsburgh Press.

Simonton, D. K. (1995). Foresight in insight? A Darwinian answer. In R. J. Sternberg & J. E. Davidson (Eds.), *The nature of insight* (pp. 465–494). Cambridge, MA: MIT Press.

Sio, U. N., & Ormerod, T. C. (2009). Does incubation enhance problem solving? A meta-analytic review. *Psychological Bulletin, 135*, 94–120.

Snyder, A., Mitchell, J., Ellwood, S., Yates, A., & Pallier, G. (2004). Nonconscious idea generation. *Psychological Reports, 94*, 1325–1330.

Strick, M., Dijksterhuis, A., Bos, M. W., Sjoerdsma, A., Van Baaren, R. B., & Nordgren, L. F. (2011). A meta-analysis on unconscious thought effects. *Social Cognition, 29*, 738–762.

Varendonck, J. (1921). *The psychology of daydreams*. New York: Macmillan.

Wallas, G. (1926). *The art of thought*. New York: Harcourt Brace.

Weisberg, R. W. (2006). *Creativity: Understanding innovation in problem solving, science, invention, and the arts*. New York: John Wiley & Sons.

Zhong, C.-B., Dijksterhuis, A., & Galinsky, A. D. (2008). The merits of unconscious thought in creativity. *Psychological Science, 19*, 912–918.

16 Brain-Based Bounded Creativity

Riccardo Viale

Bounded rationality is a concept that was introduced by Herbert Simon (1955, 1957) to represent the way human rationality is limited or bounded when making social choices.[1] As in the scissors metaphor, bounded rationality lies between the blade of the limits of the human mind's computational capacity and the blade of environmental complexity. The subjective blade of bounded rationality became a genuine research program that, over the course of several years, led to the identification of various empirical anomalies in the processes of reasoning, judgment, and decision making. These anomalies were identified in comparison to the canonical models of rationality in deductive logic and in probability calculus. Seeking to explain these anomalies and biases, heuristics were identified, namely psychological procedures that provide ready, but often inaccurate, solutions to the problems of choice. As an alternative to the study of *Heuristics and Biases* (led by Daniel Kahneman and Amos Tversky), a program was developed that adhered more closely to Simon's original inspiration in order to analyze both blades of the scissors of bounded rationality. The ABC Research Group (led by Gerd Gigerenzer) focused on *Fast and Frugal Heuristics* and the adaptive toolbox. The analysis of human choices in relation to environmental complexity led to the introduction of the concept of ecological rationality.

Kahneman and Tversky and their co-workers had above all centered their attention on the rational value of conscious cognitive judgment and decision-making mechanisms and had regarded the intuitive emotional sphere as being mainly responsible for rationality pathology. Gigerenzer's program, on the contrary, reversed this model. It is precisely the contribution of the intuitive and emotional sphere that gives an individual the ability to adapt as well as his or her own ecological rationality. The unconscious, intuitive, and emotional dimension of human choice is the factor that links the Gigerenzer program to studies on human creativity. As in the choices we make in everyday life when we lack both time and information, so too when we solve complex problems in art and science, the creative leap toward originality and innovation often occurs in unconscious, intuitive-type conditions, with a strong affective and emotional component. In the same way that rationality is bounded in everyday choices by these types of factors, creativity is also bounded when one is resolving artistic and scientific problems by the implicit, intuitive, emotional, and perceptual characteristics of the human mind.[2] In the same way that rationality cannot be Olympic, that is, it cannot satisfy the canonical requirements of logic and probability calculus, so creativity too cannot be divine. Although this was something ardently desired by the nineteenth-century Romantics, creativity cannot be set apart from Earth's gravity and human bounds. These constraints and the limits of human thought have been analyzed in the past above all from a psychological point of view. Instead, in more recent years there has been a strong awareness of the need to combine psychological tests with neurocognitive research. As is clear from the results of the new neuroeconomics and neuroaesthetics programs, we need above all to study the cerebral dimension if we wish to understand the characteristics of *bounded rationality* and *bounded creativity*. Indeed, it is the biological structure of the brain, as it has developed through adaptive evolution, that sets the computational bounding parameters for the style of human thought.

A subdivision that might be interesting to introduce into the neurocognitive analysis of bounded creativity is that between "procedural" creativity and "substantive" creativity. In a similar manner to bounded rationality, this subdivision marks two aspects of creativity. On the one hand are the neurocognitive processes that supervise problem solving, surpassing the initial starting conditions. On the other is the result itself, which can be defined as creative because it satisfies concomitant requisites and values at a historical level. Based on this separation a process can be deemed creative even if the result, although innovative with respect to the starting conditions, is not accepted as an innovative solution in a particular historical period. In this sense a number of differences emerge between scientific and artistic creativity. Scientific problems generally draw closer to the ideal of the "well-structured" problem to which Herbert Simon (1973) refers. Therefore, solutions that surpass the initial starting conditions are rarely rejected on the basis of sociocultural prejudices, above all when they concern puzzle-solving activities (Kuhn, 1970) inside the same research program (Lakatos, 1970) or the same disciplinary matrix (Kuhn, 1970). The same cannot be said of creative solutions that introduce a radical theoretical change, albeit one that is not recognized by the scientific community at a particular historical moment. In that case years will pass before the solution is accepted and recognized as a discovery. At first sight the case of artistic creativity is very different. In this instance creativity is much more subject to historical variables but also to the idiosyncrasy of individual aesthetic tastes. Nevertheless, in figurative art, for example, there are periods during which standards of judgment have been applied to the progress made (Viale, 2013, pp. 198–204).[3]

From the beginning art has been conceived as something created for a very precise purpose. Architecture, sculpture, and painting were judged by practical rather than aesthetic criteria. In the primitive age the meaning of a work of art lay not in its beauty but in its "influence" and the possibility of having the desired magical effect. Artists worked for members of their tribe who knew the meaning of every form and every color exactly and did not expect novelty, "but only wanted to dedicate themselves to the execution of the work with all possible skill and ability." As Gombrich pointed out (1950, Italian trans. 1989, pp. 42–51), in Ancient Egypt the reliefs and paintings that adorned the walls of the tombs blended "geometrical eurhythmy and an acute observation of nature." This art was only meant to be seen by the dead man's soul. The most important thing was accuracy, and to satisfy it rules and models had to be followed, such as including everything in the human form that was considered important. This explains the lack of perspective and the flatness of the figures, which satisfies the criterion of completeness, based on magical concerns: How could a man with his arm foreshortened by perspective or cut off bring or receive the required offerings to the dead? Progress was judged not on the basis of originality but according to the degree to which it complied with a number of very strict laws.[4] The criteria change, but the adhesion to precise aesthetic canons remains a constant of the passage to Greek art. From the seventh century BC to the first century AD, although there was a certain continuity with Egyptian art, new criteria emerged in Greece regarding the production and evaluation of artistic work. Natural forms and the foreshortened view were discovered. Artists looked for new techniques and new ways to depict the human figure. They tried to reflect the inner life and travails of the soul, as Socrates advocated, first through the movements of the body and later through the very expression of the face. The artist became aware of his possibilities and his freedom to express beauty, and the comparison and judgment were transferred in an assessment of the extent to which the form and style were capable of representing beauty. Even in the Renaissance, it was essential to respect set models. Vasari (1986) wrote about artistic progress: "… I judge that it is a property and a particular nature of these arts, which can slowly improve from a humble start, finally reaching the peak of perfection." It is a typical form of incremental progress, which he considered tied to the prevalence of a naturalistic attitude and to the use of tenuous colors and rudiments of perspective. It was very different from the seamless progress that takes us through Picasso's *Les Demoiselles d'Avignon* (1907) into the Cubist period, through the decomposition of the natural form, which is present in the preceding Impressionism, in its structural levels and planes.

Where art is concerned, then, judgments about progress stem from a particular tradition, cultural context, and aesthetic viewpoint, so they are relative and not absolute. This is valid both for assessments within a given period or for those comparing different periods. For example, in the passage from Egyptian art to Greek art, most art historians based their assessment of the progress on aesthetic and humanistic criteria (naturalism, the depiction of inner life, the balance and harmony of forms, etc.), set free from magical and religious concerns, which would not have been admitted or understood by Egyptian culture. However, above all with reference to the art of the last century, the rules for judgment gradually fragmented, becoming dispersed and individualized as the idiosyncratic taste of each art critic, art historian, and gallery owner. Yet, procedural creativity can also be identified in art from the success or otherwise of the work of art. It could be defined on the basis of neurocognitive processes capable of generating new solutions that surpass the starting conditions (definable, in this case, as styles already present on the market, whether historicized or otherwise).

In this chapter I try to analyze some of the neurocognitive procedural characteristics that bound the process of creativity. Among the various expressions of creativity, figurative art will be preferred. It seems to involve more parts of the brain and psychological functions than other forms of creativity. In particular, the chapter analyzes the tendency to assign responsibility for creative thought to the right hemisphere, and we try to understand why this is so. I explain how a structure of the brain known as the *default mode network* produces one of the characteristics of human creativity: that of being enhanced, above all, during periods when the mind wanders. Both results highlight a particular dynamic of creative thought linked to the unconscious spreading activation of neural networks. This neurobiological finding seems to correspond to the results of psychological research on incubation. In conditions where the mind wanders or is engaged in other tasks, there is an empirically controlled increase in creative capacity with regard to the target problem. This appears to be caused by an unconscious spreading activation of semantic networks. Last, the characteristics of bounded creativity linked to the implicit and unconscious dynamic of neurocognitive and perceptual limits are highlighted by examining the emotional and visual dimension of creativity in figurative art. In the concluding paragraph I show how the unique structure of the brain has constrained the emergence of particular artistic styles and expressions.

Cosquer and Chauvet; Nadia and Leonardo da Vinci

When he analyzed the horses painted in the caves at Cosquer and Chauvet in France (figure 16.1), full of admiration and amazement, Ernst Gombrich in 1996 exclaimed, quoting from the Latin: "*Magnum miraculum est homo*"[5] (quoted in Kandel, 2012, p. 490). His amazement when he observed these paintings, created 30,000 years ago, was such that he felt they demonstrated an evolutionary leap in man's psychological development. Their fantastic naturalistic pictorial and evocative capacity could be taken as strong evidence of a more sophisticated mind, capable of communicating at a symbolic level and therefore able to use language. The beauty of the painted horses really is amazing. It speaks volumes in favor of graphic skills and considerable stylistic elegance. However, one question we must ask ourselves is whether this mastery in naturalistically depicting a herd of horses is evidence of a linguistic evolution in the handling of symbols at a verbal level, or whether the two skills are simply not linked.

As Eric Kandel underlines (2012, p. 488), Nadia is a 5-year-old autistic savant child who can draw a horse as shown in figure 16.2. If, as Ramachandran (2004) suggests, we

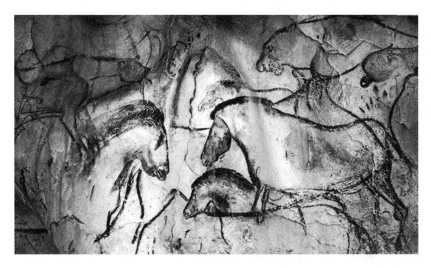

Figure 16.1
Painted horses from the Chauvet caves.

Figure 16.2
Horses by Nadia.

compare her drawing skills with those of a child of eight, we are amazed. Her capacity to dynamically grasp the horse's movement is comparable with that represented in Leonardo da Vinci's drawing. Nadia is autistic, and, like 10% of autistic people, she is a savant. Such people precociously develop an extraordinary ability to focus their attention on the task they have to perform. They often have better mnemonic, sensory, and manual skills. They tend to excel in music, art, calculations, and mechanical and spatial abilities. How can we explain this capacity? Various scholars feel that it seems to reflect a dysfunction in the left hemisphere of the brain, the one dedicated to language and analytical thought. This damage could cause an increase in the activity of the right side of the brain, which controls associative and creative thought. Ramachandran maintains:

Perhaps many or even most of Nadia's brain modules are damaged because of her autism, but there is a spared island of cortical tissue in the right parietal. So her brain spontaneously allocates all her attentional resources to the one module that is still functioning, her right parietal. (2004).

Autistic children like Nadia have great drawing skills but great deficits in communications and symbolic language. They have no empathy in relation to other people's emotional situations. The *simulationist theory* explains this as a deficit in the structure of the *mirror neurons* at a cortical and subcortical level. As an alternative, *theory theorists* point to a deficit in the *theory of mind,* which makes autistic people unable to *mind read* and therefore unable to interpret other people's actions and emotions. Starting from these considerations some scholars overturn the theory sustained by Gombrich, who was impressed by the beauty of the horses in Cosquer and Chauvet and felt that these paintings demonstrated the emergence of a new mind capable of communication and symbolic language. For one, psychologist Nicholas Humphrey (1998) wondered whether the similarity between Nadia's horses and those in Cosquer and Chauvet did not demonstrate the opposite theory. In other words there is no evolutionary leap toward a capacity for communication and symbolic language. Both Nadia and the artist of the horses in Cosquer and Chauvet reveal a naturalistic ability to represent movement photographically. However, this does not demonstrate that this ability is correlated to linguistic and symbolic skills. The theory at this point is that language is an evolutionary change, subsequent to the capacity to represent external reality in a figurative way. Pictorial ability therefore precedes linguistic ability. What is more, this example, like those of other autistic savants, seems to demonstrate not only that a language skill is not necessary, but also that it is possible to draw objects from the real world in a refined, precise way, even without an empathetic emotional life. As we know, empathy develops above all in relation to our peers. Autistic children lack it. Our ancestors in Cosquer and Chauvet probably also lacked it. In fact, at that time only animals were depicted, never men. The first drawings to depict humans, which were found in the caves of Lascaux, were produced 13,000 years later. We can perhaps speak of the emergence of a new mind capable of empathy and perhaps even of symbolic language at that moment. In any case, the absence of empathetic skills does not seem to influence the ability for naturalistic depiction. For Nadia, and for the men of Cosquer and Chauvet, the beauty of these images is a given fact. We cannot even say that it is a cold beauty. This distinction between a perceptual-representational capability and an empathic emotional capability seems to confuse us considerably. As Oliver Sacks asked himself, regarding Stephen, another famous autistic savant whom he studied, "Was not art, quintessentially, an expression of a personal vision, a self? Could one be an artist without having a 'self'?" (Sacks, 1995).

The answer seems to be that artistic skill, seen as a representational technique suited to the external image, can exist without the emotional and affective dimension. The perceptual dimension is sufficient. In fact even a camera is able to represent objects from the world. But the question of artistic creativity is another matter. It is difficult to think of a creative intelligence that does not merge perception and emotion with the cognitive dimension. In fact if we consider another category of language disturbance, *dyslexia,* we can find answers to this problem. In dyslexia we have a language deficit, particularly at a level of reading and writing, without emotional or empathetic deficits. In many cases the emotional and affective sphere is actually accentuated. Dyslexia also seems to be linked to a deficit in the left hemisphere associated with a dominance of the right side. In particular, Wernicke's area, which is linked to verbal comprehension, seems to be compromised. This combination slows down learning to read and the translation of sounds into letters. Sally and Bennett Shaywitz have discovered that when dyslexic children learn to read, an area on the right side of the brain thought to be associated with visual-spatial thinking takes over for the word-forming area on the left side. Dyslexia is not associated only with the technical pictorial skills of autistic savants but also with genuine artistic creativity. There are numerous cases of famous dyslexic artists, from Chunk Close to Robert Rauschenberg, Malcolm Alexander, and apparently even Leonardo da Vinci. And there are famous dyslexics in other creative and innovative fields: Henry Ford, Walt Disney, John Lennon, Winston Churchill, Agatha Christie, Mark Twain, and William Butler Yeats.

The first question at this point is this: should we continue to assert such a clear-cut separation between the right and left hemispheres with regard to creativity? And if the answer is yes, then what neurocognitive mechanism in the right hemisphere can explain creativity?

Right Brain and Creativity

Today, neurocognitive studies tend to represent knowledge and thought like the structure of an iceberg. The part above water—the conscious part—only represents a minimal part of human cognitive activity. But an iceberg is "driven" by its submerged part. For example, in various studies of decision-making, Dijksterhuis (2004) seems to demonstrate that conscious thought becomes overwhelmed by the unconscious in complex qualitative decisions with many possible alternative solutions, such as when one is choosing a car to purchase, changing careers, or evaluating a painting. Gerd Gigerenzer reached the same results in various rational tasks, demonstrating the importance of heuristics based on unconscious intuition (Gigerenzer, 2007). The same is true for the discussion of the role of the so-called System 1 of the mind, which is implicit and intuitive in accelerating and boosting decision-making processes. The diffuse resources of unconscious thought processes are better suited to thinking that involves several variables. Conscious thought works from the top down, is guided by expectations and internal models, and is hierarchical. Unconscious thought works from the bottom up, is not hierarchical, and may therefore allow more flexibility on finding new combinations of ideas and new reconceptualizations. On the other hand the superiority of the unconscious mind over the conscious mind is confirmed in many aspects of psychic life other than thought.

Many experiments, starting with the 1970 experiment of Benjamin Libet, who replicated the discoveries of German scientist Hans Kornhuber in 1964, have demonstrated that in voluntary movements like that of raising an arm, the nervous readiness electrical potential in the brain occurred a little less than 1 second before the voluntary movement. In Libet's experiments he also discovered, surprisingly, that the very will to move a finger, rather than occurring just before the electrical brain activity linked to movement, was preceded by the readiness potential by all of 300 milliseconds. In other words the nerve network responsible for moving the finger is activated before the person consciously decides to lift his or her finger. Libet proposed that the process of initiating a voluntary action occurs rapidly in an unconscious part of the brain but that just before the action is initiated, consciousness, which is recruited more slowly, exerts a top-down approval or veto of the action (Libet, 1985).

According to Wegner (2002), the conscious will is thus the compass of the mind: it is a course-sensing mechanism that examines the relationship between thoughts and action and responds with "I willed this" when the two correspond appropriately. The same also applies to perceptual experience. Dehaene and Changeux (2011) found that conscious awareness of an image emerges relatively late in the course of visual processing: as much as one-third to one-half of a second after visual processing begins. At the start there is an unconscious activation for 200 milliseconds only of the areas of the primary visual cortex. Immediately afterward, there is a burst of simultaneous broadcasting of neuronal activity to widely distributed regions of the brain. "Our conscious ability to report a perceptual experience derives from synchronous activity in the cerebral cortex that emerges some time after a stimulus has been presented and is then broadcast globally to critical areas in the prefrontal and parietal cortices" (Kandel, 2012, p. 464). This top-down amplification seems to be conveyed in the parietal and frontal regions by a network of pyramidal neurons. The same model also applies to the senses of touch and sound.

The prevalence of the unconscious mind over the conscious mind is therefore pervasive from decision to action and perception. Its role in artistic creativity was first introduced by Ernst Kris (1952) and Ernst Gombrich (1960) and then developed by Kihlstrom (2007) in 1987 and Epstein (1994). By reiterating Freudian concepts we can break the components of creative activity into three. The conscious part inferentially controls the informative material that is developed by the *cognitive unconscious*, which is linked to the *dynamic unconscious*. The cognitive unconscious has two main components: the *procedural unconscious*, which is assimilable to *tacit knowledge* above all where the unconscious memory of motor and perceptual skills are concerned, and the *preconscious unconscious*, which always corresponds to *tacit knowledge* (Viale, 2013) in the aspects most closely linked to cognitive and inferential rules. On the other hand, the dynamic unconscious is the one that contains conflicts, sexual striving, and repressed thoughts and actions. In creative activity there is controlled management of communications between the conscious and the unconscious part. The artist regresses consciously and voluntarily to his unconscious dimension, which is above all dynamic, and brings the force of unconscious drives and desires into the forefront of his images.[6] Because the cognitive unconscious is mainly analogical, freely associative, characterized by concrete images, and guided by the pleasure principle, it explains the importance of the phenomenon of the wandering mind and of incubation in the development of creative solutions to problems. Stimulating the unconscious dimension facilitates the emergence of moments of creativity that promote new combinations and permutations of ideas. When someone comes to a dead end in the solution to a problem, incubation and letting the mind wander generate a set shift, that is, an unconscious transition from a rigid convergent perspective to an associative, divergent perspective, which may represent the solution to the problem. Obviously, at the very moment that a new idea emerges, the conscious mind takes action to structure the solution so that it fits the problem. The top-down conscious cognitive processes monitor the creative activity and update and hone the memory repertoire of knowledge and cognitive skills for creative acts.

Let us return to the problem posed at the end of the first section: To which part of the brain should responsibility for creative activity be attributed?

Roger Sperry was awarded the 1981 Nobel Prize in Physiology and Medicine for his research on the split brain. In his view the two parts of the brain play different roles in cognitive functions. The left hemisphere is dominant in all activities involving language, calculation, and analysis. The right hemisphere is superior in functions involving spatial awareness, such as understanding maps and recognizing faces. However, as Kosslyn and Wayne Miller remind us (2013), Sperry warned that "experimentally observed polarity in right-left cognitive style is an idea in general with which it is very easy to run wild. ... It is important to remember that the two hemispheres in the normal intact brain tend regularly to function closely together as a unit" (Sperry, 1984). Indeed, according to Kosslyn and Wayne Miller, both hemispheres play similar and connected roles. For example, in language it is necessary to understand the syntax and structure of sentences (for which the left hemisphere is responsible), the meaning of changes in tone (right hemisphere), and how to decipher the meaning (undertaken by both hemispheres). When we look at an object, the left hemisphere allows us to see its parts and to categorize them, while the right hemisphere perceives the outline and the distance between the parts.

In other words, both hemispheres form part of a single system. It is true that small areas of the brain are specialized in different ways. For example, one area of the frontal part of the left hemisphere controls the movements of the tongue, lips, and vocal cords during talking, whereas the symmetrical area on the right controls the same movements during singing. In other words, as Kosslyn and Wayne Miller affirm (2013), it is not possible to talk about a logical versus an intuitive brain, or an analytical versus a creative brain. Both hemispheres play a role in logical and intuitive thought and in analytical and creative thought. If anything, it is more interesting, in their opinion, to posit a vertical rather than a horizontal division of the cognitive functions. The top parts (occipital and temporal lobes) and the bottom parts (the parietal and most of the frontal lobes) of the brain have different functions. Researchers have shown that the two parts of the brain play specialized roles in functions as diverse as visual perception, memory, decision making, planning, and emotion. For example, the top brain formulates and execute plans, whereas the bottom brain classifies and interprets incoming information about the world. The two halves always work together, and the top brain uses information from the bottom brain to formulate its plans. Kosslyn and Wayne Miller summarize the result of an unprecedented analysis of the scientific literature on the functions of the top and bottom brain:

The bottom-brain system organizes signals from the senses, simultaneously comparing what is being perceived with all the information previously stored in memory and then uses the results of such comparison to classify and interpret the object or event that gives rise to the input signals.

The top-brain system uses information about the surrounding environment (in combination with other sort of information, such as emotional reactions and need for food and drink) to figure out which goals to try to achieve. It *actively* formulates plans, generates expectation about what should happen when a plan is executed, and then—as the plan is being carried out—compares what is happening with what was expected, adjusting the plan accordingly. ... (Kosslyn and Wayne Miller, 2013, p. 14)

This proposal would seem to rule out any clear-cut separation between the right and left lobes with regard to creativity. Indeed, characterization of the four cognitive modes in which there is a respective predominance of either the bottom brain or the top brain appears to attribute the greatest propensity for creativity to the *Stimulator mode*, which involves a predominance of top-brain activity. This thesis should undoubtedly be taken into account in order to refute oversimplified theories regarding the right and left hemispheres. The brain is made up of highly interconnected and sometimes integrated subsystems. This fact cannot preclude our appreciation of the presence of specialized areas in the right brain that play an important role in creative activity. Indeed, the Kosslyn and Wayne Miller model is focused on the cortex alone and does not examine the possible asymmetric role of the subcortical areas.

We have seen that in autistic and dyslexic people, their linguistic handicap seems to be associated with a dysfunction of the left lobe of the brain. Simultaneously, their greater propensity for artistic activities seems to depend on heightened activation of the right lobe, linked to a decrease in the inhibitive activity of the dysfunctional left lobe over the right lobe. A study of children's musical abilities by Hughlings Jackson, the pioneer of neurology, revealed that in children with acquired aphasia due to disorders in the left hemisphere, their musical ability, governed by the right hemisphere, did not decrease but increased. Narinder Kapur (1996) used the term *paradoxical functional facilitation* to describe the unexpected improvement in creative capacities following a lesion in the left hemisphere. Howard Gardner (2006) analyzed numerous cases of artists who, after a stroke on the left side, significantly increased their novelty seeking and unconventional thinking and stepped up their divergent thinking.

Why do we presume that the right lobe is responsible for this propensity for creativity? It is because numerous studies indicate that it performs an associative function. In particular the right anterior superior temporal gyrus and the right parietal cortex seem to be active in experiments that aim to solve problems requiring creative insight. Goldberg and Costa (1981) suggested that this function of the right hemisphere is possible because it continuously processes loose or remote associations between the elements of a problem. This also happens during the resting stage. The associative capacity is strengthened when the mind is wandering and during the relaxation and incubation phase. Brain imaging revealed that the anterior superior temporal gyrus becomes particularly active when the participants experience an Aha! moment (i.e., the moment of creative insight) and during the initial effort to solve the problem. This differential capacity for associative thought, which is crucial for creative ability, combines with the different cognitive functions of the two hemispheres, as I pointed out earlier. The left uses logic as its language; it is oriented to detail, it deals with facts, rules, and language, above all from a connotative viewpoint, and is specialized to process routine and familiar information. On the other hand, the right hemisphere tends to use fantasy, images, imagination, figurative mental models, and symbols; it is responsible for the denotative features of language and is specialized to process novel information. This cognitive difference of the right hemisphere therefore seems to explain why, in patients with lesions to the left side, and therefore with reduced inhibitory activity on the right, this boosts the potential to think in images, reinforcing the creative capacity of figurative artists. The same could be said of scientific creativity, which, as numerous studies in the field of the history, philosophy, and psychology of science have pointed out, seems to be guided more by figurative mental modeling than linguistic and verbal modeling (Giere, 1988; Viale, 1991, 2013).

Although the right anterior superior temporal gyrus and the right parietal cortex seem to be responsible for the associative and unconscious dimension of creativity, the prefrontal cortex is involved in working through creative insights using secondary processes and logical thinking (Miller & Cohen, 2001). "Once a person has arrived at a creative solution, the prefrontal cortex becomes active, focusing not only on the task at hand but also on figuring out what other areas of the brain need to be engaged in order to solve the problem" (Kandel, 2012, p. 476). The prefrontal cortex corresponds to the conscious part of creative activity. The conscious mind has the task of processing and structuring the solution that derives from insight. In the scientific field, this component above all is fundamental to reach well-articulated, supported, and shared solutions.

Mind Wandering and Creativity: The Role of the Default Mode Network

It is known that a vital aspect of the creative experience takes the form of unconscious creativity in moments of incubation and when the mind is wandering. What data exist in psychological research, and how can this phenomenon be explained from a neurocognitive point of view? In other words, how do some particular neurocognitive features of the brain enable the wandering mind to become one of the preferred conditions for creativity?

Before attempting to answer this question it is worth mentioning two great creative spirits of the past who seem to emphasize the role of mind wandering:

The first is Wolfgang Amadeus Mozart, quoted in Andreasen (2005, p. 40):

When I am, as it were, completely in myself, entirely alone, and of good cheer—say traveling in a carriage, or walking after a good meal, or during the night when I cannot sleep; it is on such occasions that my ideas flow best and most abundantly. Whence and how they come, I know not; nor can I force them. All this fires my soul, and, provided I am not disturbed, my subject enlarges itself, becomes methodized and defined. … All this inventing, this producing, takes place in a pleasing lively dream.

The second is Arthur Schopenhauer (1851/1970, pp. 123–124):

One might almost believe that half of our thinking takes places unconsciously. … I have familiarized myself with the factual data of a theoretical or practical problem; I do not think about it again, yet often a few days later the answer to the problem will come into my mind entirely of its own accord; the operation which has produced it, however, remains as much a mystery to me as that of an adding machine: what has occurred is, again, unconscious rumination.

Creativity seems to be linked to moments in our psychic life that are not characterized by consciousness and awareness. As Jonathan Schooler pointed out (Schooler et al., 2011), big ideas, moments of creative insight, often seem to come not when people are hard at work but when they are sidetracked: going for a walk, taking a shower, thinking about something else. All of a sudden, ideas that were previously isolated come together, and people see connections that had escaped them before. Creativity involves removal of inhibition on novelty seeking. Novelty seeking encompasses capacities such as the ability to think unconventionally, to use divergent thinking in open-ended situations, and to be open to new experiences. Creative people are inclined to wonderment, independence, nonconformity, flexibility, and the capability of relaxation. The creative mind depends on abilities such as constructing metaphor, reinterpreting data, connecting unrelated ideas, resolving contradictions, and eliminating arbitrariness (Kandel, 2012, pp. 457–458). Creativity therefore depends on the ability not to be prisoners of mental patterns, of dominant hypotheses, or, to use Kuhn's metaphor, of the paradigm. These patterns act in a top-down manner on our search for solutions, limiting novelty seeking. When we consciously tackle the solution to an artistic or scientific problem, we are trapped in previous solutions and guided by models linked to these previous solutions that originate from our background knowledge. Conscious attentional focus on the problem space blocks the mind's capacity for free association, the connection of unrelated ideas, and the reinterpretation of data. The attention of the conscious mind can examine only a small number of possibilities at a time. On the other hand, letting the mind wander bypasses the conscious mind's conceptual grids and the limits of attention processing, and it allows the unconscious mind to develop its associative potential.

What is the empirical evidence of this relationship between creativity and the wandering mind? In the past few years various studies have been undertaken on the role of the incubation period in improving creative capacity (see Gilhooly, chapter 15, this volume). In the first place, as highlighted by numerous meta-analyses, the incubation, namely that period when attention is withdrawn from problem solving and the problem is not consciously addressed, undoubtedly has a beneficial effect on creative thinking (Sio & Ormerod, 2009; Strick et al., 2011). How can this facilitation be explained? As we have noted repeatedly in these pages, and as seems likely on theoretical grounds, this benefit can probably be explained by free associations at an unconscious level. Even Poincaré (1913) introduced the term "unconscious work" as early as 1913 to explain the creative facilitation in mathematical problem solving at times that can be assimilated to incubation periods. However, this explanation is not shared by all. A significant alternative hypothesis is that proposed by Herbert Simon (1966), among others, which refers to an attentional shift and a forgetting mechanism. When problem-solving activities hit a wall or are blocked by inadequate conceptual schemas or misleading answers, the possibility of focusing on other images or forgetting the problem for a while offers a fresh boost when returning to the problem, as if it were new or being tackled for the first time. This releases the person from the old, inadequate schemas and allows her to explore new ones.

Although this provides a plausible explanation of the incubation effect, the hypothesis is undermined by empirical tests. In particular, this mechanism did not produce beneficial effects in immediate incubation tests (Dijksterhuis & Meurs, 2006) when the interpolated task was performed immediately after instructions for the target problem and before the subject could consciously devise possible solutions. Instead, the results of tests using immediate incubation appear to show a beneficial effect of incubation even in these cases, thereby belying the "forgetting" or "attention-shifting" hypothesis. The same kind of test appears, on the other hand, to support the thesis that benefits of incubation are linked to unconscious-type processes. What form might this unconscious work take? There are two main hypotheses. The first is that the unconscious work could be like conscious work, that is, rule governed, but carried out without conscious awareness. This hypothesis seems unreasonable especially in view of the structure of cognitive activity. Rule-based processing, which is typical of problem solving in psychological tasks, requires high cognitive activation and is manifest in working memory and therefore is, by definition, conscious. Without this passage, it is does not seem possible to resolve even elementary numerical problems of addition or subtraction. Therefore, it seems paradoxical to suggest a duplication of the same type of mental activity, one with and the other without awareness. Moreover, as I explain later, when analyzing the *default mode network* theory, two distinct areas of the brain are activated in conscious problem solving and during incubation and mind wandering. This rules out any superimposition of the two different types of activity. The most credible hypothesis therefore remains that of unconscious work. During incubation, parallel spreading activation seems to occur through semantic networks that allow original and unforeseeable associations to be implicitly generated. This association is usually random or quasi-random. As Dijksterhuis and Nordgren (2006) state in their Unconscious Thought Theory, unlike conscious thought, this activity is parallel, bottom-up, inexact, and divergent.

How can this phenomenon be explained at a neurocognitive level?

By observing people at rest during brain scans, neuroscientists have identified a "default mode network" that is active when people's minds are especially free to wander. When people then embark on a task, the brain's executive network lights up to issue commands, and the default network is often suppressed (Mason et al., 2007). Why do these moments when the mind is allowed to wander seem to be the most fruitful for the development of creative solutions?

The default mode network (DMN) was discovered fortuitously during a study using neural imaging of the activation of brain areas following external task-based experiments. These experiments generally compared activation during a psychological task to control situations in which no external task was used, that is, during resting, mind wandering, and idle situations. Task-based neuroimaging studies tend to use a resting baseline that is assumed to be a neutral comparison to task-induced brain activity. In this way it is hoped to isolate the neural networks responsible for the psychological task. However, researchers began to notice that during this control condition a constellation of regions, often on the medial surface of both the hemispheres, sometimes exhibited heightened activation relative to external tasks (Buckner, 2012). Although these regions were deactivated during the active phase of the experiment, they unexpectedly became activated during resting time. In 2001 Marcus Raichle and colleagues described the baseline resting state as the "default mode" of function that is suspended during specific goal-directed behaviors (Raichle et al., 2001). Two years later Greicius and colleagues (2003) coined the functional anatomic term "DMN," arguing that it "accounts, in large part, for the phenomenon of task-related decreases in brain activity." It includes the medial prefrontal cortex (MPFC), posterior cingulate cortex (PCC), temporoparietal junction (TPJ), lateral temporal cortex (LTC), superior frontal gyrus (SFG), and the hippocampus. Patterns of spatial correlation measured in the absence of direct tasks (resting-state fMRI) support this network structure and suggest that the DMN is composed of midline hub regions (MPFC, PCC) and two subsystems.

What is the function of DMN? Two sets of data seem to address the answer. First, a growing body of results shows that DMN is also activated in tasks that do not involve perceptual-type external stimuli. Second, the anatomy of DMN appears to limit its possible functions. For example, it does not include motor or sensorial regions but does include areas associated with the medial temporal lobe memory system (Buckner et al., 2008). Therefore, the main possibility is that DMN supports internal mentation separated from any external stimuli. In particular, DMN appears to play a crucial role in the realization of "dynamic mental simulations based on personal past experiences such as used during remembering, thinking about the future, and generally imaging alternative perspectives and scenarios to the present" (Buckner et al., 2008, pp. 17–18).

This hypothesis seems to be confirmed by a series of studies that show how the network is activated during forms of internal mentation, such as autobiographical memory (Svoboda et al., 2006), theory of mind (Amodio & Frith, 2006), and envisioning the future (Schacter et al., 2008). The common denominator of this internal mentation activity is precisely the simulation of alternative scenarios to the actual situation. In both, the goal of the simulation processes is to imagine events that go beyond the present and immediate reality. This type of activity uses counterfactual conditionals (If *A* had been the case, then *B* would have been the case) or subjunctive conditionals (If it were *A,* then it would be *B*). All the simulations refer to autobiographical mnemonic material and use the self as a barycenter. The subsystem linked to the mnemonic part is based in the PCC and hippocampus. The part linked to the simulation and self-referential judgments and exploration is based in the MPFC. In theory of mind, the possible behavior of a person can be predicted by simulating in first person the relation between the hypothetical premises and the behavior deriving from them. Or, on the contrary, an attempt is made to interpret the reasons for the behavior observed by simulating the potential causes that generated it. In autobiographical memory, past episodes are recalled by simulating the said episodes. When envisioning the future, future scenarios are imagined by simulating the possible factual details in first person. In all these cases it is the simulator's ego, with his mnemonic history and emotional and affective characteristics, that provides inferential activity. No content comes from external stimuli.

How can DMN, which is activated during wandering mind episodes, be connected to the phenomenon of creativity? One of the characteristics of DMN is its opposition-competition with the other mode, the *external attention system*: "as activity within the default network increases, normalized activity in the external attention system shows activity decreases" (Buckner et al., 2008, p. 25). This finding appears to confirm the thesis that there is slippage from one mode to another in relation to the presence or otherwise of external sensorial-type stimuli. "One mode, marked by activity within the default network, is detached from focused attention on the external environment and is characterized by mental explorations based on past memories. The second mode is associated with focused information extraction from sensory channels" (Buckner et al., 2008, p. 25). Therefore, DMN activity is manifested when the mind is not engaged by external stimuli and can wander in mental space made up of memories in the form of propositions, images, and relative emotions.

In the past some psychologists had already highlighted the importance of this phase. "Left without an immediate task that demands full attention, our minds wander, jumping from one passing thought to the next." William James (1890/1983) had called this process "stream of consciousness." "When absorbed in intellectual attention we become so inattentive to outer things as to be 'absent-minded,' 'abstracted,' or 'distracted'" (James, 1890/1983). In scientific creativity simulations using subjunctive and counterfactual conditionals above, all play an important role in problem solving under incubation. Take, for example, the role of analogies and metaphors that serendipitously inspire simulations in a scientist's mind. In these instances of creativity the autobiographical and personal dimension is maintained above all at a cognitive level and less at an affective and emotional level. Instead, in art, it is the affective and emotional sphere that is fundamentally important. In artistic creativity DMN appears to be a powerful medium. The artist transfers his or her most intimate and hidden self into the creative process. It is the emotional and affective synthesis at an autobiographical level that guides the writer's pen, and even more so the artist's brush. It is self-simulation in others and in external social relations that gives such power to figurative art (and particularly portraiture, as we will see). The wandering mind represents the moment when artists can move away from traditional schemas, express their emotional and cognitive biography, and try to be emphatic by placing themselves in other people's shoes, whether other individuals or social groups. In this way, through themselves, artists transfer others into their paintings or novels. And the most successful artists are precisely those who involve the viewer or reader in this relationship of private and intimate reflection.

A demonstration of the importance of this personal dimension transferred into figurative art is provided by a study by Vessel and colleagues (2013). These researchers analyzed whether DMN was activated during the aesthetic experience of viewing a series of paintings from different periods and in different artistic styles. The surprising finding was that although those paintings not deemed to be interesting failed to generate any DMN activation, those that were most attractive led to a dramatic reduction in DMN deactivation and its complete absence in the MPFC. In the first case the external attention system mode was activated and DMN deactivated. In the second, on the contrary, both were activated. The strong activation in the MPFC area associated with self-evaluation and exploration emphasizes the stimulus of those paintings that most appeal to the viewer's autobiographical component.

The Perceptual Limitations of the Work of Art

All figurative art is a matter of visual perception. Even when early twentieth-century art, from Cubism onward—and Cezanne before that—attempted to break down phenomenal reality into its basic layers and forms, it did so above all from a perceptual point of view. Moreover, this viewpoint was radically different from that used in Impressionism—which sought to capture the immediate perceptual impression of external reality—and Fauvism—which sought to separate an object's color from its form. Abstract art, too, in all its myriad ramifications, is based on the means of perception. It denies any sort of perceptual-like representation of reality. Instead, representation becomes conceptual, metaphysical. The artist's message—and this is true of Mondrian's squares, the lines and splashes of color used by Kandinsky and Klee, Pollock's drips and splashes, and Calder's moving solids—is constructed through a code of perception. This code attempts to gauge the impact of the brushstroke on canvas or the moving object or video image on the eye-brain structure. Much of the time artistic creativity is unaware of the laws of neuroperceptual dynamics. As in the case of empathy, the perceptual limitations of the human brain set the bounds of artistic creativity. This is why we can also talk about bounded creativity in this case. Some examples, taken mainly from Zeki (1999) will serve to illustrate the neural limitations of artistic creativity more clearly.

When we look at Jan Vermeer's *Girl with a Pearl Earring* (figure 16.3), we are bombarded by an enormous number of mixed emotions. The painting talks to us as if it were alive. The girl entices, but at the same time she withdraws. She is submissive but also dominating. She seems happy but also melancholy. She looks as if she is about to move, while at the same time staying still. In terms of the meaning attributed to the painting, this ambiguity (Zeki, 1999) highlights an effect that is well known to neuroscientists, namely the crucially important role played in the brain's organization by recognition of the human face. The brain has an entire region of the cortex dedicated to the recognition of faces. Even minute changes in facial expressions are registered and interpreted by the brain. It is the face, rather than the limbs or trunk, that offers the brain this specialized information because of the evolutive role of the face compared to other parts of the body. Interpreting the face allows a person to anticipate other people's behavior, thereby increasing adaptive fitness. From a neuroperceptual point of view, there are two face-related properties that correspond to two different parts of the

Figure 16.3
Johannes Vermeer, *Girl with a Pearl Earring*. Mauritshuis, The Hague, The Netherlands (http://www.mauritshuis.nl).

brain. These properties have been highlighted through the study of pathologies such as prosopagnosia—the inability to recognize faces.

The part of the brain dedicated to the recognition of unfamiliar faces corresponds to the posterior fusiform gyrus (figure 16.4); on the other hand, the part dedicated to the recognition of familiar faces is the anterior fusiform gyrus in association with the frontal lobes. Other parts of the brain are also able to interpret affective-type facial expressions. An example is the response of the amygdala in interpreting facial expressions of fear. Indeed, patients with amygdala lesions are able to recognize faces but not fearful expressions. In this respect the reading of the face, comparable in some respects to mindreading, involves the system of mirror neurons and all the nervous structures affected by the empathic dimension of the emotions. Artistic creativity is above all manifest in portraiture precisely because of the extraordinary wealth of possibilities offered by the brain. An artist's ability to represent complex aspects of the human soul,

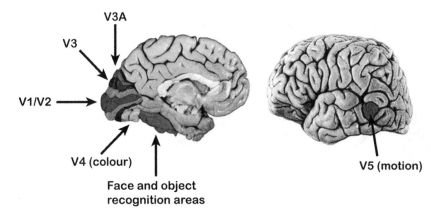

Figure 16.4
The visual brain consists of multiple functionally specialized areas that receive their visual input largely from V1 and an area surrounding it known as V2. These are the best-charted visual areas but not the only ones. Other visual areas are being continually discovered (Zeki, 1999. Italian translation, 2003).

including states of uncertain and ambiguity, stems precisely from his or her mastery of the *bounded codex* of the face-brain relationship.

Take, for example, Titian's *Portrait of Gerolamo (?) Barbarigo* (The National Gallery, London: http://www.nationalgallery.org.uk) in which the head faces the one way while the eyes look in another, gazing obliquely at the viewer to produce the effect of a scornful and arrogant personality (Zeki, 1999), or the *Portrait of a Venetian Gentleman* (Fondazione Zeri, Bologna: http://www.fondazionezeri.unibo.it), or the *Portrait of a Musician* (figure 16.5). The brilliance of Titian or Vermeer, not to mention Leonardo da Vinci, lies in knowing how to exploit the perceptual limitations of the brain in original and surprising ways. Leonardo's *sfumato* is another example of how artistic genius succeeds in using the neuroperceptual limitations of the brain, at an unconscious level, in order to emphasize certain expressive traits. When we look at the Mona Lisa, we realize how her ambiguity, her enigmatic qualities, as well as her sweetness and submissiveness are the result of this new technique that blurs the contours of the figures, the clearly marked boundaries and demarcations using subtle variations of light and color that blend together imperceptibly. When the eye fixes on a point on the contour of the face or the curve of the lip, cells in areas V1, V2, and V3 of the visual cortex (figure 16.4) are stimulated to analyze the visual scene piece by piece.

Each small segment of visible shading excites the cells depending on its direction. Given that these cells are in homogeneous orientation-selective clusters (as David Hubel and Torsten Wiesel discovered in 1959), their activation is perturbed whenever the contour or curve tends to blur, as in the case of Leonardesque *sfumato*.[7]

Figure 16.5
Tiziano Vecellio, *Ritratto di Musico*, Galleria Spada, Roma (http://galleriaspada.beniculturali.it).

This reference to *sfumato* introduces the whole question of the receptive field into the visual system. Each cell in the visual system can only be activated by visual stimuli that correspond to part of the visual space. For example, a cell can be stimulated by a red square present in its field, or by a light of one color in the background of a light of another color, while remaining insensitive to stimulation with white light. Others can show preferences for lines running in a particular direction. As their orientation gradually changes and moves away from that direction, the cells' reaction decreases progressively. Last, other cells react to moving stimuli in their receptive field depending on direction while not reacting to stationary stimuli. In short, receptive fields show three key characteristics: position, form, and specificity (Zeki, 1999).

Some aspects of the dynamics of contemporary artistic creativity can be understood through an analysis of the limitations of the various receptive fields. The building blocks for neural form processing are the cells that react selectively to straight lines.

A   B

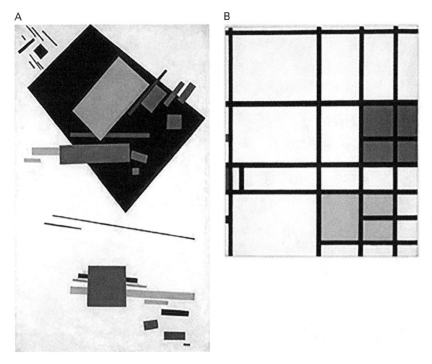

Figure 16.6
(A) Kasimir Malevich, 1915, *Suprematist Painting*, Stedelijk Museum Amsterdam, The Netherlands (http://www.stedelijk.nl). (B) Piet Mondrian, *Composition n. 11, 1940–42-London with Blue, Red, and Yellow, 1940–42*, Albright-Knox Art Gallery, Buffalo, New York (http://www.albrightknox .org).

The same principle underlies the conceptual revolution of Mondrian's and Malevic's abstract art (figure 16.6).

They too saw the straight line as being the universal shape, the basis for constructing the most complex forms. Mondrian thought that all shapes could be reduced to a multiplicity of straight lines, which then form rectangles through oppositions. These forms built from straight lines represent the most universal aesthetic dimension. Rectangular shapes are neutral, surpassing the figure and creating nonfigurative art. An analysis of the shape of receptive fields in the visual brain, particularly in areas V1 and V4, shows they are square or rectangular. The cell reacts only when the appropriate visual stimuli are projected into these square or rectangular receptive fields. New forms of abstract art consisting of lines, squares, and rectangles, seem to be fully able to stimulate cells of this kind. This does not mean that the stimulation of this type of highly specialized cell necessarily leads to artistic creation but only that this experience is "bounded to"

Figure 16.7
André Derain, *Charing Cross Bridge, London 1906*. National Gallery of Art, Washington. John Hay Whitney Collection (http://www.nga.gov/content/ngaweb.html).

these particular cells. If we did not have them then we would not be able to experience any aesthetic reaction to art works of this kind, and therefore this art would not exist (Zeki, 1999).

Last, there is one further example of the unexpected ways in which art and neuroscience interact. The aim of Fauvism—an early twentieth-century artistic movement that included such renowned artists as Henri Matisse, André Derain, Maurice de Vlaminck, and Kees van Dongen—was to liberate color from form. The solution chosen by these artists was to paint everyday objects in unnatural color combinations (figure 16.7).

Abstract art later adopted a more radical solution by using colors in straight lines, squares, or rectangles with no reference to phenomenal reality. An interesting finding discovered by neuroscientists (described in Zeki, 1999) is that activation differs when an observer examines a naturalistic painting compared to a Fauve painting. In the first case in addition to V1 activation, which serves to evaluate the spectral composition of each point of the image, the V4 complex located in the fusiform gyrus is also activated. This appears to be involved in the construction of abstract color, not linked to any particular object. Together with these areas, activation is also observed in a structure in the temporal lobe, the hippocampus (linked to the memory function), and in an area located inside the inferior frontal gyrus of the right hemisphere. Compare this to what happens when a viewer looks at Derain's painting, *Charing Cross Bridge* (figure 16.7).

Apart from the areas V1 and V4, there is no other activation in the areas observed in the previous experiment. The hippocampus, for example, is not involved, and activity in the frontal cortex is in the center, not in the inferior zone.[8] What does this mean? The involvement of cerebral areas—the temporal and frontal cortex and the hippocampus—in the naturalistic experiment serves to attribute color to the objects and to check that the attribution is correct, based on memory of these objects. This does not happen in the Fauve experiment. The impossibility of separating form from color is the result of constraints on the organization of the visual brain. We see the objects in different light conditions. A red apple in the evening has a spectral composition of reflected light that is different from that of a red apple in full sunlight. In spite of this we preserve color constancy when attributing the color red, albeit in different shades, to the apple. It is the brain that makes the transformation, creating constancy where none exists in reality (Maffei & Fiorentini, 2008). A comparison is made by the brain between the spectral composition of the light reflected by the area being observed and that of the light reflected from surrounding areas. The apple's color therefore also derives from the spectral composition of light reflected by the areas surrounding the apple. By making these comparisons the brain can subtract the "illuminant," overlook the changes, and preserve the object's color constancy (adaptive trait). At this point it becomes clear that the attempt made by the Fauve movement to separate form from color was an impossible feat. The relationship between contiguous areas of lighting is highlighted by a boundary, namely a line. It can be straight or curved, but it remains a contour that defines a shape. Therefore, the visual brain itself sets limits on the impossibility of separating form and color. Form itself, corresponding to the outline of contiguous areas of illumination, is the condition for the brain's construction of color constancy. So form is indissolubly linked to color (Zeki, 1999).

Conclusion: The Multimodality of Aesthetic Experience

The third section of this chapter highlighted the importance of the emotional and autobiographical dimension in art and aesthetic perception. In figurative art both the perceptual dimension and the empathic and emotional one are crucial aspects of creativity, and together they differentiate it from other forms of creativity. Literary creativity also has a strong emotional and empathic component, but the perceptual component is less evident. Creativity in science has a perceptual component but a reduced emotional and empathic one.[9] On the contrary, there is an intrinsic correlation between perception and empathy in figurative art, and both play a powerful primary role in the creative act and aesthetic experience. This role is both enabled and limited by the biological characteristics of the human brain. For example, in a 2007 study, Freedberg and Gallese highlighted how vision is a multimodal process that entails the activation of sensory-motor, visceral-motor, and affective brain circuits.

Empathy in the work of art is multifaceted. It is achieved at the moment that an artist creates a person's portrait and tries to identify with his psychology or when an artist portrays a landscape and his interpretation reflects the sensitivity of other known people or of the social environment of his time. Empathetic identification obviously also occurs in the attentive visitor to an exhibition, when he tries to interpret what lies at the base of a work of art. There are two ways in which empathy works in this case: the first concerns empathetic feelings deriving from the simulation of actions, emotions, and sensations contained in the work of art; the second concerns the relationship between this simulation and the visible traces of the artist's expressive gestures, for example, Fontana's cuts or Van Gogh's brushstrokes. In the production of a work of art and its aesthetic perception, empathy and the related embodied simulation (Gallese, 2005) seem to be conveyed by the system of *mirror neurons* discovered by Giacomo Rizzolatti et al. (2001).

As Kandel underlined (2012), referring to the Modernist period and to such artists as Klimt, Kokoschka, and Schiele, the empathetic identification of painters leads them to exaggerate certain traits of the person portrayed and to a deliberate overstatement of certain characteristics peculiar to that person. As Ramachandran (1999) maintained in his *peak shift principle*, the artist tries not only to capture the essence of a person but to amplify it and thus to activate more powerfully the same neural network that would have been triggered by the real person. This ability seems to correspond to what the visual and auditory systems are doing: to discard redundant and insignificant information and to concentrate on the relevant features. The ability of figurative artists is to allow the viewer to stimulate and read emotions according the limitations of our brain. The limited empathetic capacity of our brains sets the bounds on artistic creativity. This *bounded creativity* is evident in the Austrian Expressionists. They were very successful in emphasizing the feelings and instinctual striving that are submerged deep in an individual's ego. They used caricature and the exaggeration of form and color to achieve these results. The *peak shift principle* is evident in "Klimt's overstatement of the erotic and destructive power of women, in Kokoschka's exaggeration of facial emotions, and in Schiele's exaggerated, anxiety-driven bodily postures and in his depictions of the modern person struggling simultaneously with the life force and the death wish" (Kandel, 2012, p. 448).

Many other artists have tried more or less successfully to stimulate these empathic feelings. But what significance can this analysis have in relation to contemporary abstract and conceptual art? The goal of a Kandinsky, Pollock, Burri, Fontana, or Mertz is certainly not the same as that of the Austrian Expressionists or of Caravaggio. These works of art have perceptual and cognitive goals but, in general, not empathetic goals. They are often conceptual projects and networks thrown onto the world without any need to represent it from an emotional viewpoint. So one would also expect spectators not to be stimulated at a level of the centers charged with empathetic identification. Counterintuitively, recent research (Umiltà et al., 2012) seems to demonstrate the contrary. In their experiment this group of scholars showed some people high-resolution reproductions of Lucio Fontana's torn canvases, alternating with a stimulus control that was a picture in which the cut was replaced by a line. The cortical motor system was activated in the subjects, conveyed by mirror neurons. This does seem to confirm the theory that suggested that the signs of the artist's hand on the canvas trigger in the spectator the motor areas that control the execution of the gestures that produce these same images. And this stimulates the activation of the mirror neurons so that they become central to the perception of even an abstract work of art. However, in this case, we can say that this is an embodiment of motor action plans rather than a real process of emotional empathy. In other words, the spectator is prompted to simulate the cutting of the canvas as a motor action. This seems different from the emotional empathy the spectator feels in front of a work by Caravaggio or Kokoschka, where the profound psychological dynamics of the ego are experienced and reproduced.

To conclude, in the recent past, the neurosciences have made huge advances in understanding the biological features of artistic bounded creativity, in particular with regard to the complexity of our perceptual experience. In short, sensorial and motor systems appear to respond to various sensorial modalities. As was recently underlined by Vittorio Gallese (Gallese, 2014; Gallese & Ebisch, 2013), the processing of tactile sensitivity is not limited to the somatosensory cortex, but other areas of the brain, such as vision and hearing, are involved in processing tactile stimuli. The frontal and parietal areas of the motor system also contain neurons that respond to visual, auditory, and somatosensorial input. And all of these areas work together with those involved in the empathetic experience. Mirror neurons are the most striking example of the multimodal complexity of the embodied simulation of perception.

These findings are radically changing our concepts of the aesthetic experience and creativity. Neither the experience of the viewer nor that of the creator can be understood without reference to the relationship between a particular individual body-brain and the world. In other words, we must use a neurohermeneutical methodology. "Studying the brain-body system means trying to understand how our experience of images—including those laden with symbolic meanings—is generated by the functional states of our central nervous system, autonomous nervous system, and the integration of both with the cardiorespiratory and muscular apparatus of *an individual with a personal history*. Experimental aesthetics involves trying to understand the *individual* physiology of creating and experiencing the symbol" (Gallese, 2014, p. 65).

Notes

1. The first and second paragraphs are based on Viale (2013, chapter 8).

2. The analogy between bounded rationality and bounded creativity is fuzzy. Rationality is bounded compared to a precise idea of rationality characterized by the canonical rules of probability calculus and of inductive and deductive logic. Bounded rationality is characterized by the limitation of the human mind in applying these rules in order to interact with a highly complex environment. For creativity, on the other hand, the situation is less clear-cut. There is no canonical model to which human creativity can be compared in order to set its limits. The concept itself is controversial and subject to semantic slippage of various kinds.

3. As Salvatore Settis states (Maffei & Fiorentini, 2008, VI–XIII), the idea of a progress in art and an evaluation of quality appears in the Renaissance and then again in the late eighteenth century in Germany when art history became an academic discipline. In the ancient world, and in Greece in particular, the artistic was never perceived as a value, or art as something distinct from nonart.

4. To summarize Gombrich's brilliant interpretation (1950, It. trans. 1989, pp. 4–51), Egyptian artists resembled geographical cartographers rather than painters in the way they precisely followed the dictates of the time. Because the head was most easily drawn in profile, they drew it sideways. But if we imagine the human eye, we think of it as seen from the front. Accordingly, a full-face eye was planted into the side view of the face. The top half of the body, the shoulders and chest, are best seen from the front, for then we see how the arms are hinged to the body. But arms and legs in movement are much more clearly seen sideways. They found it hard to visualize either foot seen from outside. They preferred the clear outline from the big toe upward. So both feet are seen from the inside, and the man looks as if he had two left feet. The Egyptian style comprised a set of very strict rules: seated statues had to have their hands on their knees; men had to be painted with darker skin than women; Horus, the sky god, had to be shown as a falcon or with a falcon's head; Anubis, the god of funeral rites, as a jackal or with a jackal's head, and so forth.

5. "Man is a great miracle."

6. By contrast, regressions to earlier, more primitive psychological functioning that occur in psychotic episodes are involuntary.

7. The importance of codifying the contours of figures and the influence of this visual characteristic in the development of art history is highlighted by Maffei and Fiorentini (2008, pp. 60–67). The images present on the fundus are reduced in the visual cortex to outlines or segments of these. This symbolic activity is universal and underlies the symbolic paintings made by our forebears in the caves at Cosquer and Chauvet. The capacity to symbolically represent external reality using outlines is an example of bounded creativity. It is the brain that imposes its symbolic rules and neural signs. Ontogenesis and phylogenesis appear to trace a transition from figurative expressions, based on the typical outlines drawn in childhood and the earliest period of art history (up to the Egyptian era), to depictions that gradually fill with details, color, depth, chiar-

oscuro, and become less symbolic and more realistic, as we move into adolescence and from the Greek and Roman eras onwards. For reasons that are cultural in nature, contemporary art then sees a return to the symbolic dimension and to increasing levels of abstraction.

8. It still has not been explained why the hippocampus, specialized in memory, is not activated by a Fauve painting. Indeed, even the view that asserts the noncorrespondence of the object's color to its natural color needs to be backed up by the memory of the object with which the color is associated (Zeki, 1999).

9. The empathetic component of scientific creativity is minor but present. Processing an experiment or drafting a scientific paper is also guided by empathetic elements. For example, we identify with the person who will read the publication, whether a friend, colleague, or impersonal public. We identify with the colleague or teacher who first attempted the experiment. All this shapes and influences the choices we make when drafting the publication, focusing it on the understanding of our hypothetical readers and on the fine-tuning and conduction of the experiment, unconsciously imitating the style of our mental references.

References

Amodio, D. M., & Frith, C. D. (2006). Meeting of minds: The medial frontal cortex and social cognition. *Nature Reviews. Neuroscience, 7*, 268–277.

Andreasen, N. C. (2005). *The creating brain: The neuroscience of genius.* New York: Dana Press.

Buckner R. L. (2012). The serendipitous discovery of the brain's default network. *Neuroimage. 62*(2), 1137–1145.

Buckner, R. L., Andrews-Hanna, J. R., & Schacter, D. L. (2008). The brain's default network: Anatomy, function, and relevance to disease. *Annals of the New York Academy of Sciences, 1124*, 1–38. doi:10.1196/annals.1440.011.

Dehaene, S., & Changeux, J. P. (2011). Experimental and theoretical approaches to conscious processing. *Neuron, 70*(2), 200–227.

Dijksterhuis, A. (2004). Think different: The merits of unconscious thought in preference development and decision making. *Journal of Personality and Social Psychology, 87*(5), 586–598.

Dijksterhuis, A., & Meurs, T. (2006). Where creativity resides: The generative power of unconscious thought. *Consciousness and Cognition, 15*, 135–146.

Dijksterhuis, A., & Nordgren, L. F. (2006). A theory of unconscious thought. *Perspectives on Psychological Science, 1*(2), 95–109.

Epstein, S. (1994). Integration of the cognitive and psychodynamic unconscious. *American Psychologist, 49*, 709–724.

Freedberg, D., & Gallese, V. (2007). Motion, emotion and empathy in aesthetic experience. *Trends in Cognitive Sciences, 11*(5), 197–203.

Gallese, V. (2005). Embodied simulation: From neurons to phenomenal experience. *Phenomenology and the Cognitive Sciences, 4*, 23–48.

Gallese, V. (2014). Arte, corpo, cervello: Per un'estetica sperimentale. *Micromega, 2*, 49–67.

Gallese, V., & Ebisch, S. (2013). Embodied simulation and touch: The sense of touch in social cognition. *Phenomenology & Mind, 4*, 269–291.

Gardner, H. (2006). *Five minds for the future*. Boston: Harvard Business School Press.

Giere, R. N. (1988). *Explaining science*. Chicago: University of Chicago Press.

Gigerenzer, G. (2007). *Gut feelings*. London: Viking.

Goldberg, E., & Costa, L. D. (1981). Hemisphere differences in the acquisition and use of descriptive systems. *Brain and Language, 14*, 144–173.

Gombrich, E. (1950). *The story of art*. London: Phaidon.

Gombrich, E. (1960). *Art and illusion*. Princeton, NJ: Princeton University Press.

Greicius, M. D., Krasnow, B., Reiss, A. L., & Menon, V. (2003). Functional connectivity in the resting brain: A network analysis of the default mode hypothesis. *Proceedings of the National Academy of Sciences of the United States of America, 100*(1), 253–258. doi:10.1073/pnas.0135058100.

Humphrey, N. (1998). Cave art, autism, and the evolution of the human mind. *Cambridge Archaeological Journal, 8*(2), 165–191.

James, W. (1983). *The principles of psychology*. Cambridge, MA: Harvard University Press. (Original work published 1890, New York: Holt).

Kandel, E. (2012). *The age of insight*. New York: Random House.

Kapur, N. (1996). Paradoxical functional facilitation in brain-behavior research. *Brain, 119*, 1775–1790.

Kihlstrom, J. F. (2007). Consciousness in hypnosis. In P. D. Zelazo, M. Moscovitch, & E. Thompson (Eds.), *Cambridge handbook of consciousness* (pp. 445–479). Cambridge: Cambridge University Press.

Kosslyn, S. M., & Wayne Miller, G. (2013). *Top brain, bottom brain: Surprising insights into how you think*. New York: Simon & Schuster.

Kris, E. (1952). *Psychoanalitic exploration in art*. New York: International Universities Press.

Kuhn, T. (1970). *The structure of scientific revolutions*. Chicago: University of Chicago Press.

Lakatos, I. (1970). Falsification and the methodology of scientific research programmes. In I. Lakatos & A. Musgrave (Eds.), *Criticism and the growth of knowledge*. Cambridge: Cambridge University Press.

Libet, B. (1985). Unconscious cerebral initiative and the role of conscious will in voluntary action. *Behavioral and Brain Sciences, 8*, 529–566.

Maffei, L., & Fiorentini, A. (2008). Arte e cervello (2nd ed.). Bologna: Zanichelli Editore. (Original work published 1995.)

Mason, M. F., Norton, M. I., Van Horn, J. D., Wegner, D. M., Grafton, S. T., & Macrae, C. N. (2007). Wandering minds: The default network and stimulus-independent thought, *Science, 315*(5810), 393–395.

Miller, E. K., & Cohen, J. D. (2001). An integrative theory of prefrontal cortex function. *Annual Review of Neuroscience, 24*, 167–202.

Poincaré, H. (1913). *The foundations of science.* New York: Science House.

Raichle, M. E., MacLeod, A. M., Snyder, A. Z., Powers, W. J., Gusnard, D. A., & Shulman, G. L. (2001). A default mode of brain function. *Proceedings of the National Academy of Sciences of the United States of America, 98*, 676–682.

Ramachandran, V. S. (1999). The science of art: A neurological theory of aesthetic experience. *Journal of Consciousness Studies, 6*, 15–51.

Ramachandran, V. S. (2004). *The emerging mind.* London: Profile Books Ltd.

Rizzolatti, G., Fogassi, L., & Gallese, V. (2001). Neurophysiological mechanisms underlying the understanding and imitation of action. *Nature Reviews. Neuroscience, 2*, 661–670.

Sacks, O. (1995). *An anthropologist on Mars.* London: Picador.

Schacter, D. L., Addis, D. R., & Buckner, R. L. (2008). Episodic simulation of future events: Concepts, data, and applications. *Annals of the New York Academy of Sciences, 1124*, 39–60.

Schooler, J. W., Smallwood, J., Christoff, K., Handy, T. C., Reichle, E. D., & Sayette, M. A. (2011). Meta-awareness, perceptual decoupling and the wandering mind. *Trends in Cognitive Sciences, 15*, 319–326.

Schopenhauer A. (1970). *Essays and Aphorisms.* London: Penguin Books. (Original work published 1851.)

Simon, H. A. (1955). A behavioural model of rational choice. *Quarterly Journal of Economics, 69*(1), 99–118.

Simon, H. A. (1957). *Models of man.* New York: Wiley.

Simon, H. A. (1966). Scientific discovery and the psychology of problem solving. In R. Colodny (Ed.), *Mind and cosmos.* Pittsburgh, PA: University of Pittsburgh Press.

Simon, H. (1973). The structure of ill-structured problems. *Artificial Intelligence, 4*, 181–201.

Sio, U. N., & Ormerod, T. C. (2009). Does incubation enhance problem solving? A meta-analytic review. *Psychological Bulletin, 135*, 94–120.

Sperry, R. W. (1984). Consciousness, personal identity and the divided brain. *Neuropsychologia, 22*(6), 661–673.

Strick, M., Dijksterhuis, A., Bos, M. W., Sjoerdsma, A., Van Baaren, R. B., & Nordgren, L. F. (2011). A meta-analysis on unconscious thought effects. *Social Cognition, 29,* 738–762.

Svoboda, E., McKinnon, M. C., & Levine, B. (2006). The functional neuroanatomy of autobiographical memory: a meta-analysis. *Neuropsychologia, 44,* 2189–2208.

Umiltà, M.A., Berchio, C., Sestito, M., Freedberg, D., & Gallese, V. (2012). Abstract art and cortical motor activation: An EEG study. *Frontiers in Human Neuroscience, 6,* 1–9.

Vasari, G. (1986). *Le vite de' più eccellenti architetti, pittori, et scultori italiani, da Cimabue insino a' tempo nostri.* Torino: Einaudi.

Vessel, E. A., Starr, G. G., & Rubin, N. (2013). Art reaches within: Aesthetic experience, the self and the default mode network. *Frontiers in Neuroscience, 7,* 258.

Viale, R. (1991). *Metodo e società nella scienza.* Milano: Franco Angeli.

Viale, R. (2013). Methodological Cognitivism. (Vol. 2). *Epistemology, Science, and innovation.* Heidelberg: Springer.

Wegner, D. M. (2002). *The illusion of conscious will.* Cambridge, MA: MIT Press.

Zeki, S. (1999). *Inner vision: an exploration of art and the brain.* Oxford: Oxford University Press. [It. Trans. *La visione dall'interno. Arte e cervello.* Torino: Bollati Boringhieri Editore, 2003.]

Closing Thoughts

17 A *Psycho-rhetorical* Perspective on Thought and Human Rationality

Giuseppe Mosconi

Editors' Note: Giuseppe Mosconi (1931–2009) was the founder of the psychology of thinking in Italy. An outstanding scholar, he was one of the leading Italian psychologists. His psycho-rhetorical perspective reveals his clear intention to recover and restore to the research on the psychology of thinking, the wealth of observations and reflections regarding the intimate connection between language and thought (*logos*), which rhetoric has accumulated over the centuries. His rich scientific output, characterized by this originality of vision, developed along closely interrelated themes of research that ranged from studies on language to studies on thought, both in the fields of reasoning and of problem solving.

The Editors are grateful to the Department of Psychology, University of Milano-Bicocca for a grant to support the translation of the present chapter. We thank also Frances Anderson for her accurate translation. And, of course, we thank Giuseppe Mosconi, who inspired us all.

17.1 The Greeks Said *Logos*[1]

The Greeks said *logos* where we would say *word, speech, oral communication* on the one hand and *reason, calculation, evaluation* on the other; for example, *dià lógon* meant "orally," and *lógos estí* "it is said that"; *katà lógon* meant "as a logical consequence," and *lógos aireî* "reason teaches us." One single word thus serves a complex and articulated reality that was in fact experienced as a whole and is able to point out and express its various aspects.

The Greeks were aware of thinking-speaking, as "faculty,"[2] as an organic and distinct part of our psychic existence, not merely in natural language but also in learned language. Isocrates expressed this very clearly:

With this faculty we both contend against others on matters which are open to dispute and seek light for ourselves on things which are unknown; for the same arguments which we use in persuading others we can dispute debatable matters and explore what is yet unknown: the arguments which we use in persuading others when we speak in public, we employ also when we deliberate in our own thoughts; and, while we call eloquent those who are able to speak before a crowd, we regard as sage those who most skillfully debate their problems in their own minds. (Isocrates, *Nicocles: Prologue.* In Jaeger, 1959)

This passage clearly states the fundamental homogeneity of *thinking-speaking*, both when we are arguing with ourselves and with others. The acknowledgment of a distinct cognitive activity that has common traits in its diverse modalities emerges from Isocrates's words.

In modern Italian too there are still traces of the awareness of the genuine psychological wholeness of *thinking-speaking*. Consider the verb *ragionare* (to reason)[3]: it can mean a thought process but also to talk, or more precisely, it expresses two meanings with one word, a particular way of *thinking-speaking*. Today, however, it is mainly used in dialects, although in the early thirteenth century it was much used by Dante and his contemporaries (Cortellazzo & Zolli, 1979). This organic area of experience has lost its identity in our contemporary culture and its "unitariness" is obfuscated, having been divided by the prevailing segmentation in language and thought. In psychological conceptions and terminology this distinction has been taken to the extreme, almost to the point of being a separation. If you look up "reasoning" in a dictionary of psychological terms, all it gives is: "a more or less systematic thought process, the results of which are evaluated on the basis of logical principles" (Dalla Volta, 1974).

This is not the place to investigate how and why *distinction* prevailed over *unitariness* and resulted in separation. Suffice it to say that it has been neither an improvement nor an enrichment in knowledge. As a consequence it is more difficult for us to be aware of and to talk of that complex, articulated, psychological entity, which the ancient Greeks called *logos* whereas we no longer have an equivalent term for it, although we continue to experience it just as strongly as they did.

Talking about the ways of using language, Noam Chomsky said, "if I were forced to choose, I'd have to say something quite classical, and rather that language serves mainly to express thought" (1977, p. 90).

Language also serves for this. Take for example the meaning we attribute to the saying "express his *own* thought," where what is meant is that the fully formed thought, the concept, is externalized in the written or spoken language. But language is much more than a vehicle for expressing thought: it serves to think. This is not to say that language is necessary to thought or that thinking is always of necessity *thinking-speaking*. What I mean to say is that one of our cognitive activities is *thinking-speaking*, whether externalized to others or simply to ourselves, and that this activity is recognized, autonomous, organic. We are not talking about language *and* thought, or the relationship *between* thought and language, but about that activity or area of cognitive experience for which this distinction is not relevant, or only superficially so. A Greek would have said, quite simply, "I'm talking about *logos.*"

The idea of language as something to be added to thought, which serves as a vehicle to transmit thought more or less satisfactorily, is quite widespread, almost commonplace. When we use expressions such as "I can't find the word," "it is on the tip of my tongue," or "I don't know how to say this," what we really mean is that we are having difficulty in expressing the thought in words to communicate it; it certainly does not mean that the notion we want to communicate is missing or has still to be defined (this of course is not the case where the difficulty arises from a concern regarding how to express the notion most effectively). This theory can be found both in a popular usage of language and also in a long and erudite tradition advocated by great philosophers such as Locke (1689/1975).

With regard to the term "disc(o)ursive thought," my intention is to transmit the concept that, far from expressing thought in the manner we have just seen, language serves to think, and, even more, thought comes into being through discourse. The expression "discoursive thought" does have at least one small merit: it does not bring to mind, or only bring to mind, the notion of "logical thought." In fact, it is important to keep research free, so to speak, of distinctions such as logical/illogical and correct/incorrect (even though sometimes these distinctions can be very useful), taking into consideration every occurrence and every form of discoursive thought, as it is not a question of reconstructing logic from or of looking for logic in what people say (or maybe of pointing out how little logic there is!) but of studying how the mind really functions. Discourse, in the fundamental aspect we are considering here, does not come after, nor does it spring from or emerge distinct from thought, of which it is a translation or a phonographic materialization, in other words, a sort of "packaging for export." When we say "I can't find the words," as mentioned earlier, what we really mean is that the thought has not yet acquired a form that can be communicated. It is still embryonic, nebulous, it is still only a wish or a need.

The intimate connection between language and thought is to be found in all the conceptions of thought as inner speech or internal dialogue. Whitehead (1938) said that "language has two functions: conversing with others and conversing with oneself." The starting point can be language or thought; what matters is that it is not language *and* thought or the relationships *between* thought and language but an activity or area in our cognitive experience that is not involved in such a distinction. The expression *thinking-speaking* or *discoursive thought* emphasizes the need to recompose this unjustified separation, which has occurred and individuates, identifies, the discoursive nature of thought as the core of this psychologically organic unit.

From this point of view even the distinction between talking to oneself and talking to others loses its relevance. The rules that govern "conversing with another person" and "conversing with oneself" share a common basis and are greatly similar.

With regard to the rules for effective communication and those for the functioning and the organization of logical thought (which could be defined respectively as rules of communication and rules of logic), a distinction may be made, almost in the sense that these rules do not match perfectly, as some rules are specific to communication.

In other words, if a complete system of rules of logic were to be established, it would contain only some of the rules of communication, whereas the converse is true—a complete system of rules of communication would contain all the rules of logic.

If a communication is not sufficiently clear, it is difficult to understand and even more difficult to remember without some form of distortion. This depends mostly on how the communication has been thought out, but sometimes the problem lies in how it has been communicated (for example, the omission of certain elements that are necessary to the listener).

In my view it is reasonable to hypothesize that conversing with others and with oneself, public thought and private thought, are not governed by different rules, apart from the consideration that the rules disciplining public thought are more circumstantiated, analytically determined, more "bureaucratic." The *corpus* of rules governing thought destined for public use is more extensive and, shall we say, pedantic: in this case the *thinking-speaking* is subject to more conventions.

If what I have said until now should appear on the whole acceptable and reasonable, then the use of the term *"psycho-rhetorical"* would be justified.[4] Rhetoric has considered and studied discourse (both discourse in general and, in particular, persuasive discourse) from the point of view of its acceptability and effectiveness.

Although rhetoric research is sometimes belittled, considered mainly as a set, a machinery of possible stratagems and strategies for the persuader, it concerns the conditions, the means, and all that regards the production of a discourse to ensure that it will be acceptable to the listener. Assuming the point of view of the speaker, rhetoric has studied speech in relation to the listener and therefore has taken into consideration how acceptable it will be to him, how comprehensible and how effective. I believe it could be said that rhetoric, within the sphere and limits of persuasion, has studied discourse "psychologically." It is worthwhile adding that, contrary to what theories regarding rhetoric might lead us to believe, the rich heritage of knowledge produced by this discipline is not exclusively related to persuasive discourse but to speech in general. What might have been considered to be results relating to the purposes of rhetoric become results, information, knowledge, material, and to a certain extent a starting point for research on discursive thought or, in general terms, the functioning of the mind. We certainly do not have the intention of establishing rules regarding how to speak effectively, but we cannot ignore or even simply neglect the "rhetorical dimension" of discourse, which is an essential aspect of *thinking-speaking* and a fundamental purpose of psychological research. When I talk of a "psycho-rhetorical" approach or point of view, I intend to refer primarily to the need to consider the rhetorical dimension in the study of discoursive thought.

17.2 Logic as the Model of Reasoning[5]

As a consequence of logic being the normative schema in research on reasoning, experimental tasks tend to be identified as exercises in logic, presented to subjects in a popularized version, concealed as it were behind a "real-life" situation.

In fact, generally the experimental situation (or more precisely the communication that realizes in the experimental situation) tends to occur on two levels: a deep or secret structure, the structure that the "omniscient experimenter" as Simon (1979) calls him has in mind, and a surface or manifest structure that is revealed to the subject in natural, common language. Another consequence is that the answers given by the subjects are evaluated according to the rules of logic, which may deviate significantly from those of everyday discourse, that is, the conversational rules (despite being derived from them). This produced no end of misunderstandings, *qui pro quo,* and serious problems. Quite often communication between the experimenter and the subject resembles a conversation between two deaf people. The experimenter conceives the task as an exercise in logic to be solved with the codified rules of logic. The subject, on the other hand, actually solves a different task, decodified or interpreted according to different rules, those of natural discourse. Finally, the experimenter evaluates the subject's responses according to the rules of logic, those rules that apply to the deep structure with which he has organized the experiment (if not generally, at least in some cases, especially in the area of research dedicated to the syllogisms, this mechanism, which produces *qui pro quo,* misunderstanding was considered with regard to the meaning to be attributed to certain terms, such as *few/some,*[6] or to the requirement that conclusions necessarily derive from premises, and so on).

A particularly illustrious example of this appears in the errors that arose in Wason's selection task (Wason, 1966). These errors were in fact due to the mechanism we have just described, which led the subjects to ignore the pragmatic, *psycho-rhetorical* existence (which in this case would have been decisive) of the *plausibility* of the rule to be falsified.

Psychologically speaking, falsification is a *second* operation that presupposes verification. In stark contrast with previsions based on the *bias* of verification, the subjects have no difficulty assuming a falsification approach, and they employ it when the rule or proposition to be verified is plausible, that is, has already been verified or assumed to have been verified.

Although attention for pragmatic, *psycho-rhetorical* aspects and their influence on the outcome of experimental research has intensified, in the psychology of reasoning this progress does not seem to have undermined either the presupposition of the legitimate predominance of the logical structure underlying the text of the experiment or, therefore, the presupposition that it is legitimate to expect that the subjects will apply a logical code to interpret this text rather than giving it a pragmatic interpretation based on the rules of natural discourse and the psychological requirements this implies. This is the presupposition that has led experimenters to consider those responses that conform to the logical code correct and all the others to be erroneous.

The misunderstanding or mismatch between the *logical* code (used by the experimenter in designing the task and in evaluating the subjects' performance) and the *natural* or *psycho-rhetorical* code used by the subjects in performing the task can even have a completely different effect from that explained previously; in other words the solution to the logical task provided by the subjects may seem to be appropriate, but in reality it has a totally different significance and has little or nothing to do with that logical task.

Logic Tasks with Realistic Materials

Realistic materials have been used to improve performance on selection tasks, the best results being obtained by Johnson-Laird, Legrenzi, and Sonino Legrenzi (1972). The participants in their experiment were told to imagine that they were Post Office workers whose job was sorting letters and ensuring that they conformed to the following rule:

"If a letter is sealed, then it has to have a 50-*lire* stamp on it."[7]

Instead of the four cards used in Wason's task, the participants were given:

- The back of a sealed envelope (*p*)
- The back of an open envelope (*non-p*)
- The front of an envelope with a 50-*lire* stamp (*q*)
- The front of an envelope with a 40-*lire* stamp (*non-q*)

Twenty-one out of 24 subjects made the correct choice: the back of the sealed envelope (*p*) and the front of the envelope with the 40-*lire* stamp (*non-q*), while only 2 out of 24 subjects solved the same problem in the symbolic condition (A, B, 2, 3). The authors attributed this dramatic improvement to the sense of reality and familiarity created by the use of the materials.

Examining the problem in depth, we have taken into consideration the information effectively transmitted to the subjects through instructions and the use of realistic materials:

- The subjects were told to imagine that they were Post Office workers sorting letters, which implied that they had to decide whether a surcharge was to be applied.
- A surcharge was to be applied to letters stamped with a lower value than required.
- Letters with a stamp of the correct value or a higher value than required are admitted.
- A sealed envelope needs a stamp of a higher value than an open one.

In fact, there are two cases that can infringe the Italian postal regulations: a sealed envelope stamped with less than 50 *lire* or an envelope without any stamp at all. To summarize, *a sealed envelope with a 50-lire stamp or more does not constitute an infringement.*

In our opinion the participants in the experiment of Johnson-Laird et al. solved a real-world problem, and the correct responses were erroneously interpreted as a demonstration of deductive reasoning. As we see it, if the "logic" of the real situations did not coincide with the logic of the falsification of a rule, then this level of accuracy would not be maintained, even using realistic materials. We tested this hypothesis by running the following experiments.

Experiment 1. Twelve people were given the same instructions, postal rule, and realistic material as used in the Johnson-Laird et al. experiment, with the only difference being the value of the stamp in the *non-q* condition.

- The back of an open envelope (*non-p*)
- The front of an envelope with a 50-*lire* stamp (*q*)
- The front of an envelope with a 100-*lire* stamp (*non-q*)

It is important to keep in mind that, this time, *non-q* is represented by an envelope with a stamp of a higher value than required.

None of the participants chose the envelope with the 100-*lire* stamp (*non-q*); two thirds of them chose the sealed envelope (*p*) and were able to give a very clear explanation of their choice—that the Post Office would only check sealed envelopes carrying a stamp with a value below 50 *lire*. When asked if it might have been important to check the envelope with the 100- *lire* stamp, all participants said that there was no point because the Post Office is not interested in stamps with a value of more than 50 *lire*.

It was clear that the participants were not acting simply on the basis of a literal interpretation of the instructions given but were considering the situation in its entirety, including the implicit information. In the logical verification of an implication, each case that differs from the consequent represents *non-q* and so must be selected. Actually, there are values other than *q* that can be irrelevant or even, as in the case of the 100-*lire* stamp, *above-q*, in that they are perfectly acceptable, at least as far as the real world is concerned.

As in the Johnson-Laird et al. experiment, the idea of pretending to be a Post Office worker made their participants select the envelope with the 40-*lire* stamp (*non-q*), in our experiment it led our participants to neglect the envelope with the 100-*lire* stamp (*non-q*). Logic has little or nothing to do with these selections, which were still perfectly reasonable and adequate.

Experiment 2. We ran a further experiment to test our hypothesis that people act in accordance with realistic norms in real situations and that interpreting their behavior in purely "logical" terms can be rather arbitrary. A different group of twelve people were presented with the same task designed by Johnson-Laird et al., but with a different rule:

"If an envelope is left open, it must have a 40-lire stamp."

Here the open envelope represents *p,* a sealed envelope represents *non-p,* an envelope with a 40-*lire* stamp represents *q,* and an envelope with a 50-*lire* stamp *non-q.*

In this version of the experiment none of the participants chose the open envelope (*p*) and the envelope with the 50-*lire* stamp (*non-q*), and the majority of the subjects chose the sealed envelope (*non-p*) and the envelope with the 40-*lire* stamp (*q*). From a logical point of view this is an "aberration," but it is perfectly reasonable and comprehensible in the real world. In fact, the selected envelopes are those that could have constituted an infringement of the postal regulations: the sealed envelope could have had a stamp of a lesser value than 50 *lire,* and the envelope with the 40-*lire* stamp could have been sealed. In fact, the participants explained that it was pointless taking the other two cases into consideration, as these conformed to the postal regulations.

Final Considerations

The research on the selection task wound its complicated and tortuous way to finally reach a purely banal conclusion, that in order to understand what subjects do and how they proceed, it is essential first of all to find out "what they have in mind" regarding the problem, or in other words, what is the *actual message* they are working on (in the case of experiments this is the message they receive, although in everyday situations the message can even be self-produced). Observance of this principle does not identify with the solution of psychological issues, but it does eliminate or reduce the risk of having to deal with poorly conceived questions. In fact, the core issue regarding the effects produced by abstract versus thematic materials—as in general it has been studied and has evolved—is indeed an example of a poorly conceived question. An appropriate psychological analysis (which I consider to be *psycho-rhetorical*) of "abstract" and "thematic" or "concrete" problems could reveal (and indeed has been shown to do so) the basic psychological faultiness or impropriety of the presumption of fundamental identity among problems that are interpreted as having the same logical structure. This presumption has played an essential role in much of the research that it has influenced (although not to any great advantage), and it has also indirectly had an effect on the "new ideas." Accepting that not only *p* and *not-q,* but also *not-p* and *q or p, not-p, q and not-q* (and possibly others) can be "correct," is to conclude a research cycle (the complicated and tortuous one) and suggests an approach based on a direct reference to conversational rules rather than to those of logic (although I personally would not exclude an "instrumental" use of logic).

17.3 On Problems[8]

Tasks and Problems

According to Simon and his collaborators, Newell and Shaw, "the maze provides a suitable abstract model for most kinds of problem-solving activity" (Simon, Newell, & Shaw, 1979, p. 148).

If it is true that *mazedness* (the essence of the labyrinth) is an essential or at least typical characteristic of problems, and that the labyrinth is an appropriate abstract model for problem solving, then the subject facing a problem will move in the problem space just as she would move in a labyrinth, searching for the right path. Each time she comes up against a dead end, she will retrace her steps and even return to the beginning; she will form and apply a strategy of sorts, implementing a selective search, progressing step by step toward the goal. This is more or less what the manuals teach us now. This conception of the problem or this model of *problem solving* is in general well suited to problems such as the well-known *Missionaries and Cannibals* problem in which:

Three missionaries and three cannibals are on the left bank of a river with a boat that can carry a maximum of two people, and they want to cross to the right bank. For obvious reasons the number of missionaries can never be less than the number of cannibals, whether on the left or on the right bank.

The task consists of finding a sequence of crossings that will leave all six persons on the right bank, respecting the condition that the boat can only transport two people at a time and that the missionaries cannot be less in number than the cannibals. Quite often people think that this is a complex task and that its solution will be long and involved, and quite often they are "good prophets." Missionaries and cannibals cross and recross the river many more times than is actually necessary. Sometimes they are moved randomly, "just to try" in the hope that the right solution will turn up of its own accord; other times strategies are applied, such as the exploitation of the capacity of the boat or moves dictated by an equal ratio between missionaries and cannibals.

In general the problem is quite difficult. People tend to approach it as they would a labyrinth, applying a trial-and-error process; they try a few moves, then go back because they have finished up in a dead end, or they think that is the case; they try strategies or adopt criteria, and so on. "It is so complicated you get lost," someone says, and someone else, discouraged, "you really need to be able to calculate everything." This then is in fact the heart of the matter: it is the complexity of the calculation as perceived by the problem solver that makes the task so difficult to solve and transforms it into a problem.

But it is not always like this. Sometimes the complexity of the calculation does not have a role to play in determining the difficulty of the problem: and in this case, rather than the labyrinth, the abstract model that comes to mind is the trick or the *qui pro quo*. There are problems that, unlike the *Missionaries and Cannibals*, seem quite easy to solve due to both the limited amount of data and the apparent brevity of the series of operations to be performed. In reality, problems of this type are not at all easy to solve.

Sometimes the answer is given very rapidly, and often or almost always it is wrong. An example of this is the *Four Balls* problem (Petter, 1974):

Four balls look the same, but one is a different weight. Find this ball, using a pair of scales only twice.

Practically everyone attempts this problem with one of the two following procedures (the situation is purely imagined): in the first procedure, two balls are placed on both plates of the scales in the expectation that one side will "move downward." The second move is to take the two balls of the plate that moved downward and put one ball on each plate of the scale in the expectation that it will create new imbalance between the two plates. The subjects expect that the ball with a different weight will be on the plate which moves downward. In the first move of the second procedure, a single ball is placed on each plate. If the scales are balanced, then the second move is to repeat the first, using the balls that have not yet been weighed; the subjects expect that this will necessarily cause an imbalance. Once again they expect that the ball that is a different weight will be on the plate that moves downward. Neither of these procedures is satisfactory as the ball of a different weight could be heavier or lighter, but all the subjects performed as if they knew already that the ball which is a different weight will be heavier.

The *Horse-Trading* problem (Maier & Janzen, 1969) presents a similar twist:

A horse trader bought a horse for £70 and sold it for £80; then he bought it back for £90 and sold it for £100. How much did he make?

Once again, as Steiner (1982) says, nearly everyone gives the answer straight away but only 40% get it right. The typical reasoning that leads to the erroneous response of £10 is "When the trader sells the horse for £80, he makes £10 on the transaction but loses this sum when he buys it again for £90. His final sale of the horse for £100 brings him £10."[9]

Sometimes the subjects find themselves in an impasse. The question seems to be totally incompatible with the data provided and the described situation, as in the well-known *9 Dot* problem, or it may appear to necessarily, though inexplicably, contain a contradiction as in the case of the popular problem used here in a version we will call the *Restaurant Bill* problem:

Three friends have dinner together in a restaurant. The bill comes to €60. Each places a €20 bill on the table, but they protest and ask for a discount. The owner gives them back €10, and they leave €4 for the waiter and take €6. So, each has paid €18, which multiplied by 3 makes a total of €54. The waiter has received €4. This makes a total of €58. What has happened to the missing €2?

Many subjects are intrigued and are aware that there must be some sort of trick, but they are also conscious of the fact that they have all the relevant information and have followed and verified their reasoning step by step. These subjects have added the €4 tip to the €54 paid for the dinner, whereas they should have realized that the €54 includes the tip. The fact that this operation is clearly suggested in the text of the problem is not sufficient to justify the error, as can be seen below.

Taking a closer look at these examples, we can see that at least one distinction can be drawn. In some of the situations the difficulty for the subject lies in the complexity of the calculations involved, in the length of the operations that have to be performed, in the quantity of data to be managed and remembered, and so on.

In other cases the impasse into which the subjects find themselves is not so much a result of the difficulty of the calculations as of one or two critical points which they come up against; a *qui pro quo*, an interpretation that is incompatible with the solution.

Moreover, in this case the subject is not just faced with difficulties, nor is he external to the case but, in a certain manner of speaking, actually collaborates in creating the difficulty. Take the *Restaurant Bill* problem, for example. If the subject does not interpret *pay* as *pay the bill* but rather as *shell out*, he will not add the €4 tip given to the waiter to the overall amount of €54 and so will thwart what, in his view, is a clumsy trick.

The same applies to the *Four Balls* problem. If the subject does not interpret *different* weight as *heavier*, he will not make the error explained above.

In these cases the action is in the hands of the subject himself. Through his interpretation he *transforms* the original message, *the given message*, into the *actual message* on which he works. In fact, for the problem to be a problem at all, the subject has to accept the message, to which he himself has contributed, as acceptable discourse. This does not happen only in cases where he is misled without realizing it (such as the *Four Balls* or the *Horse-Trading* problems) but also when (as in the *Restaurant Bill* problem) the message on first sight appears to be definitely contradictory. In fact, even in this case the subject does not perceive any incongruencies in the text of the problem, the discourse sounds flawless; of course the difficulty for those subjects for whom the problem is formed consists precisely in the fact that a message that is acceptable initially leads to an unacceptable situation or conclusion.

Even where the *qui pro quo* or the contradiction is already de facto present in the text, as in the case of the *Restaurant Bill* problem, this is not sufficient to explain the behavior of the problem solvers. What has to be understood and explained is that fact that problem solvers actually accept the "misleading" message. Each time a subject accepts a *discourse-problem*, it is as if he himself had produced it.

If the message had not been formulated in accordance with the same rules the subject uses to process information in his mind, he would not have accepted it as it was. This is particularly evident in the *Restaurant Bill* problem. The version of this problem we have illustrated above is misleading: it prompts the strange way of making the calculations that led to the final contradiction. However, even if another version is used, eliminating the misleading section with the biasing suggestion that the tip be added to the total amount paid (and changing the final question), the subjects still make the same mistake. Some will even repeat the problem to themselves and in so doing will reproduce the first "dishonest" version. This implies that the suggestion in itself is not sufficient to explain the subjects' behavior. The fact that an indication has been supplied, irrespective of whether it is erroneous or not, does not guarantee that it will be accepted. Acceptance of a suggestion implies that it is congruous with the receiver's cognitive system: the "way of thinking" or the rules according to which it has been produced must be the same as those of the receiver. The fact that under certain conditions the "suggestion" can be produced autonomously is proof of this.

Naturally what we have said about accepting or rejecting a suggestion or a discourse in general, considered as the reproduction of that discourse, is valid not only for problems of the type we are considering here but also for others such as the *Missionaries and Cannibals* problem, and for problems in general. However, it is worth keeping in mind that in situations of the *Missionaries and Cannibals* type, the fundamental difficulty depends on the complexity of the calculations or the performance required; therefore, the *discourse-problem* actually is only an introduction to the situation and has no function other than that which it would have in any communication. Instead, in the type of problems we are considering here, the *discourse-problem* itself creates the difficulty, and therefore its acceptance and reproduction are decisive.

To conclude, in this first section I intended to suggest an opportunity, or rather indicate the necessity of distinguishing between problems and tasks or, if we want to use current terminology, between problems of type A and type B.

A criterion to distinguish tasks from problems could be found in the sensitivity of the problem to a not misleading reformulation.

In the case of *task*, the message that passes between the experimenter on the one hand and the participant on the other is univocal in both the explicit and implicit (given as being said) parts; in other words, its meaning is understood in the same sense by both parties unless there are fortuitous errors of comprehension.

Problems, on the other hand, have a characteristic that a casual observer might think is an accidental misunderstanding. In fact, the problem is posed by the experimenter in terms that are different from those that are commonly understood.

Contrary to what happens in tasks where the subject is faced with an "objective" external difficulty, in true problems the subject plays a decisive role in *forming* the problem itself or in creating the difficulty. The traditional and current problem defini-tions (which only include the notions of the starting point, of the aim or final state, and of the obstacles or difficulties in passing from one state to another) do not take this aspect into consideration and erroneously communicate a notion of homogeneity in the ill-defined area of problems.

Consequences of the Distinction between Tasks and Problems

... The predominant theory not only considers tasks and problems as being homoge-neous, it also treats them as such, claiming that "there is no qualitative separation" in the problems, but in fact it does favor the tasks, both implicitly and explicitly, attempt-ing to eliminate the specificity of the *problems* and treating them all as if they were *tasks*. Models and methods that are typical of tasks are proposed as being typical and appropriate of problems in general. The emphasis laid on the importance *in the prob-lem solving* of the limits of the memory, from the reduced capacity of the short-term memory to the "slowness" in the storage and retrieval from the long-term memory, are all coherent with this trend. With regard to problems in the true sense of the term, their characteristics, and the nature of their difficulty, it is reasonable to judge this emphasis as excessive and more limited, or rather, less specifically relevant than what was thought, the role of memory and above all the importance of its limits in problem solving (while of course there can be no question of the importance that the memory has *also* in *problem solving*).

... When [one is] dealing with *problems*, the activity of *problem solving* is only a part of what happens and of what the subject does. The distinction between *tasks* and *prob-lems* is a necessary condition to realize that by limiting our reflections and research to *problem solving,* we are skipping the first phase of the process and starting half way through the entire process with the second phase that moreover depends on and is functionally connected to the first phase. The activity of *problem solving* is subsequent and antagonist to the activity of *problem forming*. To take "a part as the whole" would be an error in any case; moreover, when the part being taken as the whole depends on the part that has been ignored and with which it is complementary, the error will certainly not be innocuous.

... The acknowledgement of the formation of the problem, the identification of *how* it has been formed and the awareness that this is a decisive phase for everything [that] follows necessarily imply to take a new look at failures; studying them should teach us just as much as studying successes in problem solving. A theory [that] has been improperly called *Problem Solving* must first of all explain how the problem has been constituted, and then how the solution is reached, but also what happens when the solution is not reached; it must explain the correct solution or the success and, and at the same time, with the same methods both the error or lack of success.

Conclusion: The Problem Not as Given but as a Fact

"There are no contradictions in Nature" (Vauvenargues, 1747/1981, maxim CCLXXXIX). Contradictions are only present in our knowledge and our theories on Nature, that is, in our thoughts. What Vauvenargues has said here regarding contradictions can be applied to problems: there are no problems in Nature. The problems are in our mind. There are no situations that are problems in themselves, and the converse of this is that every situation can be transformed into a problem. Information from the surroundings, from events and conversations, can contribute to the constitution of the problemic situation, but it is never the situation in which the subject finds himself or the information that reaches him from the outside, which in itself produces the problem.

There are many reasons to explain the fact that problems are generally assumed as *given;* starting from this assumption, research in the field has focused only on how problems are solved, how to proceed from the *given state* to the *goal,* or what happens when subjects progress toward the solution. In the course of their research experimenters usually use ready-made problems and submit them to the subjects, who then start working on these problems supplied by others (i.e., *given problems*) and attempt to solve them. It is understandable that in this scenario the problem appears as something external, something "given" that has an existence of its own, something to be used and dealt with, the starting point *par excellence.*

This conception of the problem is consolidated by the fact that frequently in our everyday life we become aware of the existence of a problem only when the problemic situation already exists, when we encounter difficulties, but often we are not aware of how the problem came into being, of our situation, or how we happened to get into the difficulty (rather like when, having taken the wrong turning, we realize that we are lost, without however having realized that we were getting lost while we were taking the wrong turning).

Our day-to-day experiences and, in particular, how the researcher experiences the experiment (both directly and indirectly through the experience of the subjects faced with the problem), explains, or at least makes it possible to understand, the assumption of the problem as something that is given, with an existence of its own, something with which we come into contact. Therefore, it also explains why in research the focus is generally on how the problem is solved, hence assuming the problem as a given. This, however, does not legitimize such an assumption as a good foundation for psychological theory.

There is a link among the conception of the problem as a given, the tendency to disregard how it is formed (and the concomitant focus on the solution or *problem solving* in the strict sense of the term), and the lack of distinction between tasks and problems. In my opinion there is a link between all this and the status of research in this field, which Johnson-Laird and Wason describe in these terms:

How do people solve problems? The short answer is that psychologists do not know. There is no comprehensive theory of problem solving, only a number of models and hypotheses about different aspects of the process. Part of the difficulty is that problems come in all shapes and sizes from nursery riddles to deep scientific puzzles, but there is no substantial taxonomy for them. There can even be a problem in recognizing a problem in the first place, whether one is trying to define the notion or is a practicing scientist looking for something worthy of investigation. It is difficult to talk about such elusive entities in general terms. ... (Johnson-Laird & Wason, 1977, p. 13)

In fact, there cannot be any "comprehensive theory of problem solving" as long as a part is confused with the whole and only the final part of the process is taken into consideration, ignoring what has gone before and on which the process depends. Indeed it would be advisable to abandon the expression *problem solving*, which reflects this fundamental theoretic deficiency, or at least to restrict its use to that part of the activity to which it correctly refers. Moreover, it is not possible to "define the notion of the problem" as long as our idea of what is to be defined is erroneous and incomplete. Nor is it possible to "recognize a problem" while there is confusion between tasks and problems and we look for something different from what we should be recognizing. Last, the lack of a clear distinction between tasks and problems makes it impossible to make any substantial taxonomy for problems (although this is certainly not the worst of the ills).

In spite of having been distorted and constrained by theoretical and empirical limitations, research on problems has made an important contribution to knowledge and holds an important place in psychological research. It is a natural point of convergence for almost all psychological knowledge (thought, language, communication, memory, perception, and learning). In order to resolve the serious "lacunae" indicated by Johnson-Laird and Wason, it is indispensable to reject the flawed approach that reduces the study of problems to *problem solving* and integrate *problem solving* with *problem forming*.

Notes

1. Excerpta from: Mosconi, G. (1978). I Greci dicevano Logos. In: *Il pensiero discorsivo*. Bologna, Il Mulino, 5–13; and Mosconi, G. (1990). Il pensiero discorsivo e il punto di vista psicoretorico. In: *Discorso e pensiero*. Bologna. Il Mulino, 7–15.

2. The term appears in the quotation reported below in the text.

3. In Italian the verb *ragionare* means both *to think* and *to speak*.

4. The term "psycho-rhetorical" identifies Mosconi's approach to the study of thought.

5. Excerpta from: Mosconi, G. (1974c). Codice logico e codice retorico. Il problema delle quattro carte. In: Mosconi, G., & D'Urso, V., *Il farsi e il disfarsi del problema*. Firenze, Giunti-Barbera, 113–129; Mosconi, G., & D'Urso, V. (1975). The selection task from the standpoint of the theory of the double code—*International Conference on Selection Task*—Trento, Oct. 1975; and Mosconi, G.

(1990). Codice logico e codice psicoretorico nel compito di selezione. In: *Discorso e pensiero*. Bologna. Il Mulino, 181–217.

6. Italian words *qualche/alcuni*.

7. The reference is to the values of the stamps in Italy in the early 1970s.

8. Excerpta from: Mosconi, G. & and D'Urso, V. (1973). Il farsi e il disfarsi del problema. In: G. Mosconi & V. D'Urso (Eds.) *La soluzione di problemi*. Firenze, Giunti-Barbera, 9–34; Mosconi, G. (1974). La formazione e la soluzione del problema secondo la teoria del doppio codice. In: Mosconi, G., & D'Urso, V. (1974). *Il farsi e il disfarsi del problema*. Firenze, Giunti-Barbera, 9–34; Mosconi, G. (1981). All'origine dei problemi: l'incompatibilità presente nella mente. In: L. Tornatore (Ed.). *Educazione alla ricerca e trasmissione del sapere*. Torino. Loescher, 181–224; Mosconi, G. (1990). Sui problemi. In: *Discorso e pensiero*. Bologna. Il Mulino, 221–243.

9. Instead, the typical explanation for the answer "£20," which is the correct one, is: "The dealer makes £10 on the first transaction (70 – 80) and £10 on the second (90 – 100)."

References

Chomsky, N. (1977). *Dialogues avec Mitsou Ronat*. Paris: Flammarion.

Cortellazzo, M., & Zolli, P. (1979). *Dizionario etimologico della lingua italiana*. Bologna: Zanichelli.

Dalla Volta, A. (1974). *Dizionario di psicologia*. Firenze: Giunti-Barbera.

Isocrates. (1959). *Nicocle (Prologo)*. In W. Jaeger (Ed.), *Paideia*. Firenze, La Nuova Italia III.

Locke, J. (1975). *An essay concerning human understanding*. P. H. Nidditch (Ed.). Oxford: Oxford University Press. (Original work published 1689.)

Johnson-Laird, P. N., Legrenzi P., & Sonino Legrenzi, M. (1972). Reasoning and a sense of reality. *British Journal of Psychology*, *63*(3), 395–400.

Johnson-Laird, P. N., & Wason, P. C. (Eds.). (1977). *Thinking*. Cambridge: Cambridge University Press.

Maier, N. R. F., & Janzen, J. C. (1969). Are good problem solves also creative? *Psychological Reports*, *24*, 139–146.

Mosconi, G. (1978). I greci dicevano logos. In *Il pensiero discorsivo* (pp. 5–13). Bologna: Il Mulino.

Mosconi, G. (1990). *Discorso e pensiero*. Bologna: Il Mulino.

Mosconi, G., Bagassi, M., & Serafini, M. G. (1988). Solutori e benrispondenti. II. Il problema della compravendita del cavallo: Discussione e ricerca di gruppo. *Giornale Italiano di Psicologia*, *XV*(4), 671–694.

Mosconi, G., Bagassi, M., & Serafini, M. G. (1989). L'antefatto del problem solving. Il problema della compravendita del cavallo. *Rivista di Psicologia-nuova serie*, *74*(3), 9–27.

Mosconi, G., & D'Urso, V. (1973). Il farsi e il disfarsi del problema. In: G. Mosconi & V. D'Urso (Eds.), *La soluzione di problemi* (pp. 9–34). Firenze, Giunti-Barbera.

Mosconi, G., & D'Urso, V. (Eds.). (1974). *Il farsi e il disfarsi del problema*. Firenze: Giunti-Barbera.

Mosconi, G., & D'Urso, V. (1975). The selection task from the standpoint of the theory of the double code. *International Conference on Selection Task,* Trento, October 1975.

Petter, G. (1974). Procedimenti euristici nel campo del pensiero produttivo. In G. Mosconi & V. D'Urso (Eds.), *La soluzione di problemi* (pp. 239–256). Firenze: Giunti-Barbera.

Simon, H. A. (1979). *Models of thought*. New Haven: Yale University Press.

Simon, H., Newell, A., Shaw, J. C. (1979). The processes of creative thinking. In H. A. Simon (Ed.), *Models of thought* (pp. 144–174). New Haven: Yale University Press. (Original work published 1962.)

Steiner, I. D. (1982). Heuristic models of groupthink. In H. Brändstatter, J. H. Davis, & G. Stocker-Kreichgauer (Eds.), *Group decision making* (pp. 503–524). New York: Academic Press.

Tornatore, L. (Ed.). (1981). *Educazione alla ricerca e trasmissione del sapere*. Torino: Loescher.

Vauvenargues (1981). In J. Dagen (Ed.), *Réflexions et maximes*. Paris: Flammarion. (Original work published 1747.)

Wason, P. C. (1966). Reasoning. In B. M. Foss (Ed.), *New horizons in psychology 1* (pp. 135–151). Harmondsworth: Penguin.

Whitehead, A. N. (1938). *Modes of thought*. New York: Macmillan.

Contributors

Maria Bagassi, Department of Psychology, University of Milano-Bicocca

Linden J. Ball, School of Psychology, University of Central Lancashire

Jean Baratgin, CHArt (PARIS), Université Paris 8 and Institut Jean Nicod, Paris

Aron K. Barbey, Decision Neuroscience Laboratory, University of Illinois

Tilmann Betsch, Department of Psychology, University of Erfurt

Éric Billaut, Centre National de la Recherche Scientifique, Université de Toulouse

Jean-François Bonnefon, Centre National de la Recherche Scientifique, Université de Toulouse

Pierre Bonnier, Ecole Normale Supérieure

Shira Elqayam, School of Applied Social Sciences, De Montfort University

Keith Frankish, Department of Philosophy, The Open University

Gerd Gigerenzer, Center for Adaptive Behavior and Cognition, Max Planck Institute for Human Development

Ken Gilhooly, Department of Psychology, University of Hertfordshire

Denis Hilton, Department of Psychology, Université de Toulouse-II

Anna Lang, Department of Psychology, University of Erfurt

Stefanie Lindow, Department of Psychology, University of Erfurt

Laura Macchi, Department of Psychology, University of Milano-Bicocca

Hugo Mercier, Cognitive Science Center, University of Neuchâtel

Giuseppe Mosconi, Professor of Psychology, Emeritus, University of Milano-Bicocca

Ian R. Newman, Department of Psychology, University of Saskatchewan

Mike Oaksford, Department of Psychological Sciences, Birkbeck College, University of London

David Over, Deparment of Psychology, Durham University

Guy Politzer, Institut Jean Nicod, Paris

Johannes Ritter, Department of Psychology, University of Erfurt

Steven A. Sloman, Department of Cognitive, Linguistic, and Psychological Sciences, Brown University

Edward J. N. Stupple, Centre for Psychological Research, University of Derby

Ron Sun, Cognitive Sciences Department, Rensselaer Polytechnic Institute

Nicole H. Therriault, Department of Psychology, University of Saskatchewan

Valerie A. Thompson, Department of Psychology, University of Saskatchewan

Emmanuel Trouche, French National Centre for Scientific Research, L2C2-Laboratoire sur le Langage, le Cerveau et la Cognition

Riccardo Viale, Professor of Behavioral Sciences and Decision Making at Scuola Nazionale di Amministrazione

Index